ICME-13 Monographs

Series editor

Gabriele Kaiser, Faculty of Education, Didactics of Mathematics, Universität Hamburg, Hamburg, Germany

Each volume in the series presents state-of-the art research on a particular topic in mathematics education and reflects the international debate as broadly as possible, while also incorporating insights into lesser-known areas of the discussion. Each volume is based on the discussions and presentations during the ICME-13 Congress and includes the best papers from one of the ICME-13 Topical Study Groups or Discussion Groups.

More information about this series at http://www.springer.com/series/15585

Carmen Batanero · Egan J. Chernoff
Editors

Teaching and Learning Stochastics

Advances in Probability Education Research

Springer

Editors
Carmen Batanero
Facultad de Ciencias de la Educación
University of Granada
Granada
Spain

Egan J. Chernoff
College of Education
University of Saskatchewan
Saskatoon, SK
Canada

ISSN 2520-8322 ISSN 2520-8330 (electronic)
ICME-13 Monographs
ISBN 978-3-030-10282-1 ISBN 978-3-319-72871-1 (eBook)
https://doi.org/10.1007/978-3-319-72871-1

Printed on acid-free paper

This Springer imprint is published by Springer Nature
The registered company is Springer International Publishing AG
The registered company address is: Gewerbestrasse 11, 6330 Cham, Switzerland

Preface

Introduction

Probabilistic competencies and knowledge are today necessary for adapting to a quickly changing society where information about random phenomena is over-abundant in all areas of human activity. Consequently, to adequately function in society citizens need to adapt their deterministic thinking and embrace chance and uncertainty in different settings. At the same time, they need to acquire strategies and ways of reasoning that help them in making appropriate decisions in everyday and professional situations where chance is present.

By including probability in the curricula at different educational levels and in the education of teachers, educational authorities in many countries have recognised a need for teaching and learning probability for everyone. Despite, however, the fact that probability is taught in school and university curricula, the research literature indicates poor probabilistic thinking among most students and adults, including those who have formally studied probability. These preconceptions (found in the literature as misconceptions) and biases contribute as well to widespread errors and misinterpretations in the application of statistical inference, which requires a well-developed probabilistic reasoning, since many inferential concepts (e.g. p-value, level of significance or sampling distribution) are defined by a conditional probability.

One explanation for the above situation, of course, is that including a topic in the curriculum does not automatically assure its correct teaching and learning. The specific characteristics of probability such as the multifaceted view of probability, the lack of reversibility of random experiments, the confusion between conditioning and causation, the number of paradoxes found even in elementary concepts and the philosophical debates around some of the most important probability or inferential concepts are not usually found in other domains; and create special challenges for teachers, students and citizens.

Research in (what is becoming known as) probability education attempts to respond to the above challenges—as shown by the many papers on this topic presented at international conferences such as the International Congresses on Mathematical Education (ICME), the European Mathematics Education Conference (CERME), the International Group for the Psychology of Mathematics Education (PME), the International Conference on Teaching Statistics (ICOTS), as well as in regional or national conferences such as the Latin-America Mathematics Education Conference (RELME) and the North American Chapter of the International Group for the Psychology of Mathematics Education (PMENA). This research is also visible in the specific books and chapters that summarise research in probability education such as, for example, Batanero et al. (2016), Borovcnik and Peard (1996), Chernoff and Sriraman (2014), Jones (2005), Jones et al. (2007), Kapadia and Borovnick (1991), and Shaughnessy (1992).

Background

This book is a consequence of and presents the results from the work in Topic Study Group 14 Teaching and Learning Probability at the International Congress on Mathematical Education (ICME13), carried out under the auspices of the International Commission on Mathematical Instruction (ICMI). The work in the group started with the appointment of the team leaders, Carmen Batanero (Spain), Egan J. Chernoff (Canada), and members Joachim Engel (Germany), Hollylynne S. Lee (USA) and Ernesto Sánchez (Mexico). This team worked collaboratively to produce a Call for Papers and a related Topical survey (Batanero et al. 2016) that served as a background of the discussions in the Topic Group sessions.

The general aim of the Topic Study Group on Teaching and Learning of Probability at the 13th International Congress of Mathematics Education (ICME-13) was to encourage *new* research in the domain. As such, the organisers welcomed diverse papers, including theoretical analyses and empirical research, in probability education whilst using a variety of research methods. A total of 41 papers and four posters were accepted for presentation after a refereeing process and were presented and discussed in the conference.

After the conference, the editors of this volume accepted the invitation to produce a monograph that covers the content of research presented in the Topic Study Group. The purpose of this monograph is compiling a collection of selected papers that represent the current variety of research related to the teaching and learning of probability. The different chapters in the monograph address a variety of theoretical, empirical and practical aspects underpinning the teaching and learning of probability, curricular issues, probabilistic reasoning, preconceptions and biases, as well as their related pedagogical implications. To achieve this purpose, the editors fixed a tentative content and a structure for the book. This structure took into account the papers presented and the discussions held at the conference and tried to assure coherence and completeness in the monograph. The Call for Papers was distributed to participants in the conference, who were encouraged to contribute with a chapter.

The first version of tentative chapters was received by October 2016, and finally, 20 papers were selected after a refereeing process in which all the tentative authors as well as additional researchers in probability education collaborated.

Structure of This Book

The book is organised into three main parts. The parts are dealing with teaching of probability, probabilistic reasoning and learning and teacher education.

In Part I, *Teaching Probability*, Manfred Borovnick and Ramesh Kapadia argue that risk and probability are twin concepts that should develop together in teaching and analyse the complementarity of different views of probability in teaching. Because objects in probability are not physically visible, humans must describe them through language, which constitute the probability register, analysed by Adam Molnar who also examined the use of specific terms. Cynthia W. Langrall reports that the majority of U.S. states have adopted a mathematics curriculum that has eliminated probability as a topic of study in the elementary grades and narrowed its focus in the middle grades and discusses the implications of this curricular decision. Per Nilsson, Andreas Eckert and Dave Pratt report on how instruction that is based on engaging students in practical experiments can create challenges and opportunities in the teaching of the relationship between classical a priori and frequentist models of probability. Katharina Böcherer-Linder, Andreas Eichler and Markus Vogel compare two visualisation tools useful for teaching probability from the psychological, educational and practical perspectives. Hollylynne S. Lee presents an analysis of the probability concepts needed for teaching inference with a repeated sampling approach. Carmen Batanero, Maria del Mar López-Martín, Pedro Arteaga and María M. Gea describe research aimed to characterise the probability problems proposed in the university entrance tests in Andalucia, Spain. Jorge Soto, Daniela Díaz-Rojas and Pamela Reyes-Santander suggest the use of the random walk metaphor to facilitate learning of probabilistic concepts. Finally, Caterina Primi, Anna Donati and Francesca Chiesi study the effect of statistics anxiety on learning probability in undergraduate students.

Part II is focused on *Students' Reasoning and Learning*. Peter Bryant, Terezinha Nunes, Deborah Evans, Rossana Barros, Laura Gottardis and Emanouela Terlektsi present an intervention study in which 75 school children, aged 9 and 10, were randomly assigned to an experimental group and two control groups in order to compare the students' learning of randomness in the different groups. Haneet Gandhi describes an open-exploratory study with eight middle-grade Indian school students (aged 13–14) working in small groups to understand their perceptions of random generators. Ana Serradó reports the results of a design-based research, where a game of chance is used to investigate 48 Spanish students' (aged 12) reasoning about decision-making under uncertainty and of risk. Ernesto Sanchez, Jaime García and Miguel explore the responses in a sample of high-school students to two tasks, one of them related to prediction and another to distribution, in order to understand how students express variability in their predictions before and after simulation activities. Pedro Landín and Jesús Salinas offer a probabilistic reasoning hierarchy for the concepts of sample space and the probability of a compound event as a tool to describe high-school students' performance when solving two problems involving binomial experiments, before and after a period of teaching. Joachim Engel and Arne Orthwein reflect on college students' intuitive strategies in a game of chance and contrast their reasoning with a normative probabilistic point of view.

Finally, in Part III, a group of papers deals with exploring the *Education of Teachers*. Egan J. Chernoff, Ilona Vashchyshyn and Heidi Neufeld contribute to the research investigating the use of logical fallacies, in particular the fallacy of composition, to account for normatively incorrect responses given by prospective teachers to relative probability comparisons. Pedro M. Huerta introduces the idea of solving problems by simulation with didactical intention. Based on this idea, he analyses the way we prepare the teachers to teach solving probability problems by simulation, in distinguishing whether the teaching is offered to prospective elementary teachers or secondary school teachers. Assumpta Estrada, Carmen Batanero and Carmen Díaz describe the development of a scale aimed at measuring primary school teachers' attitudes towards probability and the teaching of probability. They also present preliminary results with a sample of 232 prospective teachers. Susanne Podworny discusses the development of a one-semester course for preservice teachers at the university level based on intensive use of simulation software and analyses the teachers' reflections on the development of the course. Finally, Emilse Gómez-Torres, Carmen Díaz, José Miguel Contreras, and Juan J. Ortiz analyse the knowledge of ideas of sampling (population and sample, frequency, proportion, estimation, variability of estimates and the effect of sample size on this variability) demonstrated by 157 prospective primary school teachers in Spain

In summary, the articles presented in this book include research dealing with teachers and students from different ages (from primary school to university) and consider different probabilistic concepts and procedures, epistemological and curricular analysis, the role of technology, simulations, games of chance, language and visualisation, as well as use of metaphors in teaching and learning probability. We contend it offers useful information for teachers, teacher educators, researchers in mathematics and statistics education, and curricular designers. We hope this book will prove helpful towards improving the teaching of probability at school level and increasing probability literacy and probabilistic thinking of both teachers and students.

Granada, Spain — Carmen Batanero, University of Granada

Saskatoon, Canada — Egan J. Chernoff, University of Saskatchewan

References

Batanero, C., Chernoff, E., Engel, J. Lee, H., & Sánchez, E. (2016). *Research on Teaching and Learning Probability*. ICME-13. Topical survey series. New York: Springer.

Borovcnik, M. & Peard, R. (1996). Probability. In A. Bishop, M. A. Clements, C. Keitel, J. Kilpatrick, & C. Laborde (Eds.), *International handbook of mathematics education* (pp. 239–288). Dordrecht, The Netherlands: Kluwer.

Chernoff, E. J., & Sriraman, B. (2014). *Probabilistic thinking: presenting plural perspectives*. New York: Springer.

Jones, G. A. (2005). Exploring probability in schools. Challenges for teaching and learning. Mathematics Education Library (Vol. 40). New York: Springer.

Jones, G., Langrall, C., & Mooney, E. (2007). Research in probability: responding to classroom realities. In F. Lester (Ed.), *Second handbook of research on mathematics teaching and learning*. Greenwich, CT: Information Age Publishing and NCTM.

Kapadia, R. & Borovcnik M. (1991). *Chance encounters. Probability in education.* Dordrecht, The Netherlands: Kluwer.

Shaughnessy, J. M. (1992). Research in probability and statistics: Reflections and directions. In D. A. Grouws (Ed.), *Handbook of research on mathematics teaching and learning* (pp. 465–494). New York: Macmillan.

Contents

Part I
Teaching Probability

Chapter 1
Reasoning with Risk: Teaching Probability and Risk as Twin Concepts

Manfred Borovcnik and Ramesh Kapadia

Abstract Risk is a key aspect of life and probability is the mathematical tool to address risk. Our aim is to investigate how risk is embedded in probability and how probability can be used to solve problems of risk. Furthermore, we illustrate how risky situations are paradigmatic for the concept of probability, not least by the eminent role risk played for the emergence of probabilistic concepts. The analysis of risk forms the central section, which shows how risk is connected to probability. Decisions in risky situations draw from the various approaches to probability and from rational and behavioural views to decisions. The section on emerging concepts re-visits the development of probability and probabilistic thinking. Faced with the twin character of probability and risk, we argue that these concepts should be developed together in teaching. The conceptualisation of probability in terms of A Priori Theory (APT) Frequentist Theory (FQT) and Subjectivist Theory (SJT) forms the background to classify the pertinent constituents of the arguments. For understanding small probabilities/risks, we promote the ideas of a micromort (1 in a million chance of leading to death) and a microlife (a half-hour period, which is 1 in a million part of the average life span of a person aged 30 years).

Keywords A priori probability theory · Frequentist probability
Subjectivist probability · Probabilistic thinking · Risk · Uncertainty

---

M. Borovcnik (✉) · R. Kapadia
Institut für Statistik, Universität Klagenfurt, Universitätsstraße 65,
9020 Klagenfurt, Austria
e-mail: manfred.borovcnik@aau.at

R. Kapadia
e-mail: hmi.100@oxon.org

C. Batanero and E. J. Chernoff (eds.), *Teaching and Learning Stochastics*,
ICME-13 Monographs, https://doi.org/10.1007/978-3-319-72871-1_1

1.1 Introduction and Terminology

The title of this paper alludes to the origins of probability in the midst of time. In particular, the paper deals with the first theme of the ICME 13 Study Group on the nature of chance and probability, with some thoughts about the second theme on probabilistic knowledge and reasoning (see Batanero et al. 2016). We start by analysing risk as the precursor of probability (Sect. 1.2) with a commentary about both risk and probability and links to decisions. The third section contributes to a deeper understanding of probabilistic thinking, which is still—as opposed to statistical thinking (e.g., Ben-Zvi and Garfield 2004)—"a research concern" as Batanero et al. (2016, p. 8) state. We show how probability considerations are naturally embedded in the context of risk and that making risk visible in probability highlights the meaning of probabilistic concepts, before drawing brief conclusions in Sect. 1.4.

We believe that the teaching of probability does have fundamental problems partly linked to emotional and psychological considerations. This is demonstrated by our own research (based on historical studies as well as in-depth interviews) and that of Kahneman and Tversky (as summarised in Kahneman et al. 1982) and many others subsequently such as those included in Chernoff and Sriraman (2014), as well as in the proceedings of ICME and ICOTS conferences over the last forty years.

Some of the reasons that teaching probability is inescapably problematic become clear by reference to the historical sources on the emergence of the concept of probability and its different facets, which Devlin addressed when he identified as one genuine difficulty of probability that:

> We need to make it clear that those [classroom experiments] are merely motivational, and that the real power of probability theory comes from reflection on the nature of events in the world and what we can know about them based on the information at our disposal. (Devlin 2014, p. xiii)

Our aim is to present a rationale for supporting the teaching of probability as a means of addressing risk in the school curriculum. In the not too distant past, probability was included in the school curriculum as a subject underpinning statistics as well as an interesting topic in mathematics, like algebra and geometry, though its teaching was usually only to older children. We argue that modern views on risk are a new justification for the teaching of probability, a view, which is also supported in recent publications (e.g., Gigerenzer 2002; Martignon and Krauss 2009).

Devlin (2014, p. xiii) also states that "almost every actual application of probability is in the context of a one-off event, introducing the concept through atypical classroom experiments probably does more harm than good." That means while most applications use probability in a subjectivist theory (SJT) connotation, the concepts are introduced in school by experiments that embed them within a pure frequentist theory (FQT) meaning, sometimes alongside with an a priori or equiprobable theory (APT) setting that allows direct calculations for probabilities in simple cases of equal likelihood. We use the terminology of these acronyms, introduced in Borovcnik and Kapadia (2014) to delineate the key strands, which are important for the teaching of probability. Probability has to be seen "from a pluralistic perspective" as the approach of Chernoff and Sriraman (2014) implies. A comprehensive description of these terms is given in Borovcnik and Kapadia.

There is an axiomatic setting for these three approaches, which justifies the like interpretation and related methods: a reformulation of de Laplace (1812/1951) for APT; the direct characterisation of FQT by von Mises (1919), the indirect theory for FQT by Kolmogorov (1933/1956); and de Finetti (1937/1992) for SJT. Yet the historic evolution has been—and still is—signified by fierce controversies (see Hacking 1975; von Plato 1994). For each of the approaches, there are restrictions: situations with equal probabilities are artificial rather than real; in the real world it is not possible to finish a series of experiments to get the limit of the relative frequencies; the elicitation of personal probabilities always has a subjective connotation, which asks for validation.

There has been some debate in the research community as to whether APT as a priori *theory* may be confused with prior probabilities (used in the Bayes formula); Alternative suggestions brought forward were equiprobability theory (EQT) or Laplace theory (LPT). Yet, we advocate APT as it expresses the fact that within this approach probabilities are given *a priori*, which historically played a great role (one only has to know all possible cases). Our systematic notation was inspired by Çınlar (2011). We assert that our terminology encapsulates in a direct and dynamic way the key issues, which are under discussion within the field of probability education.

Two other terms are used in this paper, adopted from the work of Spiegelhalter (2012) who developed the idea of "micromort" and "microlife" to describe what low probabilities may imply in life. A micromort is linked to an event, which has a 1 in a million chance of leading to death. For example, this is estimated to be the chance of death from walking 27 miles or cycling 28 miles (on average). A microlife is 30 min period within the average life of a young person aged 30 which is around 1,000,000 half hours: a particular event may increase or reduce a life by 1 microlife. For example, a unit of alcohol a day increases life by 1 microlife, while red meat reduces life by 1 microlife, on average (Blastland and Spiegelhalter 2013).

For probability education, in line with the educational review of Hawkins and Kapadia (1984), we support the aims of the Schools Council Project on Statistical Education (1980, p. 27), "Children should become aware of, and appreciate the role of statistics in society [...] the scope of statistics. [...]" to probability and risk: Children should become aware of, and appreciate:

- That probability is a tool for modelling reality;
- That misconceptions on probability affect people's decisions in many situations;
- That the concept of risk is closely related to and dependent upon probability.

We have presented research on the first two aims in Borovcnik and Kapadia (2011c, 2014). The present paper focuses on the third aim.

1.2 Surviving Risk

We describe definitions of risk, discuss rational and behavioural approaches towards risk, and illustrate why the personal strategies to deviate from a rational approach matter for teaching. The on-going presence of risk considerations in the public do not contribute to a calibrated perception of risks and therefore influence our decisions. We conclude this section with a systematic analysis of constituents of situations under risk.

1.2.1 Defining and Describing Risk

By the analysis that follows, the reader can see how risk and probability are intertwined. Sometimes, risk involves different stakeholders and this requires using information, which includes probabilities. We start from the following definition of risk (Borovcnik and Kapadia 2011a, p. 1).

> By risk we understand a situation with inherent uncertainty about the (future) outcomes, which are related to impact. (cost, damage, or benefit)

If we operationalise the inherent uncertainty by probabilities, i.e., the probability of the various possible values of the impact, then we have a random variable. Why should this random variable be related to risk or be called risk? It depends on the context and the *choice*: the choice may be to opt for this "risky situation" and face the inherent uncertainty, or to avoid it. A simple example in the context of games is, to accept the rules of a game and play it (with specific payments as stakes) or do nothing (not to play it), or to change the roles with the other 'player'. Does it pay to accept the risky situation or is it better to do nothing? To take out an insurance policy is a risky situation with reversed signs: do nothing means accept the risk, buy the policy means avoid the risk and make a certain payment. Games and insurance were the two driving forces for the emergence of the concepts in the history of probability (see also Maistrov 1974, p. 5, 173). With this setting, we see that many situations of probability are naturally embedded in the context of risk. We see how risky situations may be described but what is risk and how is it measured?

"The common element between science and everyday [usage of risk] seems to be," according to Borovcnik (2015, p. 114), "an unwanted event that may or may not occur, often related to a low probability and severe consequences." Hansson (2007, p. 2) is specific and distinguishes five definitions of risk:

(1) An *unwanted event* which may or may not occur; […]
(2) The *cause* of an unwanted event which may or may not occur; […]
(3) The *probability* of an unwanted event which may or may not occur; […]
(4) The statistical *expectation value* of an unwanted event which may or may not occur; […]

(5) the fact that a decision is made under conditions of *known [rather than unknown] probabilities ("decision under risk" as opposed to decision under uncertainty*. [...]

The definitions of Hansson (2007) address various components of risky situations: the consequences (impact, adverse events) that may or may not occur as in (1), the probabilities of the adverse event as in (3). It is interesting that both definitions focus only on that one adverse event and model only a small part of the overall situation. (2) Relates risk back to potential causes in order to prevent the adverse outcome in the future. (4) Provides an overall summary figure for the risky situation, while (5) distinguishes between situations when such a summary figure can be calculated or not.

An unwanted event (1) reduces the basic risky situation to the outcome of uncertainty and furthermore usually reduces to only one adverse outcome, which is then called risk. This connotation of risk is meant when we state, "the surgery dissection of lymph nodes bears an increased risk of severe oedema" (see, e.g., Ki et al. 2016). Moreover, (1) does not imply a quantification of the adverse event in the sense of its probability. If the adverse event is related to a very high impact, a rational approach becomes difficult to compare the situations with and without this risk. Even if the increase of complications due to *severe* oedema has a lower probability, it may be regarded as so severe that one would be inclined to avoid this increased risk and not undergo this surgery (and not even analyse this option further).

Risk in the sense of hazard (2) as the "cause" of an unwanted event draws on an assumed link between an exposition of "something" (the hazard) and an event (the negative impact) that may or may not occur (sometimes much) later. This perception of risk requires evaluating the strength of evidence that the present exposition will cause the adverse event later. For example, "a history of sunburns, especially early in life" is listed among the risk factors for skin cancer (CDC, n.d.). Such risks are also called hazards. However, the usage of terms in an everyday context is diverse, inconsistent, and changing as may be seen from a linguistic analysis based on dictionaries. To avoid the risk/hazard here does not mean that the adverse event cannot occur. For example, a person who never exposed skin to sun can still get skin cancer.

If risk is associated with the probability of the adverse event (3), an evaluation of such a situation becomes difficult. For example, if the probability is low, one might tend to neglect it. If several options have to be compared for their risk, conception (3) of risk does not allow the ordering of the options properly as it considers only one component of the situation, namely the probability of the adverse event, and also neglects the impact of the outcome.

The way that risk is perceived in (4) meets the need to derive one figure as a characteristic of risk inherent to a choice (Resnik 1987): it uses both the impact (which could be measured by money or utility) and the probability of the various outcomes of the risky situation. It attributes an expected value to the random variable of all possible outcomes that are related to an option. This facilitates the interpretation of a single risky situation but it also enables to order several choices: i.e., to accept the risky situation or to do nothing; or to choose between several options of which all bear different consequences in terms of impact and its related probability. For example, the risk of taking out a full-coverage policy for the car can be compared to the alternative of taking no insurance (see, e.g., Borovcnik 2006).

The perception of risk in (5) makes a distinction between risk and uncertainty that goes back to Knight (1921): if the probabilities of the various outcomes are known in FQT sense, then (4) allows the evaluation of the risk by one number; if such probabilities are unknown, then such a summary figure cannot be calculated, or has to be based on SJT information (see more below).

In decision theory, when a null hypothesis is tested against an alternative hypothesis (e.g., Batanero and Borovcnik 2016; Neyman and Pearson 1928), we have two types of errors involved: to reject the null hypothesis if it in fact applies (type I) and not to reject it if in fact the alternative hypothesis applies (type II). When we speak of the risk of wrong decisions, then we often use risk in the sense of (3) if we address the size of this risk (alpha or beta); however, we switch to the meaning (1) of risk if we only encompass the event (i.e., to commit an error of that type).[1]

Borovcnik and Kapadia (2011b) mention the following constituents of risk related to health:

> The nature of the risk (and mathematical concepts like probability); the psychological matters involved [see Section 2.2]; the type of situation (treatment, prevention) and information used; the people involved, their aims (purpose) and their inherent criteria. (p. 5505)

These categories generally are suitable to describe risky situations. Once, a decision involves several stakeholders, the following question arises: Can the rational approach (4) be pursued by all of them equally, or do these personal judgements—still following a rational approach—lead to different decisions depending on the stakeholders' role in the decision? For example, it is disputed whether screening programmes for the early detection of cancer will have a positive effect overall. Of course, it cannot have a positive effect for all. Sandblom et al. (2011) evaluate the success of PSA tests and biopsies with prostate cancer.

1.2.2 Decision Making Under Uncertainty

Given that decisions made by individuals under uncertainty affect the impact of outcomes, especially when those decisions interact, a key challenge is to understand how decisions are made. Decision and risk analysis overlap. We would go beyond Bedford and Cook (2001) who state that "decision analysis seems to be the logical extension to risk analysis" as there is hardly any decision that does not involve risk

[1]For an analysis of the meaning of risk in the context of statistical inference, see Borovcnik (2015).

—at least we face the risk to benefit from the non-chosen option more than from that one chosen. Risk analysis, i.e., approaching a risk neutrally, seems much harder than the *comparison* of risks related to several options. In both cases, the future (or the unknown past) is modelled by probabilities.

There are two broad views on decision making under uncertainty—the standard, 'rational' view of 'homo economicus' (e.g., Aktipis and Kurzban 2004), and the 'behavioural' view informed by the psychology literature (e.g., Gigerenzer 2002). In practice, the applicability of each approach depends on the context. Distinguishing between these two views is important since—given that they have contrasting implications for how decisions are made and information is processed—they can yield different conclusions.

The benchmark for the rational approach to choice under uncertainty is the expected utility hypothesis, under which expected utility is maximised by rational players who follow the laws of probability.

> Expected Utility Theory (EUT) states that the decision maker (DM) chooses between risky or uncertain prospects by comparing their expected utility values, i.e., the weighted sums obtained by adding the utility values of outcomes multiplied by their respective probabilities. (Mongin 1997, p. 342)

A key point in the expected utility hypothesis (Friedman and Savage 1952) is that the uncertainty is quantifiable in the sense that the decision-maker can assign (subjective or objective) probability to each possible outcome. The pre-eminent approach to decision making under Knightian uncertainty (5), i.e., when no (or only vague SJT) probabilities can be related to the single outcomes, is to maximise utility under the worst conceivable outcome across all optional decisions (called 'maximin', Resnik 1987).

Behavioural economists and psychologists in contrast contend that people often make decisions using simplistic or "fast and frugal" heuristics rather than using expected utility in the sense of (4) because of limited time, information, and cognitive capacity (Gigerenzer 2002; Todd and Gigerenzer 2000, p. 728). This behavioural view highlights psychological traits that violate the assumptions underlying the 'rational' view. In both approaches, the impact of a decision is measured in utility and not in money terms (Resnik 1987). Since assignments of utility and probability have to be made (which both vary between people), it may be doubted that people in fact use that procedure to evaluate risk and it is difficult what such an overall figure should mean. Bermúdez (2000, p. 744) states:

> [...] the normative theories provide the standards by which practical reasoning ought to be judged (that is to say, [...] what it is rational to do, in a particular situation when one possesses such-and-such information) [...] it is not always possible in practice to assign numerical probabilities to possible outcomes, or to give them numerical desirability ratings [...].

There is indeed strong experimental evidence that people are subject to biases when collecting and processing information under uncertainty. Todd and Gigerenzer (2000, p. 740) refer to the idea of a decision-making toolkit as the product of evolution. The assumption is that the brain learns over time to make the best decision, based on previous experiences. That explains that formal processing only occurs intermittently. Conversely, people are frequently incoherent in assessing, assigning and processing probabilities, even in highly artificial settings where information is complete (Kahneman et al. 1982), and find it particularly difficult to assess probabilities which are very low or very high, partly because there is less scope for learning in these cases (Blastland and Spiegelhalter 2013). People's judgement is also influenced by the fact whether the situation is a winning or losing situation; Kahneman and Tversky (1979) show that in situations of gains, people seem to be risk-aversive, in situations of loss, they seem to be risk-prone. However, Borovcnik (2015) finds even more complexities in a rational analysis of decisions as some winning situations may also be regarded as losing (!) situations and the optimal decision to rational criteria can switch if one applies a decision only once or can repeat the decision often. In the real world when information is imperfect, such as when deciding whether to have surgery or cancer screening, people often use rough heuristics, or go on first impressions, appearances, or gut instinct or intuition. Aikman et al. (2011, p. 4803) highlight evidence of this by classifying the following deviations of rational behaviour as detected by the experiments by Kahneman and Tversky (see also Kapadia 2011):

- Perceptions of risk have been shown to be unduly influenced by recent or personal experience [Availability].
- Decisions under uncertainty have been shown to be sensitive to the way questions are presented [Anchoring].
- People tend to follow the actions of others [Herding].
- People have excessive faith in their own judgements and are subject to confirmation bias and wishful thinking [Overconfidence and Confirmation Bias].

The first point is that the availability heuristic of Kahneman and Tversky (1973) is based on personal experience such as the recall of events from memory and probabilities are judged by the ease of recall. For example, after catastrophes, people overreact immediately and then forget about bad consequences with time (Aikman et al. 2011). The second point on anchoring refers to the sensitivity of language and the formulation so that some cues get an undue importance as people's assignment of probabilities is influenced by given or perceived cues (Tversky and Kahneman 1974). Herding means that probabilities are judged according to what others believe; key players are particularly important in this dynamic and the herding effect that can generate panics. Overconfidence implies that probability judgements are prone to hope and emotions. The overconfidence in our judgement (Russo and Schoemaker 1989) mentioned in the last point means failing to collect factual information because we are too sure of our assumptions and opinions. And overconfidence is paired with a "confirmation bias" that leads us to seek and acknowledge only confirming data rather than systematically searching for interrelations and data (p. 103). The interaction between overconfidence and confirmation bias may be mutually reinforcing, for example, in the development of financial bubbles. All these effects establish a challenge for teaching.

1.2.3 Recent Scares

A key challenge for education is to understand how decisions are made. In terms of the real-life examples, there are several topical issues from this century. These cover many aspects of life, including economic and financial issues, health and pharmacological concerns, business, gambling, crime, and sport. Here we deal with three examples: flu epidemics, the problems from ash clouds, and dealing with bovine spongiform encephalopathy (BSE). Common to these situations is that there is a threat of severe impact but no one knows about the probability of the occurrence of the adverse events, i.e., risk is involved as in (1) or (5). The underlying issues are quite deep and complex, which may be seen from the analysis in Borovcnik (2015). It is hard to disbelieve experts, but one should always remember that they have their own values and prejudices, not least relating to their self-perceived expertise (Borovcnik and Kapadia 2011a, b).

With regards to flu epidemics, scares have arisen over the last few centuries. The scale of concern has increased markedly in the last few decades with increasing travel and enhanced communication, but it is not clear that the responses have improved in terms of a better understanding of probability. Indeed, it could be argued that a poor understanding of probability has led to a worse response politically, certainly in terms of the extra money spent, for example, in stock-piling medicines which turn out not to be used. This trend is perhaps encouraged by pharmacological companies, who can make large profits from "scare-mongering" (Blastland and Spiegelhalter 2013).

The ash-cloud in 2010 had a significant adverse economic effect (EuroControl, n.d.). Flights of aeroplanes were stopped and only resumed after great commercial pressure was applied by airlines. The BSE controversy in the United Kingdom led to mass culls, which some now believe were unnecessary as it might be that all positive tested BSE cattle were actually false positives (see Dubben and Beck-Bornholdt 2010, p. 64). Two underlying issues are the confusion between probabilities and impact, and the difficulties to update probabilities with new information.

The confusion between probabilities and impact, as measured by utility arises when probability relates to everyday situations and not much comparable data is available so that FQT does not give a sufficient basis for an estimate of the probability of the event of interest. Thus, a situation with a huge impact biases the judgement of risk towards (1) rather than it is done by the related probability by (3) or by a weighting procedure by (4) in the form of an expected value.

Second, updating risks by new data involves—on the rational side—Bayes formula. That means, beyond the difficulties already mentioned, a relatively complex mathematical algorithm has to be applied to evaluate the new risk. To interpret the notions of false positives and positive predictive values in the context of medical diagnosis, FQT is of no help as relevant frequentist information is missing. Mathematical statements about conditional probabilities are quite sophisticated so that people tend to oversimplify matters (Todd and Gigerenzer 2000). Knowing that there is a small probability of a positive medical test given that the disease is absent and accepting the basis of medical diagnosing, people tend to grossly overestimate the positive predictive value, i.e., the probability of the disease given the test is positive (partly by a general trust in the benefits of modern medicine).

The risks to life can be greatly exaggerated in the media, leading to public fears and concerns, with over-cautious responses from politicians. Even if the risk is very low, no politician can afford to 'do nothing' as his/her reputation would be irrevocably damaged if the highly unlikely event does indeed occur. For example, a dubious, haphazard, and overpriced purchase of breathing masks against swine flu (H1N1) by the Austrian Ministry of Health in 2005 led to an on-going dispute about the rationality of the decision (critical report of the Austrian Court of Audit in 2010). The then Minister of Health in reply to a parliamentary query to him answered (Austrian Ministry of Health 2012):

> Based on the experience with aviation flue (H1N1) it is to state that the pressure of the media and the inconsistent advice of the experts have exerted a more than critical part in finding the decision […]. (p. 2)

The major point to make is that risk is lower now than at any point in the past. Tragedies are far less common, as shown in the marked rise in life-span and general well-being, with dramatic falls in infant mortality. We note that—despite popular perceptions—life has actually become much less risky compared to even a century ago. One only has to look at the tables of micromorts of various events in Blastland and Spiegelhalter (2013) to support this. Two examples are: life expectancy has increased by 13.5 years for males between 1910 and 2010; infant mortality in England and Wales is one tenth of infant mortality in the world as a whole. Our main concern here is whether the underlying probability is well understood, and how a rational perception of the context and the evaluation of risks and probabilities can be improved.

1.3 Emergence of Probability

This section deals with the historic emergence of probability concepts re-visited from the point of risk to show how risk may be seen as driving force behind the historical development. The second part on emerging concepts is devoted to probabilistic thinking seen from the perspective of risk, which may enrich understanding and teaching of probability.

1.3.1 Historic Development Re-visited from the Point of Risk

The early historic development of probability, as discussed in Borovcnik and Kapadia (2014) is re-interpreted from the point of risk. In fact, there were some attempts to judge the likelihood of the outcomes as Cicero in his *De divinatione* (44 B.C.) expresses in the following argument on four dice:

> They are entirely fortuitous you say? Come! Come! Do you really mean that? [...] When the four dice produce the venus-throw you may talk of accident: but suppose you made a hundred casts and the venus-throw appeared a hundred times; could you call that accidental? (David 1962, p. 24)

Cicero addresses the *risk* of getting a specific result in the sense of a small probability (3). Interestingly, Cicero compares two situations related to the risk of the "adverse event" and does not attempt to give an absolute value, for example, by weighing the risk by combinatorial multiplicity. Throwing dice or casting lots were recognised as tools to decide fairly or to explore god's will for a decision. When it was difficult to decide between two persons, lots were cast for them. In modern words, risk was distributed evenly or fairly. In those times, the risk of adverse events was connected to god's will. It may have made no sense to evaluate such risks.

Cardano discusses how equally easy it is to get any of the numbers in casting one die; this may be interpreted as an attempt to an intuitive risk calculation in the sense of (4).

> One half the total number of faces always represents equality; thus the chances are equal that a given point will turn up in three throws, [...] The wagers therefore are laid in accordance with this equality if the die is honest. (Cardano 1526, quoted from David 1962, p. 58)

De Méré's problem. Two games are compared. In game 1 the player wins if there is at least one 'six' in four throws of a die; in game 2 the player wins if there is at least one 'double six' in 24 throws of two dice. For each game, Fermat and Pascal (1654) compare the risks of the two players: is it better than equal? They already evaluate the risk by its combinatorial multiplicity (more cases/reasons for the six than against the six; less for the double six). In comparing both games, de Méré uses as a criterion 4 cases of 6 in game 1 and 24 cases of 36 in game 2. The problem, however, may have been more specifically to clarify how to count cases.

Division of stakes. If Peter and Paul are competing, how should they share the stakes if a series of 11 is ended at a time when Peter has 4 while Paul has 3 points (6 points are sufficient to win the series)? This example is typical for the exchange of money and risk since both players have a risk to lose the whole series (and not as in an insurance contract where one stakeholder is completely free of risk on the outset). However, their risk differs. How to measure this and convert the differing degree of risk into money? The situation resembles today's purchase of a future at the stock market. Of all 16 cases 11 lead to a win of Peter and only 5 to a win of Paul, thus the risk to lose is 5:11 and the stakes should be divided by 11:5. As the real cases do not have the same weight, Pascal and Fermat's ingenious trick was the introduction of imaginary cases that have no real counterpart to ensure equal (weight for the) cases to apply APT probability. From our modern perspective, we would state that the modelling by the probabilities has made all other arguments about fair division of stakes obsolete.

Huygens refers to situations of equal chances and later only to having p chances. And he derives no specific probabilities but proportions of stakes or a price of the situation.

> Proposition 3: to have p chances of obtaining a and q of obtaining b, chances being equal, is worth $\frac{pa+qb}{p+q}$. (Huygens 1657, quoted from David 1962, p. 116)

In his analysis of the historical development, Shafer (1996, p. 16) explicitly refers to the non-probabilistic embedding of the emergent concepts:

> Pascal, Fermat, and Huygens were concerned with a problem of equity, not a problem of probability. They were pricing gambles, not evaluating evidence or argument. But it did not take people long to draw the analogy.

It was with Bernoulli that the evaluation of relative possibilities (as espoused by Pascal and Fermat), or the value of a situation (as suggested Huygens) was transformed to a probability scale between 0 and 1. By his Law of Large Numbers, Bernoulli also justified the use of FQT information to estimate probabilities. That justified Huygens' economic value of an uncertain situation for the calculation of risks in the insurance problem in the sense of (4), which always has been performed by cases that have actually occurred.

In the next section, we investigate how probabilistic thinking can be described as a construct, or how probabilistic thinking emerges.

1.3.2 Probabilistic Thinking and Literacy

Literacy in a general sense is defined to be the ability to read and write; statistical literacy is the ability to read and write statistics. Thinking in connection with a subject means that thinking is shaped by this subject. There is a great overlap between literacy and thinking (see Batanero and Borovcnik 2016). While the number and detail of definitions or descriptions of statistical literacy is abundant, there is little to find on probability literacy. Gal (2005) extended his ideas on statistical literacy to develop a model of probability literacy, which includes knowledge and dispositional elements: Critically evaluate probabilistic information embedded in context.

> Being literate about probability-related matters requires that a person develops some knowledge not only of relevant big ideas, ways to figure probabilities, and the language of chance, but also of the role of probabilistic processes and communications in the world. Knowledge regarding context [...]. (p. 52)

Gal includes in probabilistic literacy abilities to understand basic concepts of probability and use probability arguments properly. Beyond abilities to do that, Gal also views attitudes and beliefs, such as a critical stance towards probabilistic statements, as vital.

Probabilistic (stochastic) thinking goes back to Heitele (1975). Borovcnik (1997) built on Heitele's ideas and tried to go beyond a textbook-like listing of the principal concepts of probability by describing fundamental ideas (rather than concepts), which all deal with the peculiar type of information embedded in probability statements. In his paper on probabilistic thinking, Borovcnik (2006) illustrates this kind of thinking by a few paradigmatic examples to illustrate essential aspects of probabilistic thinking. Apart from an attempt to synthesise these ideas and complement them by higher-ordered categories of thinking by Batanero and Borovcnik (2016) and Borovcnik (2016), a convincing model of probabilistic thinking still is an open research question as Batanero et al. (2016) stated. We re-order and re-name their categories and embed the underlying ideas in the context of risk.

- Basic categories of probabilistic thinking: competing intuitions and strategies
  1. The ability to discriminate between randomness and causality;
  2. The ability to balance between psychological and formal elements of probability;
  3. The understanding that direct criteria for success in probability are missing;
  4. The understanding that criteria for reflecting on a random situation differ from those which may be applied in selecting a decision;
- Higher-ordered categories of probabilistic thinking: secondary intuitions
  5. The awareness of the theoretical character of probability (combining SJT, APT and FQT aspects) including the case of small probabilities;
  6. The awareness of conditional probability and its asymmetry;
  7. The development of concepts building probabilistic evidence (such as probabilistic dependence as conceptualised by the correlation coefficient).

1. *The ability to distinguish between causality and randomness*. This ability lies at the heart of probability. Until the last few centuries, the lack of knowledge has been seen as critical in random events (David 1962). Causal approaches have the advantage that once one recognises the mechanism of how a specific cause establishes the effect, one can predict the future. For de Laplace (1812/1951), probability was restricted to such cases where the exact causes are unknown. His famous demon in the background would know these causes and thus would not need probability (Borovcnik and Kapadia 2014). Nowadays, the opposite paradigm leads theoretical physicists to eliminate causality (Styer 2000). Most people follow a more or less intuitively impregnated causal approach to the world, based on simple interrelations between cause and effect (see Borovcnik and Bentz 1990). The judgement of hazards in the sense of (2) in Sect. 1.2, i.e., risk factors that may cause the adverse event is fragile and highly dependent on a causal link between the risk factor and the event. For example, cancer of the cervix may be caused by the HP virus. If the vaccination against HPV becomes more reliable to hinder the infection of the cervix by this virus, people will seek the vaccination

against it. That the same cancer may be induced by many other factors such as smoking or long-term hormonal contraceptive use (see WHO 2016) that can be related to lifestyle does not have the same causal connotation and thus consideration for change of lifestyle does not play a key alternative for the vaccination. Yet, it may be the better option.

2. *The ability to balance between psychological and formal elements of probability*. Balancing between psychological and formal elements is required when using a personal scale of probability. Psychological factors are related to archetypical deep-seated behaviour (see Batanero and Borovcnik 2016). Formal elements of probability and probability assignments are the result of an intellectual approach to randomness. This is difficult and remains unstable as it is hard to "see" the interrelations, which can still be forgotten. That is why people prefer short-cuts based on simplification (Todd and Gigerenzer 2000). Risk often involves negative impact (or implications) and is associated with fear (worst-case scenarios for nuclear reactors, economic crises, etc.) and measurements for prevention. The usual alternatives from psychology are fight or flight; neither is suitable for reflective consideration of the situation under scrutiny. Aikman et al. (2011) provide examples about the effect of psychological factors such as availability, anchoring, herding, overconfidence and estimation bias on decision-making under risk (see Sect. 1.2.2).
3. *The understanding that direct criteria for success in probability are missing*. It is hard to accept that there are no direct success criteria in random situations. We can, for example, come to a decision on a wrong basis (choose the numbers in the state lottery because we dreamed of them over night) and yet have success (have all six right). Who can convince that person of the (general) inadequacy of the used strategy? The usual theoretical answer is to repeat the decision and check the success of the strategy in the long run. That does not work when the situations are perceived as one-off, which is most often the case in decision making under risk. A common example is the prevalent habit to take out an insurance policy on almost every aspect of our life. Part of the success criteria is that the person would say that the adverse event did not affect him as he has taken out the policy as if the policy is a protection shield; it is like saying "it did not rain today because I took an umbrella". This is an argument that would override any monetary calculation for evaluating the rationality of the decision in favour of the insurance; it resembles the preference to enter the class with the left foot first when the student faces an oral exam to improve one's chances (Borovcnik and Bentz 1990).
4. *The understanding that criteria for reflecting on a random situation differ from those which may be applied in selecting a decision*. People tend to use different criteria depending on whether they reflect on a random situation or select a decision (Borovcnik and Bentz 1990). To come to a conclusion that the probability for an event has some numerical value on the scale between 0 and 1 does not suffice for a decision. The ranking of (adverse events) on a probability scale as is done in the definition (3) in Sect. 1.2 for risk, may not allow for a reasonable preference as it might end up in a draw if both (of two) options have the

same probability. Which side to share then (assumed that choosing the right option, one would get a prize)? The reflection does not provide a justification for a decision. In such a situation the person is free to choose and could well have strong preferences for one. In such a case, the person might be prepared to accept a further "payment" to be able to freely choose and select the preferred one (Borovcnik and Bentz 1990). The reason is that the reflection ends up with an abstract probability while the reasons for a preference for one or the other options are vivid.

5. *The awareness of the theoretical character of probability* (combining SJT, APT and FQT aspects) including the case of small probabilities. There have been many discussions on the genuine theoretical character of probability and that it distorts the concept if one reduces it to one aspect. Steinbring (1991) states,

> [We] use the term theoretical character of mathematical knowledge to denote this dependency of the concept on the theory's level of development. [...] it is only the theory which explains and develops the meaning of fundamental mathematical concepts. (p. 144)

   Spiegelhalter (2014) provides several answers to his question on the meaning of probability: "[Probability is a]'virtual' number – not directly measurable, Probability does not exist, [and] Any description is a metaphor" (p. 9); the properties named describe probability essentially as non-empirical, though this is seen as contentious by some. Devlin (2014) speaks about the negative consequences of reducing probability to the FQT aspect while many applications are one-off. Such a one-off interpretation is most appropriate in many risky situations, which are also connected to small probabilities. Usually, such values stem from other probability (risk) calculations based on assumptions (such as independence) and lack empirical foundation. Historical approaches have been to ignore probabilities smaller than a threshold (such as 10^{-6}, see Borovcnik and Kapadia 2014). Anyway, it is not possible to validate such small probabilities by data (Batanero and Borovcnik 2016).

6. *The awareness of conditional probability and its asymmetry*. Probabilistic reasoning also requires (conditional) probabilities if new information becomes available. For example, the risk to have a specific disease changes when the intermediary results of medical data are collected to prepare a decision about what to do. The probability of a positive result of the mammogram is high given that a woman has breast cancer will not be the same as the reverse conditional probability that she has breast cancer given that the mammogram is positive. Indeed in typical screening, under 10% positive mammograms are actually correct—the rest are false positives (Blastland and Spiegelhalter 2013). It may also seem paradoxical that the conditional probability for an individual may be dependent on a subgroup to which she belongs (a subgroup with a special risk such as age, gender, ethnicity, etc.; Gigerenzer 2002). The direction between events in $P(A|B)$ may be interpreted by time (earlier, later) by cause (cause, effect), or by indication (symptom, status). Depending on such connotations, the interpretation and the acceptance of the values of these probabilities may vary and be unstable in a person's perception (Borovcnik and Bentz 1990). It may be

regarded as a great step forward in probabilistic literacy to recognise that conditional probability covers all these aspects. Conditional probabilities play also an essential role in the interpretation of statistical tests (type-I or type-II errors, as discussed in Batanero and Borovcnik 2016).

7. *The development of concepts building probabilistic evidence* (such as probabilistic dependence as conceptualised by the correlation coefficient). Correlation encompasses a much weaker form of relationship than functional dependence. If a correlation coefficient between variables is high, this does not show causal interrelations between these variables. It only indicates that connections between the variables may be found by a well-designed empirical study to detect either confounding effects (other variables blurring the co-relation) or to corroborate a hypothesis between the variables under scrutiny. It is very subtle how correlation measures interrelations that are beyond causal connections. The complexity asks for simplifications such as rules of thumb (a correlation larger than 0.3 is at least intermediate, larger than 0.7 is strong). However, correlation can be increased, generated, or even changed in sign by other variables. One famous example for such phenomena is the Simpson paradox (see Batanero and Borovcnik 2016). If one has no data about such potential confounders, no such follow-up investigation is possible. Correlation and association become vital for the judgement of risks if they are related to hazards as in definition (2) of risk. Judging whether smoking is a risk factor for conceiving lung cancer depends on understanding what association really means. The discussion about potential risk factors would be much easier if correlation and association were well understood. These concepts are—besides statistical significance tests—at the core of what we call empirical evidence.

1.4 Conclusions

Probability can help to explore risky situations and make the final judgements more rational in the sense of definition (4) in Sect. 1.2 (see Borovcnik 2006; Borovcnik and Kapadia 2011c). The usual contract in games and in the insurance situation is signified by a swap of risks. One stakeholder who has no risk offers to the other to 'take over the risk' who then asks for monetary compensation. To fix the prices for such an exchange, it is essential to determine the probabilities of all (not only the adverse) outcomes—either by APT, FQT or SJT information and calculate an economic value of the diversity of outcomes. That is Huygen's value of an uncertain situation (see Shafer 1996), which is the expected value of the risky situation in the sense of our definition (4) in Sect. 1.2 for risk. Similarly, probabilistic modelling is applied to derive the Black-Scholes equations and solve them to determine the price of futures at the stock market (see Hull 2009), which forms the basement for modern stock exchange.

Historical struggles provide a further valuable orientation. While empirical research about how people think and how successful teaching programmes have been helpful to improve teaching plans, we should not lose sight of key concepts and strategies from the past. The tradition of Felix Klein (1908) and his advocacy of strong links between the historical, theoretical and philosophical ideas of mathematics and mathematics education has been renewed by a new translation of his famous Elementary Mathematics (now named “from a higher standpoint” rather than “from an advanced standpoint”), which was presented at ICME 13 in Hamburg. One key lesson from history is that probability has always been a *pluralistic* concept and has drawn its meaning from the interplay of its interpretations (Borovcnik and Kapadia 2014).

In conclusion, we repeat our plea above that multiple approaches to probability including APT, FQT and SJT are essential in the teaching of probability. We have discussed probabilistic thinking in the context of risk. Risk and emotion interfere with probabilistic thinking. Risk is a key aspect of life and probability is the mathematical tool to address risk.

References

Aikman, D., Barrett, P., Kapadia, S., King, M., Proudman, J., Taylor, T., et al. (2011). Uncertainty in macroeconomic policy-making: Art or science? *Philosophical Transactions of the Royal Society, A, 369,* 4798–4817.

Aktipis, C. A., & Kurzban, R. O. (2004). Is homo economicus extinct? Vernon Smith, Daniel Kahneman and the evolutionary perspective. In R. Koppl (Ed.), *Evolutionary psychology and economic theory. Advances in Austrian economics* (Vol. 7, pp. 135–153). Oxford: Elsevier.

Austrian Ministry of Health. (2012). Reply to the parliamentary query related to the purchase of breathing masks as precaution against the potential danger of an epidemic outbreak of swine flu H1N1 in 2005 by the Ministry of Health. *Offenes Parlament*. offenesparlament.at/gesetze/XXIV/J_12400/. Accessed 15 May 2017.

Batanero, M., & Borovcnik, M. (2016). *Statistics and probability in high school.* Rotterdam: Sense Publishers.

Batanero, C., Chernoff, E., Engel, J. Lee, H., & Sánchez, E. (2016). *Research on teaching and learning probability. ICME-13 Topical Surveys.* Cham, Switzerland: Springer online. doi.org/10.1007/978-3-319-31625-3_1.

Bedford, T., & Cook, R. (2001). *Probabilistic risk analysis: Foundations and methods.* Cambridge: Cambridge University Press.

Ben-Zvi, D., & Garfield, J. (Eds.). (2004). *The challenge of developing statistical literacy, reasoning and thinking*. Dordrecht, The Netherlands: Kluwer.

Bermúdez, J. L. (2000). Rationality, logic, and fast and frugal heuristics. *Behavioral and Brain Sciences, 23,* 744–745.

Blastland, M., & Spiegelhalter, D. J. (2013). *The norm chronicles.* London: Profile Books.

Borovcnik, M. (1997). Fundamentale Ideen als Organisationsprinzip in der Mathematik-Didaktik [Fundamental ideas as an organising principle in the didactics of mathematics]. *Schriftenreihe zur Didaktik der Mathematik der Österreichischen Mathematischen Gesellschaft (ÖMG), 27,* 17–32.

Borovcnik, M. (2006). Probabilistic and statistical thinking. In M. Bosch (Ed.), *Proceedings of the Fourth Congress of the European Society for Research in Mathematics Education* (pp. 484–506). Barcelona: ERME.

Borovcnik, M. (2015). Risk and decision making: The "logic" of probability. *The Mathematics Enthusiast, 12*(1, 2 & 3), 113–139.

Borovcnik, M. (2016). Probabilistic thinking and probability literacy in the context of risk. *Educação Matemática Pesquisa, 18*(3), 1491–1516.

Borovcnik, M., & Bentz, H.-J. (1990). *Intuitive Vorstellungen von Wahrscheinlichkeitskonzepten: Fragebögen und Tiefeninterviews.* [Intuitive conceptions of probabilistic concepts: Questionnaire and in-depth interviews.] Technical Report. Klagenfurt: Klagenfurt University.

Borovcnik, M., & Kapadia, R. (2011a). Risk in health: More information and more uncertainty. In *Proceedings of the IASE Satellite Conference on Statistics Education and Outreach*. Voorburg: International Statistical Institute. https://iase-web.org/documents/papers/sat2011/IASE2011PaperPoster2Borovcnik&Kapadia.pdf. Accessed 15 May 2017.

Borovcnik, M., & Kapadia, R. (2011b). Determinants of decision-making in risky situations. In *Proceedings of the 58th World Statistics Congress*. Voorburg: International Statistical Institute. http://2011.isiproceedings.org/papers/950138.pdf. Accessed 15 May 2017.

Borovcnik, M., & Kapadia, R. (2011c). Modelling in probability and statistics—Key ideas and innovative examples. In J. Maaß & J. O'Donoghue (Eds.), *Real-world problems for secondary school students—Case studies* (pp. 1–44). Rotterdam: Sense Publishers.

Borovcnik, M., Kapadia, R. (2014). A historical and philosophical perspective on probability. In E. J. Chernoff & B. Sriraman (Eds.), *Probabilistic thinking: Presenting plural perspectives. Advances in mathematics education* (Vol. 7, pp. 34). Berlin: Springer.

CDC. (n.d.). *What are the risk factors for skin cancer?* Whashington: Centers for Disease Control and Prevention. www.cdc.gov/cancer/skin/basic_info/risk_factors.htm. Accessed 15 May 2017.

Chernoff, E., & Sriraman, B. (Eds.). (2014). *Probabilistic thinking: Presenting plural perspectives. Advances in mathematics education* (Vol. 7). New York: Springer.

Çınlar, E. (2011). *Probability and stochastics*. New York: Springer.

David, F. N. (1962). *Games, gods and gambling*. London: Griffin.

de Fermat, P., & Pascal, B. (1654/1962). *Correspondence*, publ. 1679 in Toulouse; English translation in F. N. David (Eds.), *Games, gods and gambling* (pp. 229–253). London: Griffin.

de Finetti, B. (1937/1992). La prévision: ses lois logiques, ses sources subjectives. *Annales Institut Henri Poincaré, 7*, 1–68. Foresight: Its logical laws, its subjective sources (chapters 1–4 translated by H. E. Kyburg Jr.) in S. Kotz, & N. L. Johnson, *Breakthroughs in statistics. Vol. I. Foundations and basic theory* (pp. 134–174). New York, Berlin: Springer.

de Laplace, P. S. (1812/1951). Essai philosophique sur les probabilités. *Journal de l'École Polytechnique VII/VIII*, 140–172. English translation: *A philosophical essay on probabilities* (extended version). New York: Dover.

Devlin, K. (2014). The most common misconception about probability? In E. J. Chernoff & B. Sriraman (Eds.), *Probabilistic thinking: Presenting plural perspectives. Advances in mathematics education* (Vol. 7, pp. ix–xiii). Berlin: Springer.

Dubben, H.-H., & Beck-Bornholdt, H.-P. (2010). *Mit an Sicherheit grenzender Wahrscheinlichkeit. Logisches Denken und Zufall.* (With a probability coming close to certainty. Logical thinking and randomness.) Reinbek: Rowohlt.

EuroControl. (n.d.). *Ash-cloud of April and May 2010: Impact on air traffic*. Doc394v1.0 28/6/10. Statistics and Forecast (STATFOR). www.eurocontrol.int/sites/default/files/content/documents/official-documents/facts-and-figures/statfor/ash-impact-air-traffic-2010.pdf. Accessed 15 May 2917.

Friedman, M., & Savage, L. J. (1952). The expected-utility hypothesis and the measurability of utility. *Journal of Political Economy, 60*, 463–474.

Gal, I. (2005). Towards "probability literacy" for all citizens: Building blocks and instructional dilemmas. In G. A. Jones (Ed.), *Exploring probability in school. Challenges for teaching and learning* (pp. 39–63). Dordrecht: Kluwer.

Gigerenzer, G. (2002). *Calculated risks: how to know when numbers deceive you*. New York: Simon & Schuster.

Hacking, I. (1975). *The emergence of probability*. Cambridge: Cambridge University Press.

Hansson, S. O. (2007). Risk. In E. N. Zalta (Ed.), *Stanford encyclopedia of science*. plato.stanford.edu/archives/spr2014/entries/risk/. Accessed 15 May 2917.

Hawkins, A. E., & Kapadia, R. (1984). Children's conceptions of probability—A psychological and pedagogical review. *Educational Studies in Mathematics, 15*(4), 349–377.

Heitele, D. (1975). An epistemological view on fundamental stochastic ideas. *Educational Studies in Mathematics, 6*(2), 187–205.

Hull, J. C. (2009). *Options, futures and other derivatives* (7th ed.). Upper Saddle River, NJ: Prentice Hall.

Kahneman, D., & Tversky, A. (1979). Prospect theory: An analysis of decision under risk. *Econometrica, 47*(2), 263–292.

Kahneman, D., & Tversky, A. (1973). Availability: A heuristic for judging frequency and probability. *Cognitive Psychology, 5*(2), 207–232.

Kahneman, D., Slovic, P., & Tversky, A. (1982). *Judgment under uncertainty: Heuristics and biases*. Cambridge: Cambridge University Press.

Kapadia, R. (2011). Risk in science and society: towards new pedagogies of probability Bayes rules. *Selçuk Journal of Applied Mathematics, Special Issue,* 73–82.

Ki, E. Y., Park, J. S., Lee, K. H., & Hur, S. Y. (2016). Incidence and risk factors of lower extremity lymphedema after gynecologic surgery in ovarian cancer. *International Journal of Gynecological Cancer, 26*(7), 1327–1332.

Klein, F. (1908/2004/2016). *Elementary mathematics from a higher standpoint. Vol. I: Arithmetic, algebra, analysis, Vol. II: Geometry, Vol. III: Precision mathematics and approximation mathematics*. Berlin: Springer. New translation of the classic volumes. *Elementary mathematics from an advanced standpoint. Vol. 1: Arithmetic, Algebra, Analysis, Vol. 2: Geometry*. New York: Dover. Translation of *Elementare Mathematik von höherem Standpunkte aus* (3rd Edit. 1924–28; 1st Edit. 1908). Leipzig: Teubner.

Knight, F. H. (1921). *Risk, uncertainty, and profit*. Boston, MA: Hart, Schaffner & Marx; Houghton Mifflin Company.

Kolmogorov, A. N. (1933/1956). *Foundations of the theory of probability*. New York: Chelsea.

Maistrov, L. E. (1974). *Probability theory: A historical sketch*. New York: Academic Press.

Martignon, L., & Krauss, S. (2009). Hands-on activities for fourth graders: A tool box for decision-making and reckoning with risk. *International Electronic Journal of Mathematics Education (IEJME), 4*(3). In M. Borovcnik & R. Kapadia (Eds.), *Special issue on "Research and Developments in Probability Education"*, 227–258.

Mongin, P. (1997). Expected utility theory. In J. Davis, W. Hands, & U. Maki (Eds.), *Handbook of economic methodology* (pp. 342–350). London: Edward Elgar.

Neyman, J., & Pearson, E. S. (1928/1967). On the use and interpretation of certain test criteria for purposes of statistical inference, part I and II. *Biometrika, 20 A*, 175–240; 263–294. Reprinted as *Joint statistical papers, Nr. 1 and 2*. Berkeley: University of California Press.

Resnik, M. D. (1987) *Choices: An introduction to decision theory*. University of Minneapolis: Minnesota Press.

Russo, J. E., & Schoemaker, P. J. H. (1989). *Decision traps: Ten barriers to brilliant decision making and how to overcome them*. New York, NY: Doubleday.

Sandblom, G., Varenhorst, E., Rosell, J. Löfman, O., & Carlsson, P. (2011). Randomised prostate cancer screening trial: 20year follow-up. *British Medical Journal*, 342.

Schools Council Statistical Education Project 11–16 (1980). *Teaching statistics 11-16: Statistics in your world*. (a handbook and a series of booklets). Slough: Foulsham.

Shafer, G. (1996). The significance of Jacob Bernoulli's Ars Conjectandi for the philosophy of probability today. *Journal of Econometrics, 75,* 15–32.

Spiegelhalter, D. (2012). Using speed of ageing and "microlives" to communicate the effects of lifetime habits and environment. *British Medical Journal*, 345:e8223. doi.org/10.1136/bmj.e8223.

Spiegelhalter, D. (2014, April). *What can education learn from real-world communication of risk and uncertainty?* Invited lecture at the Eight British Congress on Mathematical Education, Nottingham.

Steinbring, H. (1991). The theoretical nature of probability in the classroom. In R. Kapadia & M. Borovcnik (Eds.), *Chance encounters: Probability in education* (pp. 135–167). Dordrecht: Kluwer.

Styer, D. F. (2000). *The strange world of quantum mechanics*. Cambridge: Cambridge University Press.

Todd, P. M., & Gigerenzer, G. (2000). Précis of Simple heuristics that make us smart. *Behavioral and Brain Sciences, 23,* 727–780.

Tversky, A., & Kahneman, D. (1974). Judgment under uncertainty: Heuristics and biases. *Science, 185,* 1124–1130.

von Mises, R. (1919). Grundlagen der Wahrscheinlichkeitsrechnung [Foundations of probability theory]. *Mathematische Zeitschrift, 5,* 52–99.

von Plato, J. (1994). *Creating modern probability: its mathematics, physics and philosophy in historical perspective*. Cambridge: Cambridge University Press.

WHO. (2016). *Human papillomavirus and related diseases report. Austria*. WHO: ICO Information Centre on HPV and Cancer. www.hpvcentre.net/statistics/reports/AUT.pdf. Accessed 15 May 2917.

Chapter 2
Language and Lexical Ambiguity in the Probability Register

Adam Molnar

Abstract Because objects in probability are not physically visible, humans must describe them through language. The descriptive words and symbols, taken from everyday life, mathematics, and other sources, form the probability register. Many terms do not have precisely the same meaning in probability as in other registers, creating confusion called lexical ambiguity. This chapter develops the probability register, starting from its basis in mathematics. Specific terms examined include random, independence, likely, probable, and other scale words. Variations for students learning probability in a non-native language are included. It is suggested that reducing lexical ambiguity requires research into misconceptions leading to better formed definitions and representations, followed by teachers' explicit attention to vocabulary.

Keywords Linguistics · Lexical ambiguity · Register · Random
Independence

2.1 Introduction

"Draw a box", a question might ask on a worksheet in a probability class. What responses might a student offer to this three-word request? If the class had just covered pseudo-random numbers drawn by computers, the student might draw a box and label the vertical edges at 0 and 1, forming a uniform distribution. A student learning about multi-dimensional probability distributions might draw a three-dimensional box. If the preceding problem had involved numbers, the student might place four lines around the answer. For a more unusual answer, imagine a student who really likes automobile racing and decides to sketch a pit box, the place where automobiles are serviced during a race, before thinking about the probability problem.

A. Molnar (✉)
Department of Statistics, Oklahoma State University, 301 Math Sciences, Stillwater, OK 74078, USA
e-mail: adam.molnar@okstate.edu

C. Batanero and E. J. Chernoff (eds.), *Teaching and Learning Stochastics*,
ICME-13 Monographs, https://doi.org/10.1007/978-3-319-72871-1_2

The point of these drawings is not to explore understandings of the box; this example introduces the study of language as applied to probability. Problem context makes a huge difference. For instance, most people would never consider a pit box, but a group of race car drivers might refer to that meaning first. Confusion about context and symbolic meaning leads to misconceptions and errors. One research question in probability education is how people understand (or fail to understand) a specific word related to probability in different contexts. In Sect. 2.2, a theoretical framework from linguistics is developed, based on the concept of *register*, a variant of language applied in a specific setting (Halliday 1975). People often apply words differently in registers for mathematics and probability than in the everyday conversation register. In Sect. 2.3, two examples of close study are presented, the important terms *random* and *independence*. Another relevant research area is learning in a secondary register, since many students study probability in a language other than their first native tongue. Studies related to learners studying probability outside their native language are discussed in Sect. 2.4. Finally, conclusions are summarized in Sect. 2.5.

2.2 Theoretical Background

As written probability includes both words and symbols, the background in Sect. 2.2.1 includes topics from the related disciplines of semiotics (study of signs) and linguistics (study of language); for simplicity, topics are combined and generally labelled as linguistics. Because solving probability problems applies many symbols and words from mathematics, the mathematics register is discussed in Sect. 2.2.2. Section 2.2.3 contains information on words and symbols used when conversing about probability, the beginnings of a probability register.

2.2.1 Language and Register

Unlike containers or pit boxes, mathematical objects are never physically visible. Humans can comprehend non-substantial objects only through signs and symbols, known as semiotic representations. One potential problem is that "the mathematical objects must never be confused with the semiotic representations that are used" (Duval 2006, p. 107). An object can be represented through any number of possible representations. For example, take the concept of *variable*, a mathematical object that can change between more than one value, dependent on circumstances that affect its value. Most readers will accept Variable or VARIABLE or Va*RiaB*lE as equivalent representations, although the last representation would be less standard. In Korean written and oral language, the object is represented with different symbols. German and Spanish use the same written symbol for variable as English, but the oral representation differs.

A structured system of representations—markings, sounds, and/or gestures—used to communicate is called a *language*. There is no consensus about whether mathematics fulfils all requirements of a language, because mathematics does not attempt to communicate about all possible events; there exists no mathematical way to represent joy or love, for instance. Pimm (1987) concluded that "mathematics is not a natural language in the sense that English and Japanese are. It is not a 'dialect' of English (or any other language) either" (p. 207). Instead of speaking of mathematics as language, Pimm followed other linguistic researchers and considered mathematics as a register. Formally defined, a *register* is "a set of meanings that is appropriate to a particular function of language, together with the words and structures that express these meanings" (Halliday 1975, p. 65). The function of a register can range from very general (like everyday conversation on a public street) to very specific (like terms for types of football kicks at goal). When working with problems in mathematics, problem-solvers use terms and symbols from their mathematics register; when working with problems in probability, problem-solvers use terms and symbols from their probability register.

Words, symbols, and structures for a register are gathered from various places, such as a native language, Greek symbols like θ, and new special symbols like = and +. Halliday (1975) gave a list of seven ways that words are introduced into a register. Table 2.1 presents the ways, accompanied by English examples. In probability and mathematics, many words are taken or created from older languages, but few objects are created in imitation of another language, and completely new words tend to come from names.

Words and symbols have entered the probability register through all seven approaches given in Table 2.1. Probability computations use algebra, a loanword from Arabic. Binomial is a term created from Latin roots; the binomial formula uses combinations, a word reinterpreted from the everyday register with a substantially different meaning. The empty set symbol ∅ is an imitation symbol adopted from Norwegian (Weil 1992); the empty set is an example of a subset—and subset is a new word created from base words in English. Compound phrases include the sample space where the empty set lies and binomially distributed to describe a variable. Finally, a few completely new words and symbols have been created, such as the word Gaussian to describe a particular distribution.

Table 2.1 Types of terms in a register

Term type	Examples
Reinterpretation of existing words	Event; random; combination
Word taken from another language	Algebra; infinite
Created from words in the base language	Output; subset
Created in imitation of another language	∅ (empty set symbol; impossible event)
Created from words in another language	Permutation; binomial
Compound phrase	Sample space; binomially distributed
Completely new word	Gaussian; Bayesian

Shifting between languages and registers often challenges people. Kotsopoulos (2007) offered a student quote that could have come from almost any classroom: "I know how to do it but the question is confusing. I don't understand the words that the question asks" (p. 301). A situation where a particular word or symbol could reasonably have more than one meaning is an instance of *ambiguity,* such as the chapter opening about a few of the dozens of possible meanings attached to the term box (Oxford University Press 2017). Confusion about words is known as *lexical ambiguity*. Structures in mathematics and probability attempt to minimize lexical ambiguity, not always successfully. Terms often change meaning between the register of everyday conversation and the mathematics register (Kotsopoulos 2007; Pimm 1987). Students who cannot manoeuvre between meanings will not be able to move forward and solve problems.

Some researchers have examined term *formations*, the explanations for mathematical objects designed to help students connect a concept with its representation in a register (Duval 2006). Teachers often present formations through structures not present in everyday conversation, such as offering sentences with one missing word (like "If the length of a rectangle increases by three, the perimeter increases by …"), and asking questions for which the questioner knows the correct answer (Pimm 1987). Advice for classroom teachers about formations generally recommends explicit attention to formal definitions. After reviewing classroom discourse between teachers and students, Kotsopoulos suggested that teachers speak less; students should have opportunities "to use mathematical language themselves to be able to see through the outwardly familiar language to the underlying mathematical meaning" (Kotsopoulos 2007, p. 304).

2.2.2 The Mathematical Register

Almost all school curricula introduce mathematics topics before probability and statistics, making mathematics the mother tongue of mathematical sciences. The probability register takes many formations from the mathematics register. One challenge is that some symbols and words shift meaning between registers, forming ambiguity. For instance, a university probability teacher might write the square root of 2 as = 1.414, using the = symbol to represent approximate equality, but a university mathematics teacher might reserve = for exact equality and write $\approx$ 1.414 instead. Another example of meaning shift is the term independence, one of the words closely studied in Sect. 2.3. In a conversation about potential misconceptions related to independence, a secondary school teacher offered what students might say about the two registers: "just getting this stuff mixed up. That's not my input. What do you mean independent? I thought independent and dependent were input and output. Domain and range" (Molnar 2016, pp. 1–5). This teacher described what her students might think when trying to move from the mathematics register into the probability register. Lexical ambiguity across problem types confuses students.

Regrettably, some aspects of registers are confusing because registers are developed by humans. Halliday (1975) recommended that communities who have the chance to develop mathematical terminology plan carefully to reduce ambiguity. Some languages have developed less ambiguous structures than others. For example, Spanish-speaking children learning to count must learn new words for 11 (once) through 15 (quince), including spelling shifts for catorce and quince. More spelling shifts appear at higher multiples of 10 such as siete for 7 versus setenta for 70. English and French also have similar complications. In contrast, children learning Mandarin Chinese count past ten with ten-one, ten-two, and so forth, in one pattern without spelling shifts. This allows Chinese children to learn arithmetic more quickly (Geary 1980). Better formations help learners.

2.2.3 The Probability Register

Many other researchers have written on aspects of the mathematics register, including comprehensive treatments like Pimm's and Halliday's. In contrast, past research on the probability register has almost always focused on specific situations. This section summarizes past studies, beginning with probability in word intensions.

One common issue is attempting to use adjectives to describe numeric probabilities. When this occurs, ambiguous words have caused potential problems from minor to country-changing. For example, during the Cold War after World War II, US intelligence analysts attempted to provide written probability estimates about Eastern Bloc nations. A 1951 intelligence paper said a Soviet attack on Yugoslavia "should be considered a serious possibility." A few days later, the authors—who had all agreed on the phrase serious possibility—disclosed their personal numeric estimates. It was a shock that "each Board member had somewhat different odds in mind and the low man was thinking of about 20 to 80, the high of 80 to 20" (Kent 1964, p. 52). This lack of uniformity caused no devastating consequences, as the United States made no major moves and the Soviet Union did not invade. Nevertheless, Kent and other analysts tried to standardize language in intelligence papers, with moderate success.

More recent researchers have also asked people to translate between words and numbers. In one set of studies, students were asked to estimate numeric levels for terms such as *not quite certain* and *possible*. Participants offered a median probability of 0.60 for possible and 0.70 for not quite certain, both moderate results near equi-probability. In general, the words possible, probable, likely, and certain can be arranged on a scale in that order, although cut-points vary from person to person (Teigen et al. 2014).

The probability register includes other topics besides numeric translations. Particularly important are word formations used when children initially encounter concepts in textbooks. Cañizares et al. (2002) compared the probability registers of three widely used school textbooks in Spain. There was little standardization. One textbook introduced probability axiomatically, while another book related probability to mathematical functions. Non-equivalent verbs were used to describe how probabilities are assigned, such as compute, estimate, and measure. Two of the books supplied different terms to represent numeric levels, and the third book avoided the issue of numeric translation by only representing probability numerically.

The Cañizares et al. study showed that textbook authors can have substantial influence. In an ideal world, the writers would have lots of research-based guidance. Unfortunately, few researchers have devoted time and resources to investigate term formations in the probability register. Two terms that have been studied are randomness and independence; they are the close studies presented in the next section.

2.3 Close Studies of Some Specific Terms

Close study of a term includes textbook analysis of intended curriculum, interviews or surveys to discover what students have learned, and task-based interventions to improve student knowledge. This process has been completed and published for the following two terms in the probability register. In Sect. 2.3.1, results are presented about the word random; Sect. 2.3.2 contains results about the word independence.

2.3.1 Random

The term *random* has lexical ambiguity; randomness changes meaning among everyday, probability, mathematics, and statistics registers. Discussions of the different meanings of randomness can be found in Batanero (2015) and Bennett (1998). In probability, a simple definition is that a phenomenon is random "if individual outcomes are uncertain, but there is nonetheless a regular distribution of outcomes in a large number of repetitions" (Moore et al. 2013, p. 261). In this definition, the process that generates outcomes is known, although the result of a trial of the process is unknown before the result is observed. Outside the mathematical sciences, the adjective random describes a process "without method or conscious choice; haphazard" (Oxford University Press 2017). In everyday life, the process that generates outcomes is not known, unlike the common usage in probability and statistics. Kaplan et al. (2014) defined the distinction visually. According to the authors, a random event in the probability register is like drawing values from a hat, while a random event in the everyday register is unexpected and haphazard, such as a person dressed in a zebra costume on a Shanghai street (Kaplan et al. 2014).

In statistics applications that build from foundations in probability, the word random is applied in many specialized situations. For example, one popular introductory college textbook defines a simple random sample as consisting of "*n* individuals chosen in such a way that every set of *n* individuals has an equal chance to be the sample actually selected" (Moore et al. 2013, p. 204). A simple random sample satisfies the general definition in the probability register, but it restricts the known process by mandating that each observation has equal probability. A student who understands randomness only as simple random sampling does not have mastery of the concept of random. To further complicate matters, the Moore et al. book (like others) includes other situations where the random process is more complex than a simple random sample, such as random digits, random digit dialling, and randomized blocks. The concept is challenging.

Random implements are ancient. Remains of chance-based games over 2000 years old have been found in Iraq, Egypt, Syria, Thebes, India, China, and North America. Decision-making based on lots appears in the Greek *Iliad,* Roman government records, and the Hebrew Old Testament (Bennett 1998). Interestingly, for most of recorded history, most people did not understand randomness as the result of a random process. Many cultures believed that chances reflected divine will, as expressed in the biblical proverb, "The lot is cast into the lap, but the decision is the Lord's alone" (Proverbs 16:33, New Revised Standard Version). Less religious people attributed "randomness" to ignorance about initial conditions; with complete knowledge, the process would be deterministic. The current generally accepted belief became common only about 100 years ago (Bennett 1998).

Perhaps due to historical belief, or the complexity of the definition, many modern people still have trouble understanding the definition of randomness. People have trouble seeing clumps or streaks of results as truly random. Falk and Konold (1994) reported on studies of the concept of randomness involving children and adults. They found that people maximized estimated randomness when a head-and-tail sequence switched 60% of the time, not the true value of 50%. Falk and Konold suggested introducing randomness through complexity in information theory, but information theory terminology would require adding several new words to the probability register and so few textbook authors have tried this idea.

To see how textbook authors tried to explain randomness, Cañizares et al. included the term in their 2002 report on Spanish secondary school textbooks. The books had multiple types of random generators such as dice, cards, urns, and spinners, with two books also including computer random number generators. The textbooks also had potentially random situations from everyday life, such as gender of births and weather forecasts. In two books, some activities did not assume classical equal probabilities, which introduced a more complex concept of random process to the students. On the other hand, the books did not include many experiences with random samples or random experiments; students would be unlikely to gain a complete conception of random processes.

Kaplan et al. (2009) began a non-textbook investigation into increasing students' understanding about randomness, as part of a project about lexical ambiguity in statistics and probability. First, they asked college students in the first week of an introductory class to provide a definition for the term random. Unsurprisingly, almost no students gave a correct mathematical definition. Over 80% of responses were in the everyday register, roughly evenly divided between random occurrences and random selection (Kaplan et al. 2009). Next, they decided to examine results at the end of college introductory courses without specific intervention. They categorized their two samples into six categories. Of the 103 students, 71 (69%) were classified into one of the first four categories that included no aspect of a known planned process. Another 24 (23%) defined randomness in terms of representative sample, which makes limited sense for statistics but not for probability. The remaining eight students (8%) provided an acceptable definition for a random sample (Kaplan et al. 2010). No students gave a definition of complete random process in the probability register.

Based on these disappointing results, the research team decided to intervene by introducing a new formation that contrasted the statistical random sample and everyday haphazard meanings. They also decided to focus on randomness as a process, not the randomness of outputs. Early in the semester, the new formation was introduced. After multiple choice questions about the colloquial and statistical meanings, the instructor visually showed photos of a person in a multi-colored zebra costume and a hat. The zebra-hat formation was repeatedly used throughout the semester to contrast the everyday and statistical meanings of the term random (Kaplan et al. 2014, pp. 13–14). Responses collected at the end of the course showed substantially more understanding than without the intervention. Of the 82 students, 40% provided the acceptable random sample definition; 11% included the word probability in their response (Kaplan et al. 2014, pp. 17–18).

Although promising, the intervention's emphasis on statistical random samples does not describe the general definition of random processes. The school textbooks were not sufficient either. Research questions remain on how to effectively communicate the complete modern definition of randomness and what benefits result from doing so.

2.3.2 Independence

Like the word random, the noun independence and adjective independent appear in multiple locations in probability and statistics textbooks. In probability, two events A and B are *independent* when "knowing that one occurs does not change the probability that the other occurs. If A and B are independent, $P(A\, and\, B) = P(A) \times P(B)$" (Moore et al. 2013, p. 309). If knowledge that one event occurs changes the other event's probability of occurrence, the events are dependent. The concept is symmetric; either event can serve as the "one" or the "other". Independent events are sufficiently common that the authors of the United States Common Core State Standards chose to introduce the definition into high school standards on rules of probability (National Governors Association [NGA]

Center for Best Practices and Council of Chief State School Officers [CCSSO] 2010, p. 82).

Studies since 2009, summarized in Edalgo and Molnar (2017), have detailed misconceptions of groups of people around the world, including college students in US introductory statistics classes, Argentine students and professors, and French pre-service teachers. One potential reason behind the trouble is lexical ambiguity; the word independence has different meanings in probability, everyday life, algebra, and statistics. According to the Oxford English Dictionary (Oxford University Press 2017), in everyday life the adjective *independent* refers to something "not depending on the authority of another, not in a position of subordination or subjection; not subject to external control or rule; self-governing, autonomous, free". In everyday language, independence and dependence are not necessarily symmetric. For instance, when choosing where to live, young children are usually dependent on their parents' decisions, but parents have independent autonomy. In algebra, variables are described as dependent or independent, as in the US Common Core standards where Grade 6 students should "write an equation to express one quantity, thought of as the dependent variable, in terms of the other quantity, thought of as the independent variable" (NGA Center for Best Practices and CCSSO 2010, p. 44). With defined sides, this mathematical definition is never symmetric, although closer to the sometimes-symmetric everyday definition than the always-symmetric probability version.

Statistics applies independence in multiple situations. Independent variables in regression are predictors that control the value of the dependent response variable. Reversing independent and dependent variables leads to an incorrect model. This definition is never symmetric, similar to mathematical variables but opposite of probability events. In inference, independent samples are necessary for some analysis techniques such as two-sample t tests. For two samples to be independent, "one sample has no influence on the other" (Moore et al. 2013, p. 467). This definition is sometimes symmetric; if observation values in sample I depend on observations in sample II, observations in sample I may or may not have influence on observation values in sample II. Although symmetry does not matter in inference, since dependence of either sample makes the test invalid, symmetry matters for lexical ambiguity purposes.

A single textbook can utilize multiple registers, creating much potential for ambiguity. For instance, introductory college probability and statistics textbooks like Moore et al. (2013) generally include problems with independent events, independent samples, and independent variables. In addition, students hear everyday applications of independence outside class. Since each definition of independence has a separate context, ideally students will always recognize the appropriate context—but students and instructors do not always have ideal un-ambiguous understandings.

The first step in avoiding linguistic problems is to ensure learners have a good initial understanding of independence in each new context. Edalgo and Molnar (2017) examined initial definitions of independent events and independent variables in nine algebra and probability textbooks and found that consistency was low.

In probability, many authors defined independent events with the formally correct multiplication rule in symbols, since for events A and B, $P(A \,\text{and}\, B) = P(A) \times P(B)$ if and only if events A and B are independent. One book introduced independence with the conditional definition that $P(A|B) = P(A)$. When $P(A)$ and $P(B)$ are greater than zero, this definition is equivalent to the multiplication rule, but using conditional probability complicates the structure. Two books did not include symbols in their initial definition, using words only. A learner who does not mentally associate a computational formula with a colloquial statement will not successfully solve problems with independence, making this formation more theoretically troubling.

Six of the nine textbooks contained a definition of independent variables. In four books, the independent variable was described as something whose value could be chosen or manipulated. The other two definitions did not say that independent variable values are chosen. This is a substantial distinction. Definition formations including choice are more restrictive, since choice implies experimental control but many physical models and observational studies have uncontrolled independent variables. A person with a restricted formation cannot apply independent variables to all potential situations, leading to trouble with some tasks.

Tasks can involve any definition or definitions. Nabbout-Cheiban (2017) asked undergraduate pre-service teachers in France and the US several questions about misconceptions in probability. One question distinguished between the everyday definition and probabilistic definition. Participants were asked to "consider the following events *A* 'it will rain tomorrow' and *B* 'Tom will study hard tonight'. Are *A* and *B* independent events?" (Nabbout-Cheiban 2017, p. 267). A participant who used the probabilistic definition would correctly say that events *A* and *B* are independent because the events are designed on separate spaces. Eight of the 29 participants gave an incorrect answer that applied an everyday notion of causality and dependence, such as one participant's answer that rainfall is independent of human actions like studying, but "the event *B* may be dependent on event *A*, since studying may be caused by the expectation of future plans" (p. 14). Incorrectly understanding the formation led to an incorrect belief in potential causality—and to potentially poor decisions if this occurred in real life.

Molnar (2016) interviewed 25 US in-service high school teachers about topics related to conditional probability and independence. In one task, taken from a 2010 AP Statistics examination, a two-dimensional table about educational levels and news sources was presented. Participants were asked if the events "is a college graduate" and "obtains news primarily from the internet" were independent. In the provided table, the two events were not independent because the probability of being a college graduate times the probability of obtaining news primarily from the internet was not equal to the joint probability of being a college graduate who obtains news primarily from the internet.

Only 3 of the 25 participants gave a correct answer with a correct explanation. Three other participants attempted to compute the answer but made a calculation error. The remaining 19 participants (76%) gave an answer with a conceptual misconception. Three participants incorrectly applied the term independence in its everyday causal understanding. Ten teachers claimed that the events were not independent because they were not mutually exclusive; a person could both have a college degree and obtain news from the internet. Six participants claimed that the events were independent because one group was not a subset of the other; as one teacher said, "there are people who received news from the Internet who are not college graduates" (Molnar 2016, pp. 1–4).

Research on independence and conditional probability is not new; the 2017 Nabbout-Cheiban paper includes references to studies about representative bias, recency effects, false causation, and confusion with mutually exclusive events. Papers have been written about task design and classroom actions. Examining linguistic formations, the pre-cursors to solving problems, can lead to new improvements in probability education. This process is ongoing.

2.4 Learning in Non-native Languages

Improving in any language register, native or secondary, is also an ongoing process. True and false cognates in one language can influence learning in another. The meaning of a word in the everyday register can also vary. For example, the Spanish-language term for independence, *la independencia*, can refer to historical actions in Mexico or Peru, but Spain does not celebrate a day of independence.

Millions of children are studying probability in a non-native language, due to government choice, immigration, refugee status, or other reasons. For instance, over 10% of students in the English-speaking United States are English language learners (ELL) with limitations that prevent mainstream instruction in English (Lesser et al. 2013). Children who speak another language outside probability class must not only deal with new terms in the probability register, but also mentally switch into the language of instruction. Much has been written about how children learn mathematics in non-native languages, but as Lesser et al. (2016) wrote, few studies exist on children learning statistics and probability outside their native tongue. The studies described in this section do not present a comprehensive view of the subject; they contain useful results and resources for further development.

A few published studies have investigated how students across different countries describe probability scale levels such as possible, likely, and certain, similar to what Kent (1964) and Teigen et al. (2014) considered. Nacarato and Grando (2014) interviewed 12 Grade 6 students in Sao Paulo, Brazil. The students were asked to order the scale levels improbable, less probable, probable, and very probable, but no group of students ordered the words in the standard adult way. The most common ordering switched probable and very probable (p. 97). In a later task discussion, some students' statements indicated that they assigned similar meanings to the words possibility, probability, and chance. A few students considered probability to only handle equi-probable or fifty-fifty events. None had the full scale expected of adult problem solvers, a reasonable result given their relative lack of experience with probability.

In the late 1970s, Wright led a team that compared probabilistic thinking between student samples in Great Britain, Hong Kong, Indonesia, and Malaysia. At that time, the surveyed Asian students adopted a less finely differentiated view of uncertainty than the surveyed British students. Further investigation revealed that problem context was crucial; in summary, "given the right task conditions, Asian respondents *are able* to think probabilistically" (Wright 1981, p. 291, italics as written).

In southern Malawi, most schoolchildren speak Chichewa as their native language but secondary school mathematics instruction is offered in English. Kazima (2006) surveyed 154 students who had yet to receive formal instruction in probability. The students were asked to provide explanations in Chichewa for the English probability scale words impossible, unlikely, likely, and certain. Since the students had not yet been taught meanings in the English probability register, their responses showed only prior understanding from their Chichewa and non-technical English registers. For all words except impossible, only a minority of participants gave the mathematical English meaning. Many students flipped likely and unlikely, perhaps because Chichewa uses double negations which would literally translate likely as "not unlikely". The complexity of translation makes it challenging for Chichewa speakers to differentiate between scale levels with modifying adjectives, such as equally likely versus not very likely. Kazima recommended that students be offered something more than teacher-spoken definitions, "instances which would help them refine their construction of meanings for the vocabulary" and how words are appropriately used (2006, p. 187).

Wright (1981) suggested something similar, that researchers should examine how people from different cultures and language registers work through tasks in probability. One such study is by Sharma (2012), who interviewed 14 secondary school students from Fiji about equi-probability and frequency estimation. The tasks were presented in English, not the students' first language. One question asked students to offer an opinion about an advertisement that claimed "I can GUARANTEE to predict the sex of your baby correctly. Just send $20 and a sample of your recent handwriting. Money-back guarantee if wrong!" (Sharma 2012, p. 67). None of the students' answers included equi-probability or numeric probability estimates. Even with prompting, students could not see that anyone could guess correctly about half the time. Sharma could not determine if the misconceptions were the result of task context, language difficulties, or cultural belief; three students said prediction was impossible because the sex of a baby depends on God, the older concept of randomness previously described by Bennett (1998).

At an English-language university near the border with Spanish-speaking Mexico, Lesser and Winsor (2009) asked two pre-service teachers who spoke Spanish as their native tongue to define terms in the everyday and statistics/probability registers. As an example, consider independence. In everyday language, one teacher gave a political definition (like independencia in Mexico) and the other referred to lack of control (like the Oxford English dictionary). When asked for a definition from probability or statistics, one pre-service teacher incorrectly defined independent events as separate and the other offered the independent variable formation (Lesser and Winsor 2009). Based on initial interviews, the researchers designed written questions and administered the questionnaire to 51 Spanish-speaking ELL and 83 English-speaking non-ELL students in an introductory statistics literacy course. ELL students found confusion about language registers and problem context more problematic than non-ELL students did (Lesser et al. 2013, p. 22). Lesser et al. (2016) also videotaped six Spanish-speaking ELLs solving problems with a bilingual coin toss applet. Working with the applet helped students correct some misconceptions, such as the meaning of the term longest run. All interviewees felt the applet was beneficial—some for language support and others for help visualizing concepts.

Suggestions about instruction for students learning in a non-native language have been offered based on research and personal experience. As noted in Bennett (1998), probability-based games vary by culture. Non-native students may need explanations of tools such as the 52-card deck of cards and the roulette wheel. For instance, Mexican coins' sides are called seal and eagle instead of head and tail in the US (Lesser and Winsor 2009, p. 17). Wood (1991) wrote about ways of teaching and adapting materials to deal with problems in vocabulary, syntax, and logic. Wood suggested that students devise their own dictionary, practice matching terms with meanings, and cooperate to solve problems where each student has only part of the required information. Lesser et al. recommended that teachers recognize synonymous terms when they exist, increase awareness of lexical ambiguity (such as the word run), and consider vocabulary activities (2013, p. 22). Activities like the applet can be developed with language support.

2.5 Conclusions

Most recommendations included increased attention to terms and definitions, because students frequently express frustrations that led to Kotsopoulos's article title; mathematics "is like hearing a foreign language" (2007, p. 301). Multiple authors suggested special terminology dictionaries. Individual efforts exist, but an organized effort to collect terms in probability (which have entered use from many sources) could support many classrooms.

Learners need explicit examples in order to connect a term or symbol with its concept. As demonstrated by authors on several continents (Kazima 2006; Nacarato and Grando 2014; Wright 1981), everyday childhood experience does not always provide sufficient context to order probability scale words. Countries such as Brazil and the United States are moving the introduction of probability into earlier grades (Nacarato and Grando 2014; NGA Center for Best Practices and CCSSO 2010), often from age 15–17 to age 11–13. Learners will have less verbal experience and require more assistance.

One particularly challenging problem is lexical ambiguity across the registers of everyday life, mathematics, and probability. Exposure to the probability register occurs after exposure to the mathematics and everyday language registers, so words and objects introduced in probability problems often already have associated concepts. When a term has substantial differences in meaning across registers, such as conditional, independent, and random, students and teachers consistently have trouble working in context. Many authors suggested identifying and discussing the lexical ambiguity, but discussion is not always sufficient. More close studies with investigations of alternate formulations, like the zebra-hat approach for random (Kaplan et al. 2014), would be helpful.

Sometimes a situation requiring probability is as large as international war; other times it's as small as drawing a box. In all situations, humans express probabilities through words and symbols. Without the probability register, none of the techniques in other chapters could be expressed—and so research on language in probability education should continue to advance.

References

Batanero, C. (2015). Understanding randomness. In K. Krainer & N. Vondrová (Eds.), *Proceedings of the Ninth Congress of the European Society for Research in Mathematics Education* (pp. 34–49). Prague, Czech Republic: Charles University in Prague: ERME.

Bennett, D. J. (1998). *Randomness*. Cambridge, MA: Harvard University Press.

Cañizares, M. J., Ortiz, J. J., Batanero, C., & Serrano, L. (2002). Probabilistic language in Spanish textbooks. In B. Phillips (Ed.), *ICOTS-6 papers for school teachers* (pp. 207–211). Cape Town, South Africa: International Association for Statistical Education.

Duval, R. (2006). A cognitive analysis of problems of comprehension in a learning of mathematics. *Educational Studies in Mathematics, 61*(1), 103–131.

Edalgo, S., & Molnar, A. (2017, February). *Textbook formations of independence*. Paper presented at The Twentieth Annual Conference on Research in Mathematics Education, San Diego, CA. http://sigmaa.maa.org/rume/crume2017/Abstracts_Files/Papers/116.pdf. Accessed June 25, 2017.

Falk, R., & Konold, C. (1994). Random means hard to digest. *Focus on Learning Problems in Mathematics, 16*(1), 2–12.

Geary, D. C. (1980). Biology, culture, and cross-national differences in mathematical ability. In R. J. Sternberg & T. Ben-Zeev (Eds.), *The nature of mathematical thinking* (pp. 145–172). Mahwah, NJ: Lawrence Erlbaum.

Halliday, M. A. K. (1975). Some aspects of sociolinguistics. In C. O. Taiwo (Ed.), *Interactions between linguistics and mathematical education* (pp. 64–73). Nairobi, Kenya: United Nations Educational Scientific and Cultural Organization.

Kaplan, J. J., Fisher, D. G., & Rogness, N. T. (2009). Lexical ambiguity in statistics: What do students know about the words association, average, confidence, random and spread? *Journal of Statistics Education, 17*(3), 1–19.

Kaplan, J. J., Fisher, D. G., & Rogness, N. T. (2010). Lexical ambiguity in statistics: How students use and define the words association, average, confidence, random and spread? *Journal of Statistics Education, 18*(2), 1–22.

Kaplan, J. J., Rogness, N., & Fisher, D. (2014). Exploiting lexical ambiguity to help students understand the meaning of random. *Statistics Education Research Journal, 13*(1), 9–24.

Kazima, M. (2006). Malawian students' meanings for probability vocabulary. *Educational Studies in Mathematics, 64*(1), 169–189. https://doi.org/10.1007/s10649-006-9032-6.

Kent, S. (1964). Words of estimative probability. *Studies in Intelligence, 8*(4), 49–65.

Kotsopoulos, D. (2007). Mathematics discourse: "It's like hearing a foreign language". *Mathematics Teacher, 101*(4), 301–305.

Lesser, L., Wagler, A., Esquinca, A., & Valenzuela, M. G. (2013). Survey of native English speakers and Spanish-speaking English language learners in tertiary introductory statistics. *Statistics Education Research Journal, 12*(2), 6–31.

Lesser, L., Wagler, A., & Salazar, B. (2016). Flipping between languages? An exploratory analysis of the usage by Spanish-speaking English language learner tertiary students of a bilingual probability applet. *Statistics Education Research Journal, 15*(2), 145–168.

Lesser, L. M., & Winsor, M. S. (2009). English language learners in introductory statistics: Lessons learned from an exploratory case study of two pre-service teachers. *Statistics Education Research Journal, 8*(2), 5–32.

Molnar, A. (2016, July). High school mathematics teachers' understanding of independent events. Paper presented at the *13th International Congress on Mathematical Education,* Hamburg, Germany. http://iase-web.org/documents/papers/icme13/ICME13_L9_Molnar.pdf. Accessed June 25, 2017.

Moore, D. S., Notz, W. I., & Fligner, M. A. (2013). *The basic practice of statistics* (6th ed.). New York: W. H. Freeman.

Nabbout-Cheiban, M. (2017). Intuitive thinking and misconceptions of independent events: A case study of US and French pre-service teachers. *International Journal of Research in Undergraduate Mathematics Education, 3*(2), 255–282.

Nacarato, A. M., & Grando, R. C. (2014). The role of language in building probabilistic thinking. *Statistics Education Research Journal, 13*(2), 93–103.

National Governors Association Center for Best Practices & Council of Chief State School Officers. (2010). *Common Core State Standards: Mathematics.* Washington, DC: Authors.

Oxford University Press. (2017). *Oxford English dictionary.* Kettering, UK: Author.

Pimm, D. (1987). *Speaking mathematically.* London, England: Routledge & Kegan Paul.

Sharma, S. (2012). Cultural influences in probabilistic thinking. *Journal of Mathematics Research, 4*(5), 53–74. https://doi.org/10.5539/jmr.v4n5p63.

Teigen, K. H., Juanchich, M., & Filkuková, P. (2014). Verbal probabilities: An alternative approach. *The Quarterly Journal of Experimental Psychology, 67*(1), 124–146. https://doi.org/10.1080/17470218.2013.793731.

Weil, A. (1992). *The apprenticeship of a mathematician* (J. Gage, Trans.). Heidelberg, Germany: Birkhäuser.

Wood, L. (1991). Teaching statistics to students from a non-English speaking background. In D. Vere-Jones (Ed.), *Proceedings of the Third International Conference on Teaching Statistics* (Vol. 1, pp. 291–297). Voorburg, The Netherlands: International Statistical Institute.

Wright, G. (1981). Cultural and task influences on decision making under uncertainty. *Current Anthropology, 22*(3), 290–291.

Chapter 3
The Status of Probability in the Elementary and Lower Secondary School Mathematics Curriculum: The Rise and Fall of Probability in School Mathematics in the United States

Cynthia W. Langrall

Abstract Probability is no longer featured as an important domain of study in the elementary and lower secondary school mathematics curriculum in the United States. The majority of U.S. states have adopted a mathematics curriculum that has eliminated probability as a topic of study in the elementary grades and narrowed its focus in the lower secondary grades. I illustrate the decline of probability via an overview of the various standards documents that have guided mathematics curriculum and instruction for the past 25 years. This situation is not unique to the United States; many other countries around the world do not include probability in the mathematics curriculum for students under the age of 11. I discuss the implications of this curricular decision and raise questions for the consideration of the research community.

Keywords Probability · Curriculum · Standards

3.1 Introduction

Probability has been recognized as an important topic in mathematics education, across the grade levels, since the 1950s (Jones et al. 2007). For example, in the United States, the authors of the Cambridge Conference report (Educational Services Incorporated 1963) described probability as a critical component of a liberal education that should be taught beginning in the elementary grades. During the era of "new math," the School Mathematics Study Group developed textbooks for probability instruction in the primary and intermediate grades (Blakeslee et al. 1966a, b). And more recently, Moore (1990) identified a set of key probability concepts to be

C. W. Langrall (✉)
Department of Mathematics, Illinois State University,
Campus Box 4520, Normal, IL 61761, USA
e-mail: langrall@ilstu.edu

C. Batanero and E. J. Chernoff (eds.), *Teaching and Learning Stochastics*,
ICME-13 Monographs, https://doi.org/10.1007/978-3-319-72871-1_3

developed beginning in the elementary grades.[1] Additionally, research on the nature of probabilistic thinking and the teaching and learning of probability has laid a strong foundation to support the inclusion of probability at all levels of the school mathematics curriculum. For reviews of this robust body of research, see Borovcnik and Peard (1996), Chernoff and Sriraman (2014), Jones (2005), Jones et al. (2007) and Shaughnessy (1992).

In 2005, Jones argued that probability had become an established strand within the mathematics curriculum, as evidenced by the inclusion of probability across the grade levels in national curriculum documents in Australia, the United Kingdom, and the United States. Indeed, a report of an analysis of state mathematics curriculum documents (Dingman and Tarr 2011) indicated that as of 2005 most U.S. states included learning expectations for probability at each grade level, albeit with considerable variability in grade level placement. However, as of August 2015, the majority of U.S. states have adopted a mathematics curriculum that essentially eliminates probability as a topic of study in the elementary grades and narrows its focus in the middle grades (Common Core State Standards Initiative 2015b).

3.2 Probability in the Mathematics Curriculum

3.2.1 Rise and Fall of Probability in the U.S.

The shift in status of probability in mathematics education in the United States can be seen in the standards that have been developed to guide the design of curriculum at state and local levels and to ultimately affect instruction in mathematics. The first of such standards documents was the *Curriculum and Evaluation Standards for School Mathematics* published by the National Council of Teachers of Mathematics (NCTM) in 1989. In that document, standards were organized according to grade bands; the elementary and middle grade levels were represented in two bands, Kindergarten–Grade 4 and Grades 5–8. For the Kindergarten–Grade 4 band, probability was addressed in a standard with statistics and was aimed at exploring concepts of chance. The emphasis was specifically on inquiry and problem solving: "The study of statistics and probability highlights the importance of questioning, conjecturing, and searching for relationships when formulating and solving real-world problems" (National Council of Teachers of Mathematics [NCTM] 1989, p. 54). An example activity highlighted the importance of exploring aspects of probability in relation to collecting and analyzing data as well as providing opportunities for children to discuss events that are likely, unlikely, or certain and to consider the notion of luck.

[1]In the United States, various terms are often used to distinguish the levels of schooling. Typically, the term *elementary* includes Kindergarten through Grade 5 (ages 5–10), *middle or lower secondary* refers to Grades 6–8 (ages 11–13) and *high school* or *secondary* refers to Grades 9–12 (ages 14–17). Unless otherwise noted, these are the designations I use throughout this paper when I refer to grade levels.

At the Grades 5–8 band, probability had its own standard. It called for students to be engaged in explorations of probability in real-world contexts and detailed five specific expectations, which are listed in Fig. 3.1. As in the earlier grades, the emphasis was on exploration and real-world problems. In fact, the document specifically stated that the study of probability "should not focus on developing formulas or computing the likelihood of events pictured in texts" (NCTM 1989, p. 109). The idea of using probability to model situations was introduced at this level: "Students should actively explore situations by experimenting and simulating probability models . . . . Students should talk about their ideas and use the results of their experiments to model situations or predict events" (NCTM 1989, p. 109).

The *Curriculum and Evaluation Standards for School Mathematics* (hereafter 1989 Standards) set the bar for educational standards and they were widely influential in the mathematics education community; their publication prompted revisions of conventional textbooks to be "standards aligned" and a program of funding from the National Science Foundation was designated to support the development and dissemination of curriculum materials that followed the content and pedagogy outlined in the standards (Reys 2008). In 2000, NCTM published an updated version of the standards, *Principles and Standards for School Mathematics* (hereafter 2000 Standards). One way in which these standards differed from their predecessor was that the same set of content standards (number and operations, algebra, geometry, measurement, and data analysis and probability) spanned the grade bands, which were restructured as Pre-Kindergarten–Grade 2 (Pre-K–2), Grades 3–5, Grades 6–8, and Grades 9–12. The data analysis and probability standard included four key components that were common across all grade bands, and each component had grade-band specific student expectations. One key component pertained to probability: "Instructional programs from prekindergarten to grade 12 should enable all students to … understand and apply basic concepts of probability" (NCTM 2000, p. 48). For the Pre-K–2 band, no expectations were listed for the probability component. However, the text describing the standard states that

> Probability experiences should be informal and often take the form of answering questions about the likelihood of events, using such vocabulary as *more likely* or *less likely* . . . . Teachers should address the beginnings of probability through informal activities with spinners or number cubes that reinforce other Standards, primarily number. (p. 114)

The specific expectations for students at the Grades 3–5 and Grades 6–8 bands are presented in Fig. 3.1.

Although probability was no longer a separate standard, it maintained its prominence across the grade bands in the 2000 Standards and was more explicitly defined for the upper elementary grades (i.e., Grades 3–5). However, there was less emphasis on modeling situations and real-world contexts than in the 1989 Standards. Probability measurements, which were evident in both standards documents at the middle grades, were also included in the Grades 3–5 band in the 2000 Standards: "Students in grades 3–5 should begin to learn about probability as a measurement of the likelihood of events. … they can begin to learn how to quantify likelihood" (NCTM 2000, p. 181). The 2000 Standards placed greater focus on the use of appropriate terminology and computing probabilities, which aligned with the expectation that students would be able to apply basic probability concepts.

Standards and Learning Expectations for Probability	Grade Levels								
	K	1	2	3	4	5	6	7	8
***Curriculum and Evaluation Standards for School Mathematics*: Probability Standards** (NCTM, 1989, pp. 54, 109)									
• Explore concepts of chance									
• Model situations by devising and carrying out experiments or simulations to determine probabilities									
• Model situations by constructing a sample space to determine probabilities									
• Appreciate the power of using a probability model by comparing experimental results with mathematical expectations									
• Make predictions that are based on experimental or theoretical probabilities									
• Develop an appreciation for the pervasive use of probability in the real world									
***Principles and Standards for School Mathematics*: Probability Expectations** (NCTM, 2000, p. 400)									
• Describe events as likely or unlikely and discuss the degree of likelihood using such words as *certain*, *equally likely*, and *impossible*									
• Predict the probability of outcomes of simple experiments and test the predictions									
• Understand that the measure of the likelihood of an event can be represented by a number from 0 to 1									
• Understand and use appropriate terminology to describe complementary and mutually exclusive events									
• Use proportionality and a basic understanding of probability to make and test conjectures about the results of experiments and simulations									
• Compute probabilities for simple compound events, using such methods as organized lists, tree diagrams, and area models									
***Curriculum Focal Points*: Connections for Grade 7 Focal Points** (NCTM, 2006, p. 19)									
• Students understand that when all outcomes of an experiment are equally likely, the theoretical probability of an event is the fraction of outcomes in which the event occurs. Students use theoretical probability and proportions to make approximate predictions.									

Fig. 3.1 National Council of Teachers of Mathematics (1989, 2000, 2006) grade level learning expectations for probability, designated by shaded cells

In 2006, NCTM published *Curriculum Focal Points for Prekindergarten through Grade 8 Mathematics: A Quest for Coherence* (hereafter Curriculum Focal Points). This document was developed in response to the concern that the U. S. mathematics curriculum was "a mile wide and an inch deep" (Schmidt et al. 1997) and aimed to address variation in the placement and emphasis of topics by grade levels in school systems across the country. Rather than referring to grade bands, focal points were identified at each grade level based on the following rationale:

> A focused curriculum would allow teachers to commit more time each year to topics receiving special emphasis. At the same time, students would have opportunities to explore these topics in depth, in the context of related content and connected applications, thus developing more robust mathematical understandings. (NCTM 2006, p. 4)

Three focal points were identified at each grade level as the content emphases for mathematics instruction. Also, several related topics were presented as possible connections to each focal point. These connections served two roles, to highlight concepts or skills that might serve as a prerequisite or follow up to the focal points or to identify "ways in which a grade level's focal points can support learning in relation to strands that are not focal points at that grade level" (p. 8). The only mention of probability in the entire Curriculum Focal Points document is as a connection to the Grade 7 focal points, which refers to determining the "theoretical" probability of an event and using it (and proportions) to make predictions (see Fig. 3.1). This connection appears to fit the latter purpose stated above. One of the three focal points for Grade 7 pertains to understanding proportionality and similarity and applying these concepts in various problem-solving contexts. The measure of the probability of an event would be considered as a comparison of favorable outcomes to the total number of outcomes (in a sample space of equally likely outcomes), thus connecting to students' understanding of ratio and proportion.

Publication of the Curriculum Focal Points marked a significant shift in curricular recommendations that essentially eliminated probability in the elementary grades. The implications of this change were far reaching. The Curriculum Focal Points were "widely used by state mathematics content developers in designing their own standards and curricula" (Achieve 2010, p. 1) and thus, influenced the most recent standards document to affect mathematics curriculum in the United States, the Common Core State Standards for Mathematics (CCSSM; National Governors Association Center for Best Practices [NGA] and Council of Chief State School Officers [CCSSO] 2010).

In contrast to the three standards documents previously described, which were initiated by NCTM, the CCSSM was a state-led initiative, directed by governors and chief state school officers (e.g., commissioners of education). According to its developers,

> The Common Core [which includes standards for mathematics and English language arts] is informed by the highest, most effective standards from states across the United States and countries around the world. The standards define the knowledge and skills students should gain throughout their K–12 education in order to graduate high school prepared to succeed in entry-level careers, introductory academic college courses, and workforce training programs. (Common Core State Standards Initiative 2015a, "About the Common Core State Standards," para. 4)

Additionally, the CCSSM was intended to address the same lack of coherence and focus in the mathematics curriculum as described in the Curriculum Focal Points. At the K–8 level, the CCSSM includes Standards for Mathematical Practice (common across grade levels) and Standards for Mathematical Content (specific to each grade level). The mathematical content standards include descriptions of three or four critical areas for study, and identify related content domains (e.g., statistics and probability) and specific standards that define what students should understand and be able to do. Grade 7 is the only grade that includes a standard for probability: "Investigate chance processes and develop, use, and evaluate probability models" (NGA and CCSSO 2010, p. 50). The specific learning expectations for this standard are presented in Fig. 3.2. Following the Curriculum Focal Points, the CCSSM has narrowed the study of probability to one grade level, essentially collapsing the learning expectations that were spread across the elementary and middle grade levels in the 2000 Standards.

Standards and Learning Expectations for Probability	**Grade Levels**								
	K	1	2	3	4	5	6	7	8
Common Core State Standards for Mathematics (NGA and CCSSO, 2010, pp. 50–51)									
• Understand that the probability of a chance event is a number between 0 and 1 that expresses the likelihood of the event occurring.									
• Approximate the probability of a chance event by collecting data on the chance process that produces it and observing its long-run relative frequency, and predict the approximate relative frequency given the probability.									
• Develop a probability model and use it to find probabilities of events. Compare probabilities from a model to observed frequencies; if the agreement is not good, explain possible sources of the discrepancy.									
• Find probabilities of compound events using organized lists, tables, tree diagrams, and simulation.									

Fig. 3.2 Common Core State Standards for Mathematics grade level learning expectations for probability, designated by shaded cells

3.2.2 Probability in Other Countries

Although my focus has been aimed at curriculum standards in the United States, there are other countries in which probability is not included in the mathematics curriculum at the elementary grade level. In the chapter on probability research in the *Second Handbook of Research in Mathematics Education*, Jones et al. (2007) highlighted the key ideas of probability addressed in the 1989 Standards as well as the curriculum frameworks of Australia and the United Kingdom. They stated the following rationale:

> We chose these latter two curriculum documents [from Australia and the United Kingdom] because they appeared at almost the same time as the NCTM Standards and, with it, seemed to set the scene for much of the curriculum development in probability and statistics worldwide. (p. 912)

It is difficult to determine the extent to which these documents actually influenced curriculum development worldwide but there is evidence that the NCTM Standards have informed the development of curriculum frameworks in other countries (e.g., Burrill et al. 2014; Mullis et al. 2015). However, the current situation in England is quite similar to that of the United States: In 1989, the national curriculum included probability across the grade levels (Department of Education and Science and the Welsh Office 1989) but by 2001, probability was no longer emphasized in the curriculum for students under the age of 11 (Kapadia 2009); this is still the case according to a recent national curriculum document (Department for Education 2014). In contrast, the mathematics curricula in Wales, Ireland, Scotland, as well as Australia do include probability at the elementary grade level (Australian Curriculum, Assessment, and Reporting Authority [ACARA] 2015; Education, Audiovisual, & Culture Executive Agency 2011).

This dichotomy is also evident on a more global level. According to the *TIMSS 2015 Encyclopedia: Education Policy and Curriculum in Mathematics and Science* (Mullis et al. 2015) almost half of the 57 countries that participated in the 2015 TIMSS assessment "mentioned introducing the concept of probability and possibility or frequencies of outcomes" (Introduction, p. 12) at the fourth-grade level. These countries are listed in Table 3.1 according to my interpretation of the curriculum topics described in each country's chapter of the TIMSS 2015 encyclopedia (Mullis et al. 2015).

Although this list is incomplete (with Central and South American countries particularly under-represented), these countries do represent a wide range of economic and geographic diversity, and the majority of these countries have a national curriculum (exceptions being Canada, Belgium, Germany, Sweden, and the United States). Thus, the implications I draw based on the current situation in the United States are relevant to mathematics education in many other countries around the world.

Table 3.1 Categorization of the 57 countries that completed the 2015 TIMSS Curriculum Questionnaire according to whether or not probability or probability related topics (e.g., chance, random outcomes, likelihood) were specifically mentioned for students under the age of 11 in the information presented in the *TIMSS 2015 Encyclopedia: Education Policy and Curriculum in Mathematics and Science*

Concepts of probability listed in the mathematics curriculum for students under the age of 11	Concepts of probability not listed in the mathematics curriculum for students under the age of 11	Unable to determine exact grade level (or age) from information provided
Australia	Argentina: Buenos Aires	Botswana (by Grade 8)
Bahrain	Armenia	Georgia (by Grade 8)
Canada: Ontario, Quebec	Belgium	Indonesia (by Grade 8)
Chinese Taipei	Bulgaria	Jordan (by Grade 10)
Cyprus	Chile	Korea (Grades 5–6)
Denmark	Croatia	Kuwait (by Grade 8)
Finland	Czech Republic	Lithuania (by Grade 8)
Germany	England	Malaysia
Hungary	France	New Zealand
Iran	Hong Kong	Norway (by Grade 8)
Ireland and Northern Ireland	Israel	Oman (by Grade 8)
Italy	Japan	Russian Federation
Malta	Kazakhstan	(Grades 5–9)
Portugal	Lebanon	
Slovak Republic	Morocco	
Slovenia	Netherlands	
Spain	Poland	
Sweden	Qatar	
Thailand	Saudi Arabia	
Turkey	Serbia	
United Arab Emirates	Singapore	
	South Africa	
	United States	

3.3 Questions for the Research Community

One could logically conclude that probability was eliminated from the elementary grades mathematics curriculum in the United States because it is not considered a necessary component of a coherent curriculum at that level of schooling. But how might that have been determined? The authors of the CCSSM document stated, "The development of these Standards began with research-based learning progressions detailing what is known today about how students' mathematical knowledge, skill, and understanding develop over time" (NGA and CCSSO 2010, p. 4). However, in the list of works consulted for the development of the CCSSM, there are no references to research on the development of students' probabilistic thinking (Mooney et al. 2014) although a significant body of research exists that includes frameworks characterizing the development of students' probabilistic thinking (Jones et al. 2007). In a plenary talk at a 2010 conference to examine curriculum issues associated with the CCSSM, Confrey commented on aspects of the standards that warranted further consideration or revision. In reference to the statistics and probability standard, she stated: "They ignore significant empirical data on children's ability to develop an understanding of variability, distributions, data use, statistical reasoning, and probability in the early grades; instead, these are added abruptly in sixth and seventh grade" (Confrey and Krupa 2010, p. 4).

It is not known why the authors of the CCSSM disregarded the research literature on the teaching and learning of probability. Their goal was to develop standards that would produce a coherent, sharply focused set of content topics across the K–12 curriculum (Dossey et al. 2016). To achieve that goal, some topics from the traditional curriculum were removed. Was the elimination of probability at the elementary level simply collateral damage; was there no one at the table with the interest or expertise to advocate for probability as an important domain of study? Perhaps the authors viewed probability solely as a tool for statistics, as it is portrayed in the *Guidelines for Assessment and Instruction in Statistics Education (GAISE) Report: A PreK–12 Curriculum Framework* (Franklin et al. 2007)—a reference that was included in the list of consulted works for the development of the CCSSM. This might explain the treatment of probability as a skill that could be adequately addressed as part of the mathematics curriculum at a single grade level, Grade 7. But the wisdom of that approach is questionable. As Greer and Mukhopadhyay (2005) have argued, probability is important in its own right, it is a way of thinking and it has applications to multiple disciplines—"probability is used to model aspects not only of the physical world, but also social and political issues of importance to students, their communities, and society in general" (pp. 308–308). Moreover, they cautioned against excessive formalization, a formulaic approach to instruction, and the use of unrealistic or uninteresting task contexts in the teaching and learning of probability.

The kind of probability modelling described by Greer and Mukhopadhyay (2005) might be viewed as too complex for students at the elementary level. Indeed, commenting on the decline of probability in the National Curriculum in England, Threfall (2004) argued that complex probability tasks are not accessible to students in the primary grades and the simple probability tasks that students experience in school are insufficient for developing a mathematical understanding of probability; he suggested that probability instruction be delayed until students are able to handle complex probability contexts. Yet, the research literature has shown that students in the primary grades are capable of learning about probability in meaningful ways and that it is important to do so (e.g., Borovcnik and Peard 1996; Chernoff and Sriraman 2014; Jones 2005; Jones et al. 2007; Shaughnessy 1992). More specifically, English and Watson (2016) have shown that aspects of probability modelling are within the grasp of Grade 4 students. They highlighted the importance of introducing probability early in the curriculum and reported that the modelling approach they used with students developed important links between probability and statistics. That is, probability served as the context within which the key statistical concept of variation was introduced (which contrasts with the notion of probability as simply a tool for statistical applications). It is noteworthy that students in the English and Watson study used TinkerPlots software to conduct probabilistic simulations.

Such technologies can provide accessibility to concepts that were once considered too advanced for students to explore. As Watson et al. (2013) stated: "As expertise in the use of technology to provide novel learning experiences increases, perhaps it is best to remain open-minded about the possibility of a return of probability to the curriculum for children below age 11" (p. 162.). I concur and encourage us to consider what this implies for the research community.

Fortunately, curriculum frameworks periodically undergo review and revision. For example, the CCSSM has been described as a "living document" that should be subjected to "short-term fixes, medium-term adjustments, and long-term review and modification, as needed, based on expert advice and empirical evidence" (Confrey and Krupa 2010, p. 9). Members of the mathematics and statistics education communities need to be proactive in calling for the review and modification of the CCSSM with regard to the placement and focus of probability; moreover, they need to be prepared and willing to provide the expertise needed to inform such work. How might probability research contribute to curricular decision making and provide research-based evidence to inform curriculum modifications? With that in mind, I close this commentary with some related questions: What might be the ramifications of an elementary mathematics curriculum devoid of probability? Will some students leave the educational system without having an opportunity to develop probabilistic reasoning? Will students' experiences representing and interpreting data (basic learning expectations that are included in the elementary grade levels in the United States and many other countries) provide sufficient opportunities for them to develop intuitions about probability and to engage in probabilistic thinking? Do students who first experience probability in the middle or secondary school curriculum develop a robust understanding of fundamental probability concepts and skills that prepare them for the formal study of probability and statistics? Because the findings of extant research cannot answer these questions, there is a need for research aimed at addressing the issues that underlie these and other related questions.

References

Achieve. (2010). *Comparing the Common Core State Standards in Mathematics and NCTM's Curriculum Focal Points*. Washington, DC: Author. http://www.achieve.org/CCSSandFocalPoints. Accessed 23 May 2017.

Australian Curriculum, Assessment and Reporting Authority (ACARA). (2015). *Australian curriculum: Mathematics (Version 7.5)*. http://www.australiancurriculum.edu.au. Accessed 4 March 2017.

Blakeslee, D. W., Bridgess, M. P., Gillman, L., Hansen, R., Hosier, M., Ingrum, R. G., et al. (1966a). *Probability for intermediate grades: Student text* (Rev. ed.). Stanford, CA: Leland Stanford Junior University.

Blakeslee, D. W., Bridgess, M. P., Gillman, L., Hansen, R., Hosier, M., Ingrum, R. G., et al. (1966b). *Probability for primary grades: Student text* (Rev. ed.). Stanford, CA: Leland Stanford Junior University.

Borovcnik, M., & Peard, R. (1996). Probability. In A. Bishop, M. A. Clements, C. Keitel, J. Kilpatrick, & C. Laborde (Eds.), *International handbook of mathematics education* (Vol. 1, pp. 239–287). Dordrecht, The Netherlands: Kluwer.

Burrill, G., Lappan, G., & Gonulates, F. (2014). Curriculum and the role of research: Report of the ICME 12 survey team. *Bolema: Boletim de Educação Matemática*, *28*(49), 682–700. https://dx.doi.org/10.1590/1980-4415v28n49a11. Accessed 1 March 2017.

Chernoff, E. J., & Sriraman, B. (2014). *Probabilistic thinking: Presenting plural perspectives*. Dordrecht, The Netherlands: Springer. https://doi.org/10.1007/978-94-007-7155-0.

Common Core State Standards Initiative. (2015a). *About the standards*. http://www.corestandards.org/about-the-standards/. Accessed 23 May 2017.

Common Core State Standards Initiative. (2015b). *Development process*. Retrieved from http://www.corestandards.org/about-the-standards/development-process/. Accessed 23 May 2017.

Confrey, J., & Krupa, E. (2010). *Curriculum design, development, and implementation in an era of common core state standards: Summary report of a conference*. http://mathcurriculumcenter.org/reports_research.php. Accessed 23 May 2017.

Department for Education. (2014). *National curriculum in England: Mathematics programmes of study*. https://www.gov.uk/government/uploads/system/uploads/attachment_data/file/335158/PRIMARY_national_curriculum_-_Mathematics_220714.pdf. Accessed 23 May 2017.

Department of Education and Science and the Welsh Office. (1989). *National curriculum: Mathematics for ages 5 to 16*. York, UK: Central Office of Information.

Dingman, S., & Tarr, J. E. (2011). An analysis of K–8 probability standards. In J. P. Smith, III (Ed.), *Variability is the rule: A companion analysis of K–8 state mathematics standards* (pp. 161–192). Charlotte, NC: Information Age Publishing.

Dossey, J. A., McCrone, S. S., & Halvorsen, K. T. (2016). *Mathematics education in the United States 2016: A capsule summary fact book*. Reston, VA: National Council of Teachers of Mathematics.

Education, Audiovisual, & Culture Executive Agency. (2011). *Mathematics education in Europe: Common challenges and national policies*. Brussels, BE: Author. https://doi.org/10.2797/72660.

Educational Services Incorporated. (1963). *Goals for school mathematics: The report of the Cambridge Conference on School Mathematics*. Boston, MA: Houghton Mifflin.

English, L. D., & Watson, J. M. (2016). Development of probabilistic understanding in fourth grade. *Journal for Research in Mathematics Education*, *47*(1), 28–62. https://doi.org/10.5951/jresematheduc.47.1.0028.

Franklin, C., Kader, G., Mewborn, D., Moreno, J., Peck, R., Perry, M., & Schaeffer, R. (2007). *Guidelines for assessment and instruction in statistics education (GAISE) report: A preK–12 curriculum framework*. Alexandria, VA: American Statistical Association.

Greer, B., & Mukhopadhyay, S. (2005). Teaching and learning the mathematization of uncertainty: Historical, cultural, social and political contexts. In G. A. Jones (Ed.), *Exploring probability in school: Challenges for teaching and learning* (pp. 297–324). New York, NY: Springer.

Jones, G. A. (Ed.) (2005). *Exploring probability in school: Challenges for teaching and learning*. New York, NY: Springer. https://doi.org/10.1007/b105829.

Jones, G. A., Langrall, C. W., & Mooney, E. S. (2007). Teaching and learning probability: A research perspective. In F. K. Lester, Jr. (Ed.), *Second handbook of research on mathematics teaching and learning* (Vol. 2, pp. 909–955). Charlotte, NC: Information Age Publishing.

Kapadia, R. (2009). Chance encounters—20 years later: Fundamental ideas in teaching probability at school level. *International Electronic Journal of Mathematics Education*, *4*, 371–386.

Mooney, E. S., Langrall C. W., & Hertel, J. T. (2014). A practitional perspective on probabilistic thinking models and frameworks. In E. Chernoff & B. Sriraman (Eds.), *Probabilistic thinking: Presenting plural perspectives* (pp. 495–507). Dordrecht, the Netherlands: Springer.

Moore, D. (1990). Uncertainty. In L. Steen (Ed.), *On the shoulders of giants: A new approach to numeracy* (pp. 95–137). Washington, DC: National Research Council. https://doi.org/10.17226/1532.

Mullis, I. V. S., Martin, M. O., Goh, S., & Cotter, K. (Eds.). (2015). *TIMSS 2015 encyclopedia: Education policy and curriculum in mathematics and science*. http://timssandpirls.bc.edu/timss2015/encyclopedia/. Accessed 23 May 2017.

National Council of Teachers of Mathematics. (1989). *Curriculum and evaluation standards for school mathematics*. Reston, VA: Author.

National Council of Teachers of Mathematics. (2000). *Principles and standards for school mathematics*. Reston, VA: Author.

National Council of Teachers of Mathematics. (2006). *Curriculum focal points for prekindergarten through grade 8 mathematics: A quest for coherence*. Reston, VA: Author.

National Governors Association Center for Best Practices & Council of Chief State School Officers. (2010). *Common core state standards for mathematics*. Washington, DC: Author. http://www.corestandards.org/assets/CCSSI_Math%20Standards.pdf. Accessed 23 May 2017.

Reys, R. (2008). The road to reform. In M. M. Meyer & C. W. Langrall (with F. Arbaugh, D. C. Webb, & M. B. Hoover) (Eds.), *A decade of middle school mathematics curriculum implementation: Lessons learned from the Show-Me Project* (pp. 3–8). Charlotte, NC: Information Age Publishing.

Schmidt, W. H., McKnight, C. C., & Raizen, S. A. (1997). *A splintered vision: An investigation of U.S. science and mathematics education*. Dordrecht, The Netherlands: Kluwer.

Shaughnessy, J. M. (1992). Research in probability and statistics: Reflections and directions. In D. A. Grouws (Ed.), *Handbook of research on mathematics teaching and learning* (pp. 465–494). New York, NY: National Council of Teachers of Mathematics and Macmillan.

Threfall, J. (2004). Uncertainty in mathematics teaching: The National Curriculum experiment in teaching probability to primary pupils. *Cambridge Journal of Education, 34*(3), 297–314. https://doi.org/10.1080/0305764042000289938.

Watson, A., Jones, K., & Pratt, D. (2013). *Key ideas in teaching mathematics: Research-based guidance for ages 9–19*. Oxford, United Kingdom: Oxford University Press.

Chapter 4
Challenges and Opportunities in Experimentation-Based Instruction in Probability

Per Nilsson, Andreas Eckert and Dave Pratt

Abstract This study reports on how instruction that is based on engaging students in practical experiments can create challenges and opportunities in the teaching of the relationship between a classical a priori and a frequentist model in estimating the probability of random outcomes. Knowledge is assumed to lie in the inferentialist relationships within the game of giving and asking for reasons (GoGAR). We report on dilemmas (challenges vs. opportunities) faced by the teachers and the researchers who co-designed the tasks: (i) whether it is effective to avoid the elicitation of deterministic reasons for random behaviour or to invite students to reflect on the lack of power of such reasons; (ii) whether the GoGAR is best served by accepting any responses from students or by challenging responses in order to clarify what is normative; (iii) whether the sample space that generates random outcomes should be revealed or not.

Keywords Classical a priori and frequentist probability · Experimentation-based instruction · Challenges and opportunities · Inferentialism · Game of giving and asking for reasons

---

P. Nilsson (✉)
School of Science and Technology, Örebro University,
701 82 Örebro, Sweden
e-mail: per.nilsson@oru.se

A. Eckert
Örebro University, Örebro, Sweden
e-mail: andreas.eckert@oru.se

D. Pratt
London University, London, UK
e-mail: david.pratt@ucl.ac.uk

C. Batanero and E. J. Chernoff (eds.), *Teaching and Learning Stochastics*,
ICME-13 Monographs, https://doi.org/10.1007/978-3-319-72871-1_4

4.1 Introduction

In the current study we examine challenges and opportunities in experimentation-based instruction in the learning of the bi-directional relationship between a classical a priori[1] and a frequentist model of probability (Borovcnik and Kapadia 2014).

Citizens would be better placed to make reasoned judgments in a modern society if they were able to assess the likelihood for the outcome of uncertain events based on the observed or reported frequencies of the event. For example, when a woman is faced with a positive result from a breast cancer test, her anxiety might be to some extent ameliorated by knowledge about the high frequency with which the testing generates a false positive. Citizens would benefit if they were able to make theoretically grounded predictions. For example, faced with the near impossibility of winning the lottery with a measurable frequency and noting that someone does indeed win the lottery each week, a citizen might see a win as purely a matter of luck—you win or you lose—with no deep sense of the astronomically high odds against winning calculated according to the number of entries.

Working practically with experimentation has been put forward as a promising activity in the teaching and learning of probability (Nilsson 2014), especially as it draws attention to the bi-directional relationship between a classical and a frequentist model of probability (Ireland and Watson 2009; Nilsson 2009; Stohl and Tarr 2002). Experimentation brings "life" to random behaviour, providing students opportunities to build an understanding of probability from experiencing and interacting with chance encounters (Fischbein 1975). In experimentation students can be challenged to make predictions and verifications of probabilities (Shaughnessy 2003) and to become engaged in formulating statistical and probabilistic questions, collecting data, analysing data, and drawing data-based conclusions and inferences (Paparistodemou and Meletiou-Mavrotheris 2008).

Experimentation is also often accepted in school practice. Activities of flipping coins, rolling dice or making inference about the outcome of throwing thumbtacks is common practice in many classrooms. However, just because students are engaged in experimentation does not guarantee meaningful learning will occur. Experimentation involves many knowledge aspects that can distract students, so hindering students in favouring probabilistic reasoning over deterministic reasoning (e.g., Makar and Rubin 2009; Nilsson 2013). For similar reasons, students might find it difficult to consider concrete random generators as fair as a computer-based random process (Konold et al. 2011), or vice versa. So, on the one hand, practical *experimentation-based instruction in probability* (EBIP) is considered promising in providing students meaningful engagement with uncertainty. But, on the other

[1]Henceforth, we will refer to a classical a priori model simply as a 'classical model'. It is an a priori approach in that it allows for modelling probabilities before any trial is made, based on the assumption of equally likely elementary events in the sample space (Borovcnik and Kapadia 2014).

hand, such experiences might raise certain instructional difficulties. The significance of the present study should be considered in relation to this dilemma in that it aims at providing further insight into the challenges and opportunities in practical EBIP.

Scanning the different contributions in Chernoff and Sriraman (2014) we note how cognitive and individualistic perspectives have been and continue to be very influential in research on probability learning. The social, situational and interactive nature of learning (Lave 1988) has gained relatively limited attention. In the present study we intend to make a small step in this direction. We will investigate and evaluate experimentation in relation to interactive, talk-based teaching (Hufferd-Ackles et al. 2004). Viewing processes of learning and understanding from an interactive and talk-based perspective follows recent calls for mathematical classrooms in which memorizing pre-defined procedures and learning rules for manipulating symbols give way to learning to communicate and to reasoning about and through mathematics (Walshaw and Anthony 2008). It gives way to a classroom practice in which talking about mathematics becomes acceptable, indeed essential, and that mathematical discussion, explanation, and the defence of ideas become defining features of a quality mathematical experience (Brodie 2011).

In the current study, we adopt the theory of inferentialism (Brandom 1994) to conceptualize meaning-making in probability in an interactive classroom practice, which is based on experimentation. Inferentialism is a philosophical theory, which stresses the social, pragmatic and inferential nature of meaning-making (Brandom 1994, 2000). Meaning-making is viewed as a construct of language use, coming to be articulated and constituted in language games (Wittgenstein 1958). Brandom uses *the Game of Giving and Asking for Reasons* (GoGAR) as a metaphor to describe how knowledge, meaning-making and understanding emerge and are constituted in the social practice of a language game, which follows an inferential logic (Bakhurst 2011; Brandom 2000).

The aim of the study is to examine challenges and opportunities in practical and interactive EBIP and address the following research question: How can practical EBIP create challenges and opportunities for students' active engagement in GoGAR that are focused on the bi-directional relationship between a classical and a frequentist approach to probability? Our goal is not to develop a prescription for the design of practical EBIP. Instead, the goal is to distinguish and characterize challenges and opportunities, aiming at sensitizing (Ruthven et al. 2009) researchers and teachers to essential aspects of practical and interactive EBIP.

4.2 Previous Research

Reviewing research containing experimentation with data, Nilsson (2014) distinguishes two overall methodological and analytical directions regarding the bi-directional relationship between a classical and a frequentist model of probability. One direction was the *mapping direction*. This is the case when researchers investigate students' understanding in situations where students are asked to reflect on how a classical model, in terms of sample space configurations, is mapped in data. The classical model is used to predict observations of frequencies (Prodromou 2012).

Nilsson (2014) refers to the other direction as the *inference direction*. This is the case when students are asked to infer a probability from data, since information about the sample space is unavailable, limited or too complex. Prodromou (2012) discusses how the variation in aggregated data for a statistical summary gradually emerges towards a predictable shape and trend when the statistics is sampled repetitively (as per the Law of Large Numbers), the probability distribution of that statistical summary. For example, a positive result from breast cancer screening might be interpreted as one result amongst many amongst which there is a large number of false positives.

Up to date, educational research on the bi-directional relationship between a classical and a frequentist model of probability can be described as *characterizing research*. Researchers have tried to distil students' mental representations (e.g., Konold et al. 2011) and characterize the state of affairs in relation to how students express understanding of basic ideas about this relationship (e.g., Ireland and Watson 2009). In research, structured in accordance to the mapping direction, three basic ideas have been particularly focused. These are students' ability to *identify the sample space as a key structure in modelling probability*, their understanding of *the role of the sample space in the distribution of data* and of *the role of the size of the sample* in the relationship between a classical and a frequentist model of probability.

Being able to identify all the possible outcomes is essential for a classical model for estimating the probability of outcomes for a random event (Horvath and Lehrer 1998). However, learning techniques to generate all possible outcomes of a sample space have little value unless students realize the importance of how the sample space regulates the likelihood of random events in this case and is reflected in data (Jones et al. 2007). For instance, in Nilsson (2009), students were challenged to infer the probability of the total of two dice. The dice were not cubical so it was clear to the students that the shape of the dice had great importance for the probability of the different outcomes. However, for many students the sample space gained limited attention. Material features of the random generator overshadowed sample space considerations in students' attempt to explain what regulated the outcomes of the random process. The sample space and the numbers of favourable outcomes of each sum was not explicitly taken into account. In Fischbein et al. (1991), students (grades one to five) were asked to compare the probability of throwing a "5" three times either by throwing one die three times or by simultaneously throwing three dice. The sample space of the two situations is the same and, hence, so are the probabilities. However, it was found that several students distinguished between the situations. The most common prediction was that by successively throwing the die, they have a higher chance to obtain a "5" three times. On the basis of follow-up interviews, it was argued that such a prediction was based on a belief that the individual can *control* the outcome of a throw. Pratt and Noss

(2002) report on two situated abstractions in probability, the *distribution resource* and the *large number resource.* The distribution resource is described as 'the more frequent the outcome in a sample space, the greater is its proportion in a sample'. The large number resource reflects an understanding of the principle that (relative) frequencies stabilize as a function of an increasing amount of data. Aspinwall and Tarr (2001) and Stohl and Tarr (2002) also show how children can develop their understanding of the role of the sample space and the number of trials in the relationship between a classical and a frequentist model of probabilities when they have the chance to run as many trials as they want and view the results in graphical representations.

The previous paragraph summarises research on how students appreciate (or not) the structure and formulation of data when they know the distribution (though they may not recognise the significance of the sample space). Research on how students infer distributions from data is not so extensive or developed (Nilsson 2014). One reason for this may be that processes of inference are perceived to be the domain of statistics rather than of probability. Making inferences involves probability generalizations that go beyond the collected data (Pratt et al. 2008). It concerns making an inference from the variation in the data to the features of the system generating that data (Makar and Rubin 2009). Research in this direction has mainly focused on two basic ideas; on the one hand, it has focused on why or why not students use data to infer to a probability distribution. In this situation it has been shown that students often seek other explanations, perhaps based on previous experiences rather than on the data, to explain variation in the data (Makar and Rubin 2009; Nilsson 2014). On the other hand, research has focused on how students evaluate information, gathered from short and long term data series. In this second situation, researchers have focused on how students evaluate and understand the role of sample size to infer an underlying probability distribution (Pratt et al. 2008).

Urn-like situations are often used to provide students with an unknown sample space. In Stohl and Tarr (2002) study, sixth grade students were challenged to use simulations and data analysis to infer information about the content of a bag of marbles. The students knew how many marbles there were in the bag (known population size), but did not know how many marbles there were of each kind. Similar to the findings of Brousseau et al. (2001), the information about the number of marbles in the bag affected the number of trials conducted. In both studies, several students drew samples of the same size as the number of marbles in the bag.

Ainley and Pratt (2002) describe purpose and utility as two key components of a learning environment. A task is purposeful when it has a meaningful outcome for the students. Utility refers to how a learning activity stimulates students to realize the power of mathematics 'to get stuff done'. In the present study we conceptualize purpose and utility in relation to inferentialism, in order to come to understand challenges and opportunities in practical and interactive EBIP.

4.3 Theoretical Background

We will use inferentialism (Brandom 1994) as our theoretical perspective to frame how students articulate understandings of chance and probability in interactive experimentation in probability. From the perspective of inferentialism, understanding moves within *the space of reasons*; people decide "what to think and do in the relevant domain in light of what there is most reason to think or do" (Bakhurst 2011, p. 136). The space of reasons refers to how people classify and label information and reflect on how and why they respond to information as they do (Brandom 1994, 2002). Understanding involves having a grasp of reasons, driven by the peculiar force of striving for more acceptable and better reasons (Bakhurst 2011). To illustrate this capacity we compare human reasoning with the capacity of a device, such as a smoke detector. A smoke detector can respond to a fire by raising the alarm. A human being can also respond differently to external stimuli. So, when he observes hot flames, for instance coming out from the kitchen, he can shout "fire!" However, what he can do that the smoke detector cannot is to understand the reasons for the hot flames and, so, the reasons for why he shouts, "fire!" He can also understand the consequences of shouting fire; that we should leave the house, call the fire brigade, not open the windows etc. He can place the meaning of an observation and the shouting in the space of reasons, which the smoke detector cannot. Things mean something to us (i.e., have content for human beings) because, by our capacity of being responsive to reasons we understand them in one way rather than another.

Brandom takes *inferences* as the primary moves of how content travels in the space of reasons. An inferential move of content means that the content can both serve as and stand in need of reasons. Grasping the meaning of an expression (understanding its content, like "fire in the house") implies grasping the inferential relations (premises and consequences) the expression is involved in or is used to articulate. In Brandom's terms:

> To grasp or understand (. . .) a concept is to have practical mastery over the inferences it is involved in—to know, in the practical sense of being able to distinguish, what follows from the applicability of a concept, and what it follows from (Brandom 2000, p. 48).

The inferential space of reasons is essentially socially articulated. Rather than considering the space in which thought moves, inferentialism suggests looking at the space of reasons from the perspective of language use (Sellars et al. 1997). On this account, Brandom adheres to the pragmatic perspective of the late Wittgenstein (1958) and his principle idea that being a meaning-maker involves being a player of a language game. Connected to the capacity of being responsive to reasons, Brandom (2000) describes the GoGAR: being a player, a meaning-maker, in GoGAR involves being able to participate in "the social, implicitly normative game of offering and assessing, producing and consuming, reasons" (p. 81).

GoGAR is structured according to the two basic normative statuses, *commitments* and *entitlements*, which, in turn, arise in a socially instituted structure of *authority* and *responsibility*. Commitments and entitlements are normative statuses in that they are acts of *judgments* made in accordance to the GoGAR and the social practice within which they are embedded. What one is doing in claiming how things are is undertaking a commitment to the things actually being that way. It is a normative act in that it implies the act of taking a stand, in terms of a belief of something or of prescribing a doing. It is also normative in terms of prescribing (obliging) one to other beliefs and doings. Such commitments, in the first place, stand in inferential relations—such as that *by* committing oneself in the probability classroom to "the probability of six (with an ordinary die) is one sixth", one thereby implicitly commits oneself (whether one realizes or understands it or not), among different things, to "the die has six sides" and "all sides have an equal chance to appear". It is such inferential relations that constitute the content of those commitments and, hence, of the concepts involved in the claims expressing the commitments. But, for such an inferential structure of commitments to count as involving assessment of appropriateness, "there must be in play also a notion of *entitlement* to one's commitments: the sort of entitlement that is in question when we ask whether someone has good *reasons* for her commitments" (Brandom 2000, p. 43). Hence, to be a knower is on this inferentialist perspective to know what one is committing oneself to by a certain claim, and what would entitle one to that commitment.

In a GoGAR, the production and consumption of commitments and entitlements are forced by the two communicative acts, *authorization* and *responsibility*. In producing a claim (and the commitment it expresses) one is first authorizing further commitments; both in relation to the consequences on the part of content (like what consequences it will have for further commitments if I claim the probability of six with an ordinary die is 1/6) and in relation to communicational consequence, that is, in how others can use an original commitment in their reasoning. In doing so, performers of claims are being responsible, "in the sense of being answerable for their claims" (Brandom 1994, p. 173). Producing claims implies undertaking a specific task responsibility. If the entitlement of a claim should be brought into question, the one who performed the claim has the responsibility to show, by producing reasons, that he or she is entitled to the commitment expressed by the claim.

With respect to teaching, the presented inferential perspective implies the importance of explorative learning environments, which invites students and teachers to express claims (expressing commitments) and to give and ask for reasons by which they can become entitled to their claims. In Brandom's (1994) words:

> Formulating as an explicit claim the inferential commitment implicit in the content brings it out into the open as liable to challenges and demands for justification, just as with any assertion. In this way explicit expression plays an elucidating role, functioning to groom and improve our inferential commitments, and so our conceptual contents. (p. 127)

Authorization implies a teaching approach that encourages students to articulate commitments to force the collective reasoning forward, in providing the possibility for others to capitalize on and undertake an original commitment, serving the role of social, person-based entitlement (e.g., "I agree with Samuel, the possibility of arriving on six is one-sixth because the six is on one of six sides of the die").

Giving an inferentialist account on purposeful tasks and utility, we say that a task becomes purposeful when students recognize its inferential significance. The task becomes purposeful when the outcome of it undertakes the role of premise or conclusion in an inference. However, a task might be purposeful without being mathematical. It can be purposeful in terms of it being rich and motivating (Ainley et al. 2006). So, we have to take into account also how students bring into the GoGAR inferences that *utilize* concepts and ideas relevant to a probabilistic perspective.

4.4 Method

We analyse data gathered from a small-scale teaching experiment with students, 11- and 12-year-olds, in a Swedish elementary school. The aim of the task was to engage students in GoGARs involving the bi-directional relationship between a classical and a frequentist model of probability.

4.4.1 Setting the Stage of the Teaching Experiment

The teaching involved two classes, one Grade 5 and one Grade 6. Two researchers (the first and second author of the paper) and the two class teachers constituted the developmental team of the teaching. Involving the two teachers in the process strengthens the practical relevance and reliability of the study and its results. There were six lessons conducted in the Grade 5 and five lessons conducted in the Grade 6. There were 22 students participating in the Grade 5 class and 24 students in the Grade 6 class. Each lesson in the Grade 5 class preceded the lesson in the Grade 6 class, and they were conducted less than a week apart. Between the lessons the team met to reflect on the previous lesson and for planning the next lesson. Hence, the lessons of the Grade 6 drew on what we learned from the proceeding lessons of the Grade 5. All lessons were video-recorded.

A characteristic of EBIP is that it calls for formulating a prototype model and making predictions based on its assumptions, and for providing students feedback on which they can evaluate and develop their initial model. The present case of practical and interactive EBIP included the random device of a bottle containing a number of small coloured marbles.[2] In the first lesson the bottle was covered. Neither the

[2]The idea of using transparent and covered bottles filled with marbles as random devices was borrowed from Brousseau et al. (2001).

students nor the teachers knew the content of the bottles. When turning the bottle, the colour of one marble was revealed while remaining inside the bottle. By this construction we created a constant, unknown sample space. The activity was presented as a competitive race in the first lesson with three contestants (blue, white or red marble) on a six-step track. As one of the three colours was observed on a bottle turn, that colour advanced one step down the track. The race finished when one colour reached the end. The students were asked to predict which colour would first appear six times during each race. Between each race the teachers allowed for reflection and discussion and the results of each race remained on the board. We wanted the students to start reflecting on the sample space and if they could infer how it was composed based on their empirical results. Hence, the challenge here was in the inferential direction, drawing on the variation in the data (the colours of the marbles appearing from the bottle observations) to make conclusions about the generating system (the unknown sample space of colours inside the bottle).

Based on the topics discussed by the students in the first lesson, the following three lessons revolved around a transparent bottle, a bottle with a visible, known sample space. When working with the transparent bottle the students were challenged to reflect on issues of randomness and the role of sample size to match the results with the content of the transparent bottle. We wanted the students to be able to map the empirical results with the now known sample space. The transparent bottle contained a uniform distribution of 2 red, 2 white and 2 blue marbles and each working group had their own bottle. More specific, the activity in lesson 2 focused mainly on issues of randomness, asking the students to predict results based on the now visible sample space. Lessons 3 and 4 used the same transparent bottle and distribution and focused on issues of sample size, aiming to establish the bi-directional relationship between a classical and frequentist model of probability. Hence, lessons 3 and 4 introduced the mapping direction but in the light of the children's experiences in the inference direction from lesson 1. The students returned to the covered bottle in the last lesson(s).

In the last lesson(s), the students were supposed to use their developed understanding of the bi-directional relationship between a classical and frequentist model of probability to infer the content (the sample space) of the covered bottle. The same unknown sample space as in Lesson 1 was used but now each working group had their own covered bottle, enabling the class to create a big joint sample from which they could infer the contents of the bottle. The Grade 5 teacher split the last lesson into two to create more time for discussions while the Grade 6 class managed in one. We use the pseudonyms Karen for the Grade 5 teacher and Tilly for the Grade 6 teacher.

4.4.2 Method of Analysis

We were interested in coming to understand challenges and opportunities in EBIP, in relation to the evolving GoGARs, conferring content on the relationship between the proportions of favourable outcomes in the sample space and the relative frequencies in samples and on the role of the sample size in understanding this relationship. To this end, we structured the retrospective analysis of the teaching experiment (Cobb et al. 2003) according to the chronological order of the lessons. Within this overall structure, we then searched for teaching instances that especially created challenges or opportunities for the teacher to shape affordances for students' active engagement in producing claims and giving and asking for reasons, relevant to the understanding of the bi-directional relationship mentioned. In our analysis we did not account for any quantification of the presence of the different findings reported on below. Instead, the empirical instances presented should be considered as representatives of our main findings. Some of the results are linked to a particular phase of the entire teaching experiment. For example, we will see how students emphasize deterministic aspects and, although such aspects appeared throughout the entire teaching experiment, they were most prominent during the first lesson. Other findings are of a more general kind, such as how the teachers acted to invite students into a GoGAR. The chronology provides the opportunity to report on findings of a more general nature along with more phase-specific challenges and opportunities obtained in EBIP.

4.5 Result and Analysis

4.5.1 Lesson 1. The Elicitation and Authorization of Deterministic Reasons

In both classes several students drew on material and person-based aspects in claiming and providing reasons for the result of a game. Reflecting on the appearance of how deterministic GoGARs appear, we see that there is a social act of authorization involved. What we see is that the first student's contribution gains authorization, shaping a kind of chain reaction. For instance, one student articulates, "it might be that there is a magnet in the bottle [the precise words are then rather unclear but the suggestion is that he means that marbles can stick to the magnet, preventing the marble from dropping down]". This suggestion then elicits reasons to shake and turn the bottle, and to consider the speed in which the bottle was turned, and then the order in which the different colours had been inserted into the bottle and differences in weight between the marbles. This kind of chain reaction we consider to be connected to an "anything-goes attitude" in how the teachers interacted with the students and responded to their contributions.

4.5.2 Asking for More Suggestions and Anything-Goes Attitude Versus Challenge and Make Explicit

In the first lesson in Grade 6, the following episode appeared after the first race:

Monica: It might be, like something in the bottle that makes it blue all the time
Tilly: It might be a thing inside that makes it blue all the time. Okay? [pointing to Oscar]
Oscar: There is more than one blue marble.
Tilly: There is more than one blue marble. Okay, Eric?
Eric: There is like, eh…, blue, like blue thing that light on [the rest is unclear]
Tilly: Okay, there is some little widget inside you think that makes…?
Eric: [Nodding]

The episode exemplifies how an experimentation-based teaching activity can support students in bringing to public scrutiny claims and reasons of different kinds. That is a good, and often necessary, starting point for inferential teaching because, to develop a holistic, inferential understanding of mathematics, there is a need to be able to capitalize on a certain range of information. However, what the episode also shows is how teaching can stop at students just providing suggestions. In a rather passive sense, the teacher just re-voiced students' contributions (cf., Eckert and Nilsson 2017). She avoided assessing or judging students' contributions and, as such, applied a kind of "anything-goes attitude", which may have made it difficult for students to come to understand which ideas were entitled and authorized in the GoGAR. What she authorized was not any particular content but the attitude that it was acceptable to come up with any suggestions. Moreover, looking at the episode above, we claim it must have been difficult for the students, particularly in the beginning of learning new content, to understand and build on others' contributions, and make connections between contributions, when the students were not challenged to take responsibility for their claims and provide reasons for them.

4.5.3 Lesson 2. Using Students' Data-Production to Challenge Deterministic Reasons

Between Lesson 1 and Lesson 2 the developmental team discussed how to deal with the dominance of deterministic reasoning apparent in Lesson 1 in both classes. Our concern was not that the students are mistaken in seeking a deterministic explanation—after all, deterministic reasons exist even if they are obscure—but that the teachers' and researchers' agenda was focused on stochastic explanations and we knew from experience that these would yield more profitable accounts of the situation. The team decided that in Lesson 2 the teachers should be more active and challenge students' deterministic commitments. It was decided that the teachers should change their interactive strategy, such that they would not always be passive and let anything go. Instead, when appropriate, they should take time and use the activity to challenge an assertion that expressed a deterministic commitment.

In Lesson 2 one of the tasks asked the students to note the longest run of one colour in a sample of 25 observations. In Tilly's class the following episode of many givings and askings for reasons appears when Tilly asked the groups why they did not arrive at the same results, even though the bottles were the same:

Harald: Maybe, because we shake it [the bottle] differently [showing a slow and a fast shake with the bottle].

Tilly reacted to this by claiming that this could not be the case. She retrieved reasons for her claim by referring to a group-discussion she been involved in during the group-work and where they came to the conclusion that it was not possible to control an outcome. The group in question gave her entitlements by humming affirmatively to what she said. She also referred to the groups' experimentations and that she had seen, when walking around in the class, that all groups were shaking their bottles pretty well. However, Harald and Albert in Harald's group were not convinced. Albert gives reasons for the idea that an outcome depends on who is shaking the bottle by referring to their observation protocol:

Albert: When I shook the bottle it became blue two times and… when Lucas… it became four times and then Harald seven.

Tilly asks Albert for reasons for how he can conclude this:

Tilly: Did you add your names to them [the observations]?
Albert: No! But we knew.

What Tilly then does is to give reasons for convincing Albert about the shortcomings of his inference, that the outcomes depend on who is shaking the bottle, by pointing to how this inference is incompatible with their observations:

Tilly: Ok, you knew since you moved it [the bottle] around in the group. But, was it the same colour? So, you did not have any long series? Didn't you say you had a long series?
Albert: Yes, five!
Tilly: Yes, but, if it had been almost the same colour when one [the same student] was shaking [the bottle], you should just have had one each time. If it [an outcome] was affected by how you were shaking.

Structuring what Tilly articulates in terms of inferences we see there are two incompatible inferences in play or, the consequences of one inference being incompatible with what is visible, a fact, in students' own series of observations. On the one hand there is the inference from "an outcome is dependent on the 'shaker'" to "there would be different colours in each observation if the 'shaker' were changed between each observation". What Tilly tries to make explicit to Albert, and the rest of the class, is that this inference is incompatible with the fact that the group actually had long series of the same colour in their sample. Being confronted with this incompatibility Albert endorsed it by claiming "You do not shake exactly the same". Later, Albert stressed this further by undertaking Tilly's reasoning when responding to a student who put forward the argument that an outcome would depend on the previous observation. The inference this student put forward moves from "an observed marble would be at the top of marbles at the bottom of the bottle when it is turned back" to "a marble located at the top of a pile of marbles at the bottom of the bottle will have an advantage to be observed when turning the bottle around". Albert made a move in the GoGAR, challenging this inference, by claiming "It is the chance that determines which [colour] arrives first". Albert also supplied his claim by shaking and turning the bottle around a couple of times. Taken together, Albert showed an undertaking (learning) of Tilly's inferential reasoning.

4.5.4 Lesson 3 and 4. Making Explicit the Role of Tasks and the Mathematics to Be Learned from Tasks

Ainley and Pratt (2002) have emphasized the importance of purposeful tasks for students' learning. To have students realize the purpose of a task and how different tasks are connected can be a challenging task for a teacher. From an inferentialist perspective, teaching is not simply a matter of a teacher telling students what to do and why. Instead, teaching should acknowledge and prompt students to be active in producing claims and giving and asking for reasons (Bakker and Derry 2011). In Tilly's third lesson it became clear how difficult it was to create such language games, when she was trying to involve the students in making explicit the purpose of the activities within the EBIP in question.

In Lesson 2 one student asked Tilly in what way they were doing mathematics. He thought they were just playing around with the bottles! Discussing this issue in the developmental team it was decided that Karen (the second teacher) and Tilly should take time to make explicit the end-goal of the teaching and how the different sub-tasks connect to this end-goal. Following Tilly's lesson we point to how difficult and challenging it can be for a teacher to involve students in the explication of such issues. We will also point to interactive strategies Tilly used in trying to overcome such difficulties.

In the beginning of Lesson 3 Tilly developed a GoGAR by inferentially positioning the task of Lesson 3 and 4 in relation to how it followed from Lesson 2 and how it can be followed, in terms of providing information for tackling the main task (end-goal) of the teaching, i.e., in drawing conclusions about the content of the covered bottle played with in Lesson 1. She introduced this enterprise by claiming, "We want to use mathematics to solve a problem" and then asked the students, "What is the problem we are going to solve?" To begin with, no student responded to the question but, after a few seconds one student said, "Why there are so many in a row!" Students' hesitation to Tilly's question and the answer provided emphases that the goal of the teaching was rather unclear to the class and that there was a need to make the aim explicit. Tilly chose to not acknowledge and give authority to the student who answered that the problem was to understand why there were so many [marbles] in a row. Instead she used the concrete random generator as a mediating tool (Sfard 2008) and the history of the class to initiate and turn the GoGAR in a direction she aimed for by holding up the covered bottle and asking the class, "Do you remember this?". The class responded that they did and Tilly continued:

Tilly: How can this [holding up transparent bottle]...because, we were a bit curious what it looked like inside here [holding up covered bottle]. How can this [holding up transparent bottle] help us to come to understand how it looks like inside this [holding up the covered bottle]?

There was no reaction from the class! During the episode Tilly switched between holding up the transparent bottle and the covered bottle as she spoke. We interpret this interactive strategy as if she was trying to show that the two random generators were equivalent. This interpretation is done in relation to her trying to clarify and justify the inferential relationship between the two bottles. In particular we claim that she aimed at the inference from "understanding the relationship between the contents of the bottle and frequencies in the case where the contents of the bottle are known" to "being able to use knowledge of this relationship to find out the content of the covered bottle". This inference is between the two bottles. However, the inference bears on another inference, that there is a relationship between the probability as represented by the sample space of coloured marbles in the bottle (whether transparent or not) and the frequentist estimation of probability, modelled from the colours generated (whether by the transparent bottle or not). This relationship has not yet been explicitly dealt with in a GoGAR and was the focus of the task of Lesson 3 and 4. By introducing the idea and purpose of this task, Tilly applied an interactive funnelling strategy (Bauersfeld 1998; Wood 1998); that is, she asked a leading question:

Tilly: Could there be some connection between how it looks like in the bottle [now holding up the transparent bottle] and how many times the different colours arrive? [The class is silent.] That can we think about!

No immediate reaction from the class this time either so Tilly continued, "If this is so, …, how could we find out that?". The students still had problems to contribute to the GoGAR. Tilly tried to develop the conversation. Tilly broke through this stalemate and so opened up the GoGAR by referring to a previous lesson in statistics (not with the bottles) where the class had analysed observations. In their talk of what they did in this previous lesson, the ideas of *observations* and *analysis* undertook a prominent position. Returning to the bottles, Tilly connects to these ideas:

Tilly: Do we have any facts?
Class: [Showing hesitation by humbling]
Tilly: Have we collected any facts?

Class: Yes! [Several students holding up their books with data from Lesson 2]

At first, the facts were not inferentially connected to observations for the students, even though they had just been talking about observations and how they had used them. However, when the term 'facts' was articulated together with 'collected' in a sentence, the students inferred that what Tilly asked about was *observations* and, in particular, the observations they had *collected* in Lesson 2. Lesson 3 then continued with the groups organizing the observations they had collected in Lesson 2.

The episode shows that EBIP did not become purposeful in some straight-forward manner. It shows that the teacher needed to pay serious attention to how the teaching might appear purposeful to the students and to be prepared to put energy into making it a purposeful and joyful activity. On a more general level, the episode also highlights a dilemma regarding questions that might facilitate progress in whole-class discussion, involving students in active participation. The dilemma is illustrated by how Tilly turned to a funnelling strategy and the use of test/factual questions when her open-ended/probing questions were unproductive (Hufferd-Ackles et al. 2004). On the one hand, funnelling and closed factual questions may leave little room for students' active participation in giving and asking for reasons. However, we know that teaching is limited in time so, on the other hand, if Tilly had not taken this active role, the students would probably have had to approach the task in Lesson 3 with less insight than they had as a result of that strategy. So, in this perspective, funnelling may have some productive consequences for moving the teaching forward. Nevertheless, the situation points to the need for developing teachers' interactive strategies to succeed in establishing productive GoGARs in the probability (mathematics) classroom.

4.5.5 Lesson 5 and 6. Task Commitment, Reasonable Answers and the Occurrence of Inferential Reasons (Inference Licensing)

The last lesson in both Tilly's and Karen's class represents the pinnacle of the lesson sequence. They both focus the whole class discussion on finding out how the sample space is composed in terms of the content of the bottle. What became a challenge here was to balance the quest for a definitive answer and engaging students in rich GoGARs involving the bi-directional relationship between a classical and frequentist approach to probability. It becomes apparent that Karen and Tilly still struggled between just asking for more suggestions (an anything-goes attitude) vs. deepening a GoGAR. In the following exchange we see how Karen tried to ask for reasons. Lisa had suggested that the sample space consisted of 1 red, 2 white and 2 blue marbles.

Karen: Did you use these, the relative frequencies when you reasoned? And saw that blue came up approximately twice as many times? 13 + 13 is 26
Lisa: I don't know
Karen: No, that's good, we are reasoning, good. Let's start from there. Others?

Karen tried to move the GoGAR forward by challenging Lisa's claim and helping her make her reasoning more explicit. Lisa, on the other hand, hesitated. The inferential norm of communicational responsibility seems not to have been very strong norm in the class. Lisa was not prepared to account for her claim! In Karen's response to Lisa's hesitation or inability to give reasons for her commitment, we see her struggle to encourage participation (by eliciting many suggestions) while also encouraging a more reflective stance, which speaks to taking responsibility of a claim. What follows is a chain of suggestions as seen in the first lesson, but this time with some connection to the relative frequencies displayed. The chain of suggestions differs from the chain appearing in Lesson 1 in two respects, which both point to how EBIP can support the learning of probability: (1) this time, the students' were committed to supply a reasonable answer, taking into consideration relative frequencies and random variation; (2) on this occasion, the students have a genuine interest in the answer.

Reasonable answer. In Lesson 1 the students' suggested one explanation after the other, seemingly only interested in participating in a game of producing suggestions. The contributions in the last lessons were more connected and focused on providing a reasonable answer, and not any answer! The patterns in Karen's and Tilly's last lesson were similar to each other, with Karen's being the more extreme case since almost all students who made a contribution gave a unique answer. The suggestions appearing in Karen's class, [122] [132] [154] [144] [222] [143] [142] in order [Red Blue White] (see Fig. 4.1), suggests that the students were now determined to provide not any answer but, according to the frequency information provided, a reasonable answer. How this interactive pattern was developed speaks to authorization and consequences of commitments: A suggestion (expressing a commitment) authorizes subsequent suggestions, leading to just small differences in further suggestions. The pattern also speaks to entitlement. Even though the suggestions were not exactly the same, they were close enough to provide for person-based entitlement for the reliability of the suggestions. However, to count as knowing, such person-based issues of authorization and entitlement depend on an understanding of random variation. For instance, in accordance to the relative frequencies the class had been calculating, an understanding of random variation would license the inference from [122] to [132] (i.e., that both claims are appropriate predictions of the content of the bottle). In the activity such licensing (reasons given) was expressed by, for instance, "it might depend on chance (randomness)" and "it is determined by chance (randomness)".

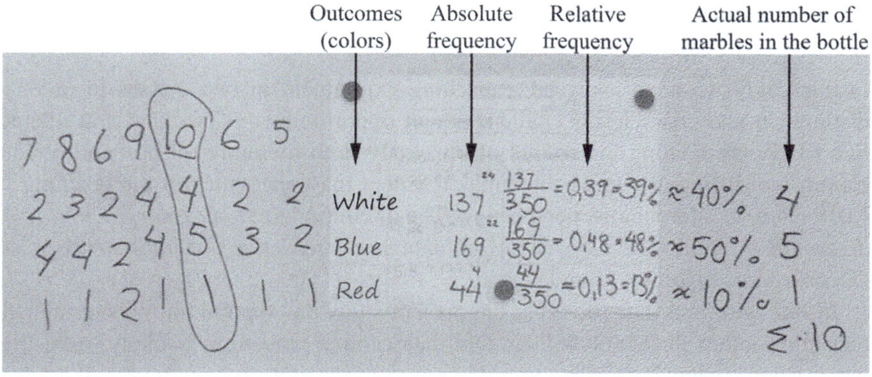

Fig. 4.1 Photo of the whiteboard (To the left on the board, students' suggestions of the content of the bottle. One suggestion is a total of 10 marbles in the bottle, distributed as 4 white, 5 blue and 1 red marble (marked by the teacher). Since, the original text was in Swedish, we added translations (white, blue, red) in the picture and clarifying text-boxes at the top) (Color figure online)

Genuine interest in the answer After some time, Lisa asked, "When we arrive at an answer, will we open it (the bottle) then?" This question sparks a discussion that became central to both Karen's and Tilly's ending of the lesson sequence; will they be allowed to reveal the hidden, unknown sample space? Compared to Lesson 1, when everyone was highly engaged in the race and who picked the winning colour, the last lesson was characterized by a genuine interest in the actual sample space, in terms of the content of the bottle. On this account, EBIP seemed to elicit and maintain a high level of situational interest (Mitchell 1993), where students experienced the task as purposeful (Ainley and Pratt 2002) in that the outcome of the tasks was perceived meaningful to them. The dilemma that both Karen and Tilly now faced was whether it was beneficial or not to reveal the mystery that had part in eliciting that interest. Their different solutions provided them with unique opportunities.

Karen chose to reveal the hidden sample space, satisfying the students' curiosity. This act enabled the class to deepen their GoGAR about the bi-directional relationship between a classical and frequentist model of probability. One student saw how the relative frequencies actually matched the now visible sample space and made this insight explicit to the rest of the class. The class also engaged in a discussion on how they could produce relative frequencies with an even larger sample, to come even closer to the proportions of marbles. Tilly chose not to reveal the hidden sample space. This act enabled her to emphasize the overall aim of giving and asking for reasons grounded in uncertainty. Tilly encouraged her students to be aware of their own learning process: No further discussions emerged but Tilly asked the students to write down in a note book about how they perceived their own development in relation to their reasoning proficiency.

4.6 Discussion

In this study we have analysed a teaching experiment in two classes in order to distinguish and characterize challenges and opportunities in practical and interactive EBIP. By relating the results of our analysis to literature on probability education, we distinguish three dilemmas teachers may encounter in the teaching of EBIP from an inferentialist perspective. The dilemmas are: *hiding versus not hiding deterministic issues*, *inviting versus challenging interactive teaching strategy* and *known versus unknown sample space*.

In line with previous research, the present study has reported how deterministic reasoning rather than probabilistic and data-centred reasoning is likely to be triggered when students are asked to predict or give account of the outcome of random processes (e.g., Fischbein et al. 1991; Makar and Rubin 2009; Nilsson 2013). This raises the question as to whether teachers should avoid practical random devices in EBIP, in order to avoid the elicitation of deterministic reasoning and give priority of probabilistic reasoning. We think not! We claim it is important that teaching does not only make use of random devices that generate uniform probability distributions, which may give students a sense that probability distributions are always uniform and reinforce an equiprobability bias (Lecoutre 1992). In addition, although equiprobability is clear when tossing a symmetric coin or a die, this assumption is rarely possible to apply to functional or everyday situations (Gal 2005). If one central goal of teaching probability is to develop rational citizens, who make reasoned judgement, teaching needs to go beyond "idealistic" game-like situations (Nilsson 2013). On the one hand, it may be appropriate for teachers to seek to reduce the elicitation of deterministic reasons in probability teaching to focus more on probabilistic accounts. On the other hand, teachers may need to make students aware of and challenge their preference for determinism while providing them with the opportunity to model probability from a theoretical, a priori point of view.

In the analysis we see two types of interactive attitudes/strategies; an inviting anything-goes attitude and a taking time and challenging attitude. Both these two strategies, or attitudes, have a place in inferentialism. An 'anything-goes attitude' is fruitful in that it is not evaluative. It invites students to contribute with claims and suggestions and so creates opportunities for teachers to provide for a holistic understanding of concepts and ideas. On the other hand, we see that such an attitude may not necessarily lead to reflection and deeper understanding. It does not challenge students' to make their commitments and entitlements explicit in the GoGAR and, thus, make it possible for teachers and students to capitalize on other people's ideas. From an inferential perspective, a balance between both interactive patterns is needed, and that is why teachers may face a dilemma in trying to develop inclusive and rich GoGARs.

Brousseau et al. (2001) suggest that teachers work with an unknown sample space and that they should avoid letting students verify whether their data-based reasoning conforms to the sample space by making the sample space known to the students. Connected to a similar bottle-situation as the present study, they argue that, if the sample space is made visible, "students would simply have "learned" a new technique: one can "guess with certainty" the unknown composition of a bottle by making or simulating a lot of drawings. But they would have had no idea of the notion which is specific to probabilistic computations/…/there will always remain a doubt" (Brousseau et al. 2001, p. 392).

Another risk of opening the bottle may be that it takes away from the students their responsibility. Differently formulated, to verify a reasoning and its conclusion by making a sample space visible may consolidate a socio-mathematical norm (Yackel and Cobb 1996) that hinders students in developing confidence and trust in their own ability and strength to reason; the final responsibility for the correct answer is handed over to something or someone else. However, from what happened after Karen opened the bottle, we can also see how this creates unique learning opportunities. Here students reflected on how and why the different suggestions differed from the proportions of marbles in the bottle and they come to bring such reflections into a GoGAR, around ideas of randomness, random variation and sample-size. So, there remains this dilemma as to whether the sample space should be made known or not to the students. However, we saw in Lesson 1 that the relationship between sample and sample space was not clear to the students. It was necessary to make this relationship visible to the students. In the present study, we have shown examples of how this can be done by systematically switching between situations with visible and invisible sample spaces, with different compositions of the sample spaces in the two cases. Connected to this, we have drawn attention to teachers needing to be alert to how students perceive the inferential significance (purposefulness) of the different tasks involved in a systematic sequence of tasks.

The present study shows the complexity in challenges and opportunities in practical and interactive EBIP. It shows the need to take a fine-grained account of the social and situational nature of how students express and develop an understanding of randomness and probability in such learning environments. On this account, we invite future research to continue offering such accounts and to develop frameworks, aimed at prescribing or sensitizing (Ruthven et al. 2009) teachers in developing effective teaching in probability, in practical and interactive experimentation environments.

References

Ainley, J., & Pratt, D. (2002). Purpose and utility in pedagogic task design. In A. Cockburn & E. Nardi (Eds.), *Proceedings of the Twenty Sixth Annual Conference of the International Group for the Psychology of Mathematics Education* (Vol. 2, pp. 17–24). Norwich, United Kingdom: International PME Group.

Ainley, J., Pratt, D., & Hansen, A. (2006). Connecting engagement and focus in pedagogic task design. *British Educational Research Journal, 32*(1), 23–38.

Aspinwall, L., & Tarr, J. E. (2001). Middle school students' understanding of the role sample size plays in experimental probability. *The Journal of Mathematical Behavior, 20*(2), 229–245.

Bakhurst, D. (2011). *The formation of reason.* London: Wiley Blackwell.

Bakker, A., & Derry, J. (2011). Lessons from inferentialism for statistics education. *Mathematical Thinking and Learning*, *13*(1–2), 5–26.

Bauersfeld, H. (1998). Interaction, construction, and knowledge: Alternative perspectives for mathematics education. In D. Grouws, T. Cooney, & D. Jones (Eds.), *Perspectives on research on effective mathematics teaching* (pp. 27–46). Reston, VA: National Council of Teachers of Mathematics.

Borovcnik, M., & Kapadia, R. (2014). A historical and philosophical perspective on probability. In E. J. Chernoff & B. Sriraman (Eds.), *Probabilistic thinking. Presenting plural perspectives* (pp. 7–34). Dordrecht, The Netherlands: Springer.

Brandom, R. (1994). *Making it explicit: Reasoning, representing, and discursive commitment.* Cambridge, MA: Harvard University Press.

Brandom, R. (2000). *Articulating reasons: An introduction to inferentialism.* Cambridge, MA: Harvard University Press.

Brandom, R. (2002). The centrality of Sellars' two-ply account of observation. In R. B. Brandom (Ed.), *Tales of the mighty dead: Historical essays in the metaphysics of intentionality* (pp. 349–358). Cambridge, MA: Harvard University Press.

Brodie, K. (2011). Working with learners' mathematical thinking: Towards a language of description for changing pedagogy. *Teaching and Teacher Education*, *27*(1), 174–186.

Brousseau, G., Brousseau, N., & Warfield, V. (2001). An experiment on the teaching of statistics and probability. *The Journal of Mathematical Behavior*, *20*(3), 363–411.

Chernoff, E., & Sriraman, B. (2014). *Probabilistic thinking: Presenting plural perspectives.* Dordrecht, The Netherlands: Springer.

Cobb, P., Confrey, J., Lehrer, R., & Schauble, L. (2003). Design experiments in educational research. *Educational Researcher*, *32*(1), 9–13.

Eckert, A., & Nilsson, P. (2017). Introducing a symbolic interactionist approach on teaching mathematics: The case of revoicing as an interactional strategy in the teaching of probability. *Journal of Mathematics Teacher Education*, *20*(1), 31–48.

Fischbein, E. (1975). *The intuitive sources of probabilistic thinking in children.* Dordrecht, The Netherlands: Reidel.

Fischbein, E., Nello, M. S., & Marino, M. S. (1991). Factors affecting probabilistic judgements in children and adolescents. *Educational Studies in Mathematics*, *22*(6), 523–549.

Gal, I. (2005). Towards "probability literacy" for all citizens: Building blocks and instructional dilemmas. In G. A. Jones (Ed.), *Exploring probability in school: Challenges for teaching and learning* (pp. 39–63). New York: Springer.

Horvath, J., & Lehrer, R. (1998). A model-based perspective on the development of children's understanding of chance and uncertainty. In S. P. Lajoie (Ed.), *Reflections on statistics: Agendas for learning, teaching, and assessment in K-12* (pp. 121–148). Mahwah, NJ: Lawrence Erlbaum.

Hufferd-Ackles, K., Fuson, K. C., & Sherin, M. G. (2004). Describing levels and components of a math-talk learning community. *Journal for Research in Mathematics Education*, *35*(2), 81–116.

Ireland, S., & Watson, J. (2009). Building a connection between experimental and theoretical aspects of probability. *International Electronic Journal of Mathematics Education*, *4*(3), 339–370.

Jones, G. A., Langrall, C. W., & Mooney, E. S. (2007). Research in probability: Responding to classroom realities. In F. K. Lester (Ed.), *The second handbook of research on mathematics* (pp. 909–956). Reston, VA: National Council of Teachers of Mathematics.

Konold, C., Madden, S., Pollatsek, A., Pfannkuch, M., Wild, C., Ziedins, I., Finzer, W., Horton, N. J., & Kazak, S. (2011). Conceptual challenges in coordinating theoretical and data-centered estimates of probability. *Mathematical Thinking and Learning*, *13*(1–2), 68–86.

Lave, J. (1988). *Cognition in practice: Mind, mathematics and culture in everyday life:* Cambridge, MA: University Press.

Lecoutre, M. P. (1992). Cognitive models and problem spaces in "purely random" situations. *Educational Studies in Mathematics*, *23*(6), 557–568.

Makar, K., & Rubin, A. (2009). A framework for thinking about informal statistical inference. *Statistics Education Research Journal, 8*(1), 82–105.

Mitchell, M. (1993). Situational interest: Its multifaceted structure in the secondary school mathematics classroom. *Journal of Educational Psychology, 85*(3), 424–436.

Nilsson, P. (2009). Conceptual variation and coordination in probability reasoning. *The Journal of Mathematical Behavior, 28*(4), 247–261.

Nilsson, P. (2013). Challenges in seeing data as useful evidence in making predictions on the probability of a real-world phenomenon. *Statistics Education Research Journal, 12*(2), 71–83.

Nilsson, P. (2014). Experimentation in probability teaching and learning. In E. Chernoff & B. Sriraman (Eds.), *Probabilistic thinking: Presenting plural perspectives* (pp. 509–532). Dordrecht, The Netherlands: Springer.

Paparistodemou, E., & Meletiou-Mavrotheris, M. (2008). Developing young students' informal inference skills in data analysis. *Statistics Education Research Journal, 7*(2), 83–106.

Pratt, D., Johnston-Wilder, P., Ainley, J., & Mason, J. (2008). Local and global thinking in statistical inference. *Statistics Education Research Journal, 7*(2), 107–129.

Pratt, D., & Noss, R. (2002). The microevolution of mathematical knowledge: The case of randomness. *The Journal of the Learning Sciences, 11*(4), 453–488.

Prodromou, T. (2012). Connecting experimental probability and theoretical probability. *ZDM, 44* (7), 855–868.

Ruthven, K., Laborde, C., Leach, J., & Tiberghien, A. (2009). Design tools in didactical research: Instrumenting the epistemological and cognitive aspects of the design of teaching sequences. *Educational Researcher, 38*(5), 329–342.

Sellars, W., Rorty, R., & Brandom, R. (1997). *Empiricism and the philosophy of mind.* Cambridge: Harvard University Press.

Sfard, A. (2008). *Thinking as communicating: Human development, the growth of discourses, and mathematizing*. New York: Cambridge University Press.

Shaughnessy, M. (2003). Research on students' understandings of probability. In J. Kilpatrick, W. G. Martin, & D. Schifter (Eds.), *A research companion to Principles and Standards for School Mathematics* (pp. 216–226). Reston, VA: National Council of Teachers of Mathematics.

Stohl, H., & Tarr, J. E. (2002). Developing notions of inference using probability simulation tools. *The Journal of Mathematical Behavior, 21*(3), 319–337.

Walshaw, M., & Anthony, G. (2008). The teacher's role in classroom discourse: A review of recent research into mathematics classrooms. *Review of Educational Research, 78*(3), 516–551.

Wittgenstein, L. (1958). *Philosophical investigations*. Oxford: Blackwell.

Wood, T. (1998). Alternative patterns of communication in mathematics classes: Funneling or focusing. In H. Steinbring, M. G. Bartolini Bussi, & A. Sierpinska (Eds.), *Language and communication in the mathematics classroom* (pp. 167–178). Reston, VA: National Council of Teachers of Mathematics.

Yackel, E., & Cobb, P. (1996). Sociomathematical norms, argumentation, and autonomy in mathematics. *Journal for Research in Mathematics Education, 27*(4), 458–477.

Chapter 5
Visualising Conditional Probabilities—Three Perspectives on Unit Squares and Tree Diagrams

Katharina Böcherer-Linder, Andreas Eichler and Markus Vogel

Abstract Two visualisations, i.e. the tree diagram and the unit square, are discussed from three perspectives of our research programme concerning possibilities of facilitating the understanding of Bayesian situations. We first regard the psychological perspective focusing on people's ability to solve specific tasks in Bayesian reasoning situations. Second, we discuss the educational perspective, where a visualisation has not only the goal to solve specific tasks but to understand the structure of a mathematical concept. Finally practical considerations for the every-day classroom constitute our third perspective. For each of these perspectives we compare the tree diagram and the unit square concerning their efficiency. Results imply that the unit square is more efficient from a psychological perspective and an educational perspective, and is on a similar level of efficiency when the practical perspective is regarded.

Keywords Visualisation · Bayes' rule · Bayesian reasoning situation
Tree diagram · Unit square

K. Böcherer-Linder
University of Education Freiburg, Freiburg, Germany
e-mail: katharina.boechererlinder@ph-freiburg.de

A. Eichler (✉)
Institute for Mathematics, University Kassel, Heinrich-Plett-Straße 40,
34132 Kassel, Germany
e-mail: eichler@mathematik.uni-kassel.de

M. Vogel
University of Education Heidelberg, Heidelberg, Germany
e-mail: vogel@ph-heidelberg.de

C. Batanero and E. J. Chernoff (eds.), *Teaching and Learning Stochastics*,
ICME-13 Monographs, https://doi.org/10.1007/978-3-319-72871-1_5

5.1 Introduction

In recent decades, psychological and educational research has proven that most people have severe problems in understanding the concept of conditional probability (e.g. Diaz et al. 2010; Gigerenzer and Hoffrage 1995; Kahneman et al. 1982). One of the most cited problems in this field of probability is the medical diagnosis test problem (e.g. Eddy 1982) that we present here without emphasising a specific disease like tuberculosis or cancer:

> The probability of a specific disease is 5% (base rate). If a person has this disease the probability is 80% that he or she will have a positive test result (sensitivity). If a person does not have the disease the probability is 10% that he or she will have a positive test result (specificity). What is the probability that a person, who received a positive test result, actually has the disease?

Research yielded one method to support the understanding of such Bayesian reasoning situations, i.e. to use natural frequencies instead of probabilities or percentages as the format of statistical information (e.g. Johnson and Tubau 2015). Based on a sample of 200 people, the statistical information given above can be expressed in the format of natural frequencies: 10 out of 200 persons have the disease (5%). Further 8 out of the 10 persons with the disease will get a positive test result (80%) and 19 out of the 190 persons without the disease will also get a positive test result (10%). Research shows that people are better able to process these natural frequencies to compute the proportion of persons having the disease given a positive test result:

$$P(disease|positivetest) = \frac{8}{8+19} = 29.6\%$$

Further, psychological research (e.g. Brase 2009; Sedlmeier and Gigerenzer 2001; Spiegelhalter et al. 2011) and educational research (e.g. Martignon and Erickson 2014; Wassner 2004) proposed several visualisations of Bayesian reasoning situations aiming to facilitate Bayesian reasoning. Although many researchers in mathematics education stated the overarching value of visualisations for understanding mathematics (e.g. Duval 2002; Presmeg 2014) the effect of visualisations in supporting the understanding of Bayesian reasoning situations seems to be ambiguous. For example, the research of Binder et al. (2015) showed that visualising statistical information in a Bayesian situation with the tree diagram or a 2 $\times$ 2-table hardly supports students' ability to solve tasks based on the Bayes' rule when probabilities are used as the format of statistical investigation. By contrast, when natural frequencies are used, visualising Bayesian reasoning situations seems to yield a further support for understanding these situations (e.g. Wassner 2004). The cited research results were focussed on solving tasks like the above given task concerning the medical diagnosis test. Thus, the mentioned research refers to the calculation of Bayes' rule which could be predominately understood as the procedure oriented aspect of Bayesian reasoning.

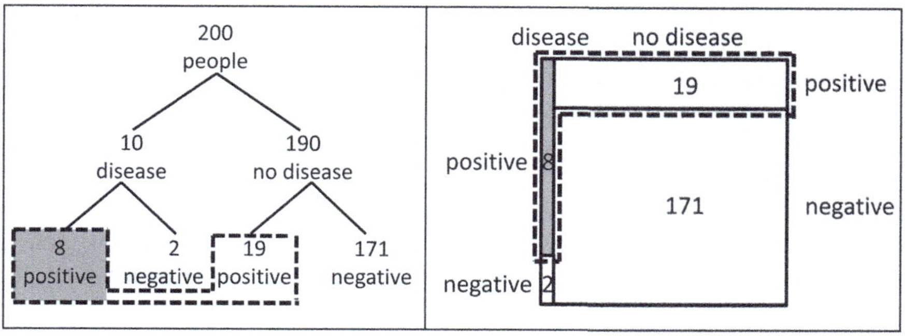

Fig. 5.1 The tree diagram and the unit square visualising the example of medical diagnosis test

For this reason, in this paper, we focus on both the procedural aspect and the conceptual aspect of Bayesian reasoning (cf. Rittle-Johnson and Schneider 2014), but we restrict our focus to two different visualisations of Bayesian reasoning situation, i.e. tree diagram and unit square. In Fig. 5.1 we present these two visualisations for the medical diagnosis test example based on a sample of 200 people: the sets and subsets being relevant for the example are emphasised by the dotted lines and the grey colour.

The rationale for the selection of the visualisations is twofold. First, the two visualisations represent different forms of visualisation that Khan et al. (2015, p. 96) call "Branch style" and "Nested style". The tree diagram representing the "Branch style" is arranged in a hierarchical way. That means, the whole sample is indicated on the top of the tree and the branches represent relations to subsets of the sample. Further, the information is given numerically. By contrast, unit squares representing the "Nested style" include no hierarchy, sets and subsets are neighboured. Further, the unit square represents the statistical information numerically and geometrically as a statistical graph (Tufte 2013). This is not the case for 2 × 2-tables. A second reason to select the tree diagram and the unit square is given by educational considerations: it is possible to use both visualisations in educational environments. Both these visualisations could be produced by learners, which is hardly the case for other elaborated visualisations that for example Spiegelhalter et al. (2011) provided in terms of communication of situations of uncertainty. Accordingly, although the tree diagram seems to be much more common than the unit square, both visualisations were used in educational contexts (e.g. Eichler and Vogel 2010; Sturm and Eichler 2014).

Given this background, our contribution has three aims concerning three different perspectives that are based on different results of our research: First, we regard tree diagrams and unit squares from a psychological perspective. The main interest from this point of view is to increase the proportion of correct answers by providing students with useful visualisations of statistical information. Thus, this line of research mostly refers to the procedural aspect of Bayesian reasoning. In psychological research two different approaches have been discussed to explain difficulties in Bayesian reasoning. These are the natural-frequencies hypothesis (e.g. Cosmides and Tooby 1996) and the nested-sets hypothesis (e.g. Sloman et al. 2003). For both approaches we discuss the visualisation by tree diagrams and by unit squares. We refer to recent research results that demonstrated the effect of both natural frequencies and nested-sets transparency for the performance in Bayesian reasoning tasks.

Second, we regard the tree diagram and the unit square from an educational perspective. From this point of view, the aim of visualising Bayesian situations is not the same as from the psychological point of view. For the educational perspective besides supporting the procedural aspect it is also important to identify visualisations that support the construction of conceptual knowledge. For the teaching of conditional probability, Borovcnik (2012a, p. 21) pointed out the need "to clarify the structure of the situation". One cited aspect is "to investigate the influence of variations of input parameters on the result" (p. 21). Thus, we focus in this section on how both visualisations reflect parameter dependency in Bayesian reasoning situations and report on the effect that the unit square and the tree diagram have on the understanding of parameter dependency.

Finally, we discuss both visualisations from a practical perspective of teaching conditional probabilities and Bayes' rule. From this point of view, one aim is the practicability in daily classroom practice for teachers and students. We discuss how both visualisations can be embedded in a spiral curriculum. In this perspective the tree diagram and the unit square are considered in a more global framework where the topic of probability is related to other topics such as fractions or combinatorics.

5.2 Psychological Perspective on Unit Squares

Two different cognitive models have been discussed to explain difficulties in Bayesian reasoning situations i.e. the natural-frequencies hypothesis (e.g. Cosmides and Tooby 1996; Gigerenzer and Hoffrage 1995) and the nested-sets hypothesis (Barbey and Sloman 2007; Sloman et al. 2003). In the following we outline these two accounts and discuss the tree diagram and the unit square from these different points of view.

5.2.1 Natural-Frequencies Hypothesis

Proponents of the natural-frequencies hypothesis depart from an evolutionary point of view: in our environment single-event probabilities are not observable. What we are observing are countable events. Hence, our mind has evolved to process natural frequencies rather than single-event probabilities: "The mind is tuned to frequency formats, which is the information format humans encountered long before the advent of probability theory" (Gigerenzer and Hoffrage 1995, p. 697).

Indeed, several studies demonstrated that the proportion of correct answers in tasks concerning the Bayes' rule increases dramatically when representing the information in the format of natural frequencies instead of probabilities (Binder et al. 2015; Cosmides and Tooby 1996; Gigerenzer and Hoffrage 1995). For the design of effective visualisations, proponents of the natural-frequencies hypothesis claim that visualisations should imitate the natural sampling situation. They propose the use of representations of "real, discrete and countable" (Cosmides and Tooby 1996, p. 33) objects as visualisations of natural frequencies. Unlike for the case of e.g. an icon array (Brase 2009) this idea is not realised in both the tree diagram and the unit square (Fig. 5.1). None of the two visualisations represents discrete and countable objects. The continuous fields of the unit square even seem to be contradictory to the idea of discreteness and countability. Nevertheless, the tree diagram and the unit square are practical alternatives to icon arrays, since the representation of a larger sample size is limited for icon arrays, most of all in an educational context.

5.2.2 Nested-Sets Hypothesis

Proponents of the nested-sets hypothesis attribute the difficulties of Bayesian reasoning to the fact that some sets of events are nested (Kahneman and Tversky 1982; Lesage et al. 2013; Sloman et al. 2003). They argue that "any manipulation that increases the transparency of the nested-sets relation should increase correct responding" (Sloman et al. 2003, p. 302). They explain the facilitating effect of natural frequencies by the fact that natural frequencies make it easier to see how many elements are in a subset and how the sets relate. The cognitive model into which the nested sets account has been incorporated is the dual process theory (Barbey and Sloman 2007): a standard probability format obscures the nested sets structure of the problem and, therefore, triggers the associative system which may lead to biases. Natural frequencies in contrast trigger the rule-based system because nested sets relations are made transparent, enabling people to reason consciously and according to the logic of set inclusion (Barbey and Sloman 2007).

For the design of effective visualisations, proponents of the nested-sets hypothesis claim that visualisations are helpful to the extent that they make the nested set structure of the problem transparent (Barbey and Sloman 2007; Sloman et al. 2003). Regarding this aspect, both the tree diagram and the unit square can be seen as nested-sets representations. In the tree diagram, the logical relations between sets and subsets are visualised by lines. In the unit square, subset relations are visualised by areas being embedded in other areas. However, when considering especially the specific subset relation that is required when calculating the posterior probability of a disease given a positive test result (see the medical diagnosis test), a fundamental difference between both visualisations becomes evident: due to the hierarchical structure of the tree diagram the set of all people tested positive is separated into two distinct parts. In contrast, the unit square is not hierarchical and the set of all people tested positive remains connected. Thus, the subset relation of the infected among those tested positive seems to be more transparent in the unit square than in the tree diagram. Following the nested-sets hypothesis, the unit square should be more appropriate to enhance correct responding in Bayesian reasoning tasks (cf. Fig. 5.1).

5.2.3 Testing the Effect of the Tree Diagram and the Unit Square on Solving Bayesian Reasoning Situations

Although there is a huge amount of research concerning the facilitating effect of visualisations on Bayesian reasoning (Binder et al. 2015), the facilitating effect of the unit square is mostly unknown (Bea 1995). For this reason, we summarise our own research results in this section.

Empirical results of our research gave evidence that for those subset relations that are not in line with the hierarchy of the tree diagram the unit square was more efficient than the tree diagram for the perception of subset relations (Böcherer-Linder and Eichler 2017). Accordingly, students' performance in Bayesian reasoning tasks was better with unit squares than with tree diagrams. In our research, this was shown for different samples with different mathematical backgrounds. For example, in Böcherer-Linder and Eichler (2017), 143 undergraduates at the Technical University of Munich (Germany) in an Electrical Engineering course were asked to complete a test including tasks involving four Bayesian reasoning situations. One task is given in Fig. 5.2.

Medical diagnosis test
In a preventive medical check-up, 1000 people are tested. The test has the following characteristic: 80% of the infected people and 10% of the uninfected people get a positive test result.
Calculate the proportion of infected people among those testing positive.

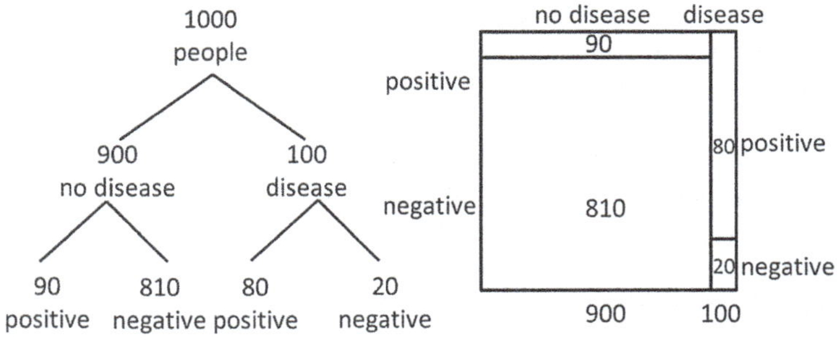

Fig. 5.2 Test item for investigating the facilitating effect of the tree diagram and the unit square concerning solving Bayesian reasoning tasks

The students got the test items only with one of the two visualisations, i.e. the tree diagram or the unit square. We used a t-test to show that for the sum over four items concerning Bayes' rule ($\alpha = .807$) with a maximum score of 4, the unit square ($M = 2.93$, $SD = 1.417$) was more effective than the tree diagram ($M = 1.72$, $SD = 1.494$), $t(141) = 4.961$, $p < .0001$, with an effect size of $d = 0.84$.

To conclude this psychological aspect of our research, the unit square seems to be more efficient than the tree diagram in supporting Bayesian reasoning regarding the procedural aspect of applying Bayes' rule.

5.3 Educational Perspective on Unit Squares and Tree Diagrams

While the main interest in psychology is how to enable people to get the correct value for a posterior conditional probability, mathematics education is also interested in how to build up conceptual knowledge of Bayesian reasoning situations. Borovcnik (2012b, p. 9) expresses this difference between psychological goals and educational goals and gives examples of what conceptual understanding of Bayesian reasoning situations consists of:

> The way of Gigerenzer (2002) – to free of mathematics by using simple absolute numbers instead of conditional probabilities – helps to find the correct solution but not always clarifies the situation (why a derived conditional probability is so small) and does not allow to check for changing the assumptions (e.g., to change prior probabilities) and is not open to an iterative application as is required often in Bayesian problems. It also gives no easy way of investigating how the posterior conditional probability depends on input parameters. Of course, the basic correct solution is a strong argument for Gigerenzer's natural frequencies. However, more investigations have to be done how the curriculum has to proceed from there.

In the following, we focus on two of the mentioned aspects, i.e. to investigate how the posterior conditional probability depends on the change of input parameters and to clarify the situation, why a derived conditional probability could be small for the diagnosis of rare events. We show how the understanding of these two aspects is supported by the visualisation with the tree diagram or the unit square respectively.

5.3.1 Change of Input Parameters

In the situation of the medical diagnosis test we have three different input parameters that impact on the posterior conditional probability P(disease|positive), i.e. the sensitivity P(positive|disease), the specificity P(positive|no disease) and the base rate P(disease). When considering potential beneficial effects of visualisations, the question arises how the visualisations reflect the change in these input parameters and how they visualise the impact of the change of the input parameters on posterior conditional probabilities.

For this purpose, we regard changes in the shape of the visualisations when input parameters are changed. In the tree diagram, only the numbers at the nodes change whereas the structure of branches and thus the shape of the tree remains unchanged. Thus, for the estimation of the influence of increasing sensitivities or increasing base rates someone has to consider the change in the relevant numbers as shown in Fig. 5.3, where the arrow up represents an increase of the natural frequencies and, the arrow down a decrease. For example, if the sensitivity would be bigger than 80%, more than 8 of the 10 people with the disease would get a positive test result. If the specificity would be smaller than 10%, less than 19 of the 190 people without the disease would get a positive test result. Finally, if the base rate would be bigger than 5% (given a stable specificity and a stable sensitivity), more people would have the disease and would get a positive test result. By contrast, there would be less people without the disease and accordingly less people without the disease and a positive test result.

In the unit square, in contrast, not only the numbers, but also the shapes of the partial areas change. Thus, statistical information is represented not only in a numerical but also in a geometrical way as shown in Fig. 5.4. This could be an advantage for the estimation of the effect of changing input parameters on conditional probabilities.

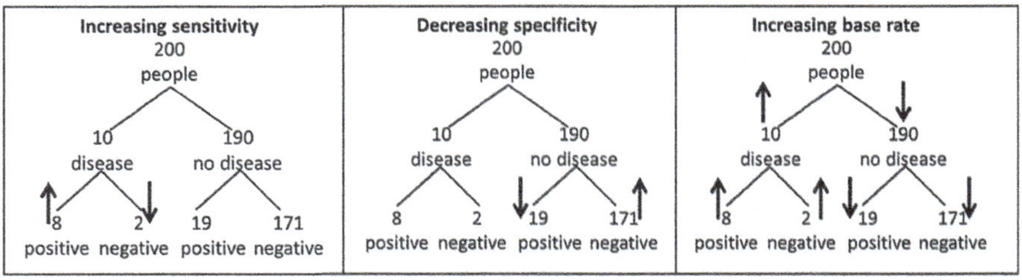

Fig. 5.3 Changes in the tree diagram when parameters in the medical diagnosis test (i.e. sensitivity, specificity, base rate) were changed

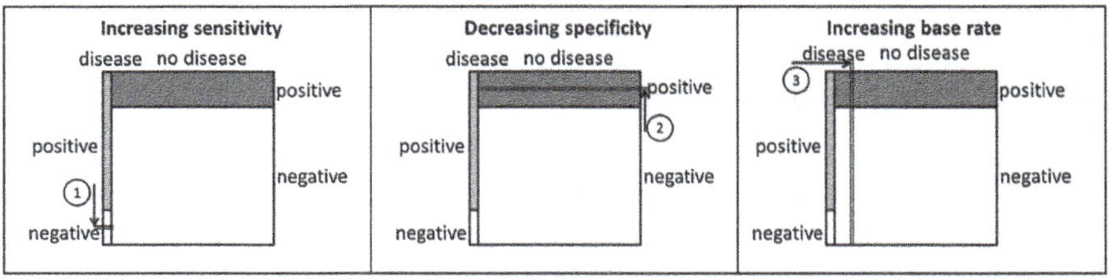

Fig. 5.4 Changes in the unit square when parameters in the medical diagnosis test (i.e. sensitivity, specificity, base rate) were changed. The arrows represent the direction of the change

One counterintuitive phenomenon in the field of the diagnosis test situation is the fact that enhancing the sensitivity of a test (Fig. 5.4, number 1) only has a small effect for diseases with small base rates if the crucial probability of

$$P(disease|positive) = \frac{P(disease \cap positive)}{P(disease \cap positive) + P(nodisease \cap positive)}$$

is considered. This phenomenon becomes transparent in the unit square: the area in the narrow rectangle (light grey representing the number of ill and positively tested people is only small compared to the rectangle representing the number of healthy and positively tested people (dark grey). If the sensitivity increases, the ratio between the light grey area (ill and positive) and the dark grey area (healthy and positive) only changes marginally. Accordingly, a change of the sensitivity in this situation only marginally changes the probability P(disease|positive). Another phenomenon in such situations is that the change of the specificity (Fig. 5.4, number 2) and also the change of the base rate (Fig. 5.4, number 3) considerably influences the posterior probability P(disease|positive). For example, the latter phenomenon becomes transparent in the unit square since the gain of area in the light grey rectangle increases considerably when the base rate increases.

The unit square also makes transparent the phenomenon of rare events and "clarifies the situation (why a derived conditional probability is so small)" (Borovcnik 2012b) referring to a medical diagnosis of a disease with a small base rate. Since the posterior probability P(disease|positive) represents the ratio of two areas (the light grey area and the sum of the light grey area and the dark grey area in Fig. 5.4), the bigger of the two mentioned areas determines the ratio. Thus, if the area representing the ill and positively tested people is small compared to the area of healthy and positively tested people, the phenomenon of rare events becomes apparent, i.e. the derived conditional probability is unexpected small. The unit square also makes one reason of this phenomenon transparent, i.e. a situation in which the base rate is smaller than the specificity of a diagnosis test.

None of the mentioned phenomena becomes transparent in the tree diagram in a geometrical sense. By contrast to evaluate the impact of a change in a parameter in a Bayesian reasoning situation, it is necessary to process the numbers representing the parameters in the situation.

5.3.2 Testing the Effect of the Visualisations on Students' Understanding of a Base Rate Change

Going beyond testing the calculation performance by applying Bayes' rule in our studies mentioned in Sect. 5.2.3, we also investigated the impact of the tree diagram and the unit square for questions concerning the aforementioned understanding of the influence of the base rate on conditional probabilities (Böcherer-Linder et al. 2017). In Fig. 5.5, we show one of our test items. Note that no calculation was required when solving the tasks and note that participants had to imagine how the numerical values and the visual appearance of the visualisations would change. As pointed out in Sect. 5.3.1, the structural relationships are more transparent in the unit square.

In two empirical studies, it turned out that the unit square was more appropriate to support the understanding of the influence of the base rate on conditional probabilities. In both studies we found a significant advantage of the unit square compared to the tree diagram. Although the practical significance was small (Cohen's d = .34 in the first study and Cohen's d = .36 in the second study), we nevertheless yielded a robust effect, i.e. a similar advantage of the unit square in each of our test items (Böcherer-Linder et al. 2017).

To conclude this aspect of our research, the unit square seems to be more efficient than the tree diagram in supporting Bayesian reasoning regarding the conceptual aspect of applying Bayes' rule.

Smoke

4000 students of a university were asked if they smoke or not. It turned out that one-third of the men smoke and one-fifth of the women smoke:

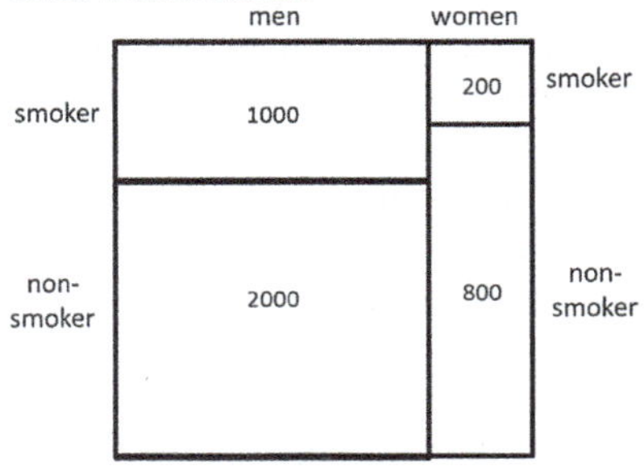

How the following proportions change if, one year later, there are more women among the 4000 students of the university and the smoking behavior of men and women is still the same?

Mark the correct solutions.

The percentage of non-smokers among the women will be bigger / smaller / constant.
The percentage of women among the smokers will be bigger / smaller / constant.
The percentage of men among the non-smokers will be bigger / smaller / constant.

Fig. 5.5 Test item for investigating students' conceptual knowledge referring the impact of a change of the base rate in a Bayesian reasoning situation

5.4 Practical Perspective on Unit Squares and Tree Diagrams

For discussing the practical perspective on unit squares and tree diagrams we refer to two different issues, i.e. the possibility of students drawing a visualisation in the classroom and the possibility of using a visualisation for different situations within a topic of school mathematics and also for different topics of school mathematics. Although there exist different names of the unit square for different purposes—for example: eikosogram (Oldford 2003), mosaic display (Friendly 1999), area model (Rathouz 2011)—we use in the following paragraphs only the term unit square for reasons of clarity.

5.4.1 Drawing Unit Squares and Tree Diagrams

A crucial practical question that concerns the selection of a visualisation for teaching mathematics (including statistics) is whether a visualisation could be appropriately drawn by students or not. The unit square provides the proportions of a Bayesian reasoning situation numerically and geometrically (Nested style). For this reason, it must be slightly more challenging to draw a unit square than a tree diagram which provides the mentioned proportions only numerically (Branch style).

However, this additional challenge of drawing a unit square becomes much smaller if only an approximate drawing draft of a Bayesian reasoning situation is demanded in contrast to an exact drawing that precisely represents given proportions of natural frequencies. We illustrate this argument in Fig. 5.6. If a student is asked to draw the unit square of the medical diagnosis situation precisely, he or she has to measure the length of the segments representing the single probabilities or the conditional probabilities and he or she has to use a ruler to produce a precise drawing. However, a rough drawing of the unit square with natural frequencies could potentially show all relevant information of a Bayesian reasoning situation and could be produced much faster than an exact drawing of the unit square.

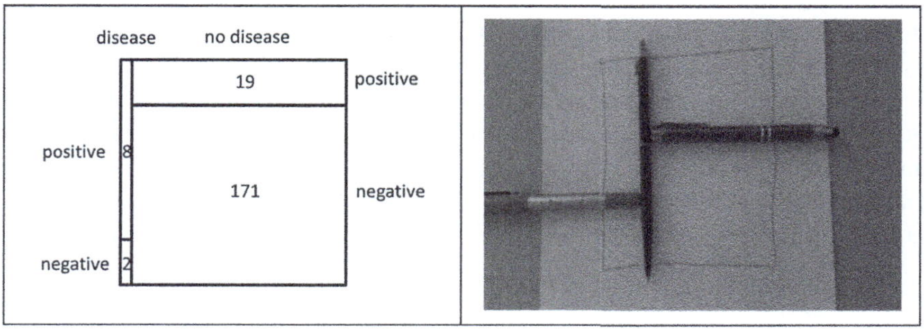

Fig. 5.6 The medical diagnosis test visualised precisely (left side) and visualised roughly (right side)

If it is further possible to use a computer applet of the unit square (e.g. http://unitsquare.leitideedatenundzufall.de/) the mentioned difference of challenges in drawing the tree diagram or the unit square disappears.

5.4.2 Using Tree Diagrams and Unit Squares in Different Situations and Topics

The difference between the two visualisations can be considered when referring to different Bayesian reasoning situations. As sometimes mentioned (e.g. Binder et al. 2015), the unit square has a limitation referring to Bayesian reasoning situations with small base rates. For example, the base rate of HIV infection is smaller than one percent in countries in Western Europe (cf. Sturm and Eichler 2014). In this case the rectangle representing the probability of being ill and infected is not visible or rather only visible as a line. Actually, every statistical graph (in sense of Tufte 2013) of a Bayesian reasoning situation that represents the different probabilities proportionally shows the same limitation. Only visualisations representing the Branch style like tree diagrams can represent extreme values.

However, to make relevant computations in Bayesian reasoning situations, the exactness of the natural frequencies seems to be sufficient (Johnson and Tubau 2015). In contrast, the geometrical shape of the unit square seems to support the identification of the relevant subsets needed, for example, to apply Bayes' rule. For identifying these subsets, a rough drawing of the unit square could be sufficient. The latter aspect is especially the case when people are asked to identify the impact of a change of one of the three relevant parameters in a Bayesian reasoning situation (see Sect. 5.3.1). To conclude, the indicated limitation is only a limitation of displaying extreme values but not a limitation of developing conceptual knowledge about Bayesian situations.

Further, the value of a visualisation could be given by its polyvalence. For example, the tree diagram represents the branching of a set into distinct subsets. This characteristic of the tree diagram could be used for multilevel random experiments and also for problems in combinatorics. Since combinatorics is not restricted to solving problems in probability there are potential uses of the tree diagram in many areas of mathematics. For example, the tree diagram could be used to examine the number of divisors of a natural number.

The unit square is appropriate for visualising nested sets in terms of normalised quantities and proportions of these quantities. To examine proportions of quantities and also proportions of proportions of quantities is a challenge when teaching and learning fractions. For this reason the unit square could serve as a visualisation for processing fractions as shown in Fig. 5.7.

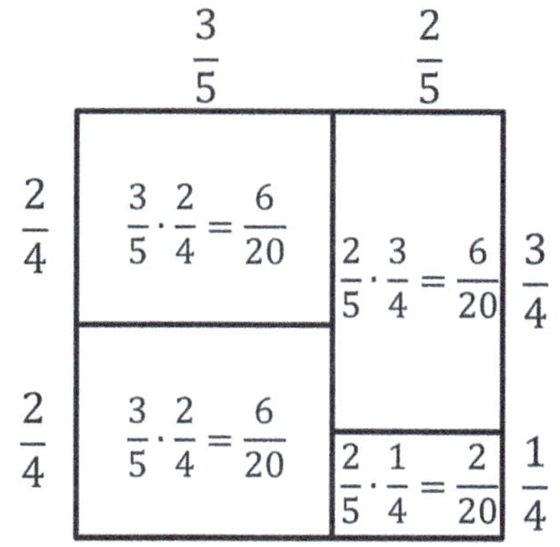

Fig. 5.7 Application of the unit square for the topic of fractions

The application of the unit square when learning fractions could further potentially lead to a connection of the topics of fractions and of probabilities and, thus, could also lead to a better understanding of conditional probabilities referring their nested sets structure.

5.5 Discussion

The question of what an ideal visualisation of a mathematical concept is, may be regarded from different perspectives. First, from a psychological perspective, an ideal visualisation should facilitate efficiently the solution of tasks related to the mathematical concept. Regarding this perspective, the unit square is more effective than the tree diagram: The unit square outperformed the tree diagram concerning different tasks, i.e. computing different relations of subsets or rather conditional probabilities, applying Bayes' rule in Bayesian reasoning situations and evaluating the impact of a base rate change. However, in our research, we only compared two visualisations. Thus, we could not exclude that a further visualisation is more efficient for solving Bayesian reasoning tasks. Nevertheless, there is some theoretical and empirical evidence that characteristics of the unit square, particularly the "Nested style", are close to ideal characteristics of a visualisation for supporting the procedural aspect of Bayesian reasoning.

From an educational perspective, an ideal visualisation has also to make the structure of a mathematical concept transparent. Regarding this perspective, we yielded empirical evidence that the unit square is more effective than the tree diagram for understanding the influence of the base rate on Bayesian reasoning situations. We further provided theoretical considerations implying that also evaluating the change in further parameters in a Bayesian reasoning situation could be effective when supported by the unit square. We hypothesise that an ideal visualisation in an educational perspective must have a Nested style and must also display areas proportionally to natural frequencies.

Finally, the third perspective in thinking about an ideal visualisation is the teaching practice. From this perspective, an ideal visualisation should be easy to use and should provide the possibility of a multiple use in different situations and mathematical topics. Regarding this perspective, both visualisations show advantages: The tree diagram is easy to draw, reflects proportions numerically and is therefore flexible concerning extreme values. The unit square can be easily drawn approximately, reflects relationships between fractions and probability and is very flexible regarding conceptual aspects.

To conclude, an ideal visualisation must integrate different features to meet the challenges of different perspectives. It needs some effort to identify an ideal of a mathematical concept. However, if the mathematical concept is important within a mathematical theory and if the mathematical concept is crucial for the understanding of real situations, it seems to be worthwhile to invest this effort. The concept of conditional probabilities and of the Bayes' rule is an important concept of probability and further a key concept for understanding risk.

References

Barbey, A. K., & Sloman, S. A. (2007). Base-rate respect: From ecological rationality to dual processes. *The Behavioral and Brain Sciences, 30*(3), 241–254. https://doi.org/10.1017/S0140525X07001653.

Bea, W. (1995). *Stochastisches Denken (Stochastical Reasoning)*. Frankfurt: Lang.

Binder, K., Krauss, S., & Bruckmaier, G. (2015). Effects of visualizing statistical information—An empirical study on tree diagrams and 2 × 2 tables. *Frontiers in Psychology, 6*(1186). https://doi.org/10.3389/fpsyg.2015.01186.

Böcherer-Linder, K., & Eichler, A. (2017). The impact of visualizing nested Sets. An empirical study on tree diagrams and unit squares. *Frontiers in Psychology, 7*(2026). https://doi.org/10.3389/fpsyg.2016.02026.

Böcherer-Linder, K., Eichler, A., & Vogel, M. (2017). The impact of visualization on flexible Bayesian reasoning. *AIEM—Advances de Investigación en Educación Matemática, 11*, 25–46.

Borovcnik, M. (2012a). Multiple perspectives on the concept of conditional probability. *AIEM. Avances de Investigación en Educación Matemática, 2*, 5–27.

Borovcnik, M. (2012b, July). *Conditional probability—A review of mathematical, philosophical, and educational perspectives*. Paper presented at the 1 International Congress on Mathematical Education, ICME-12, Seoul. https://www.researchgate.net/publication/304495166_Conditional_Probability_-_a_Review_of_Mathematical_Philosophical_and_Educational_Perspectives. Accessed May 31, 2017.

Brase, G. L. (2009). Pictorial representations in statistical reasoning. *Applied Cognitive Psychology, 23*(3), 369–381. https://doi.org/10.1002/acp.1460.

Cosmides, L., & Tooby, J. (1996). Are humans good intuitive statisticians after all? Rethinking some conclusions from the literature on judgment under uncertainty. *Cognition, 58*(1), 1–73. https://doi.org/10.1016/0010-0277(95)00664-8.

Diaz, C., Batanero, C., & Contreras, J. M. (2010). Teaching independence and conditional probability. *Boletin de Estadistica e Investigacion Operativa, 26*(2), 149–162.

Duval, R. (2002). Representation, vision and visualization: Cognitive functions in mathematical thinking. Basic issues for learning. In F. Hitt (Ed.), *Representations and mathematics visualization. Papers presented in this Working Group of PME-NA, 1998–2002* (pp. 311–336). Mexico: Cinestav—IPN.

Eddy, D. M. (1982). Probabilistic reasoning in clinical medicine: Problems and opportunities. In D. Kahneman, P. Slovic, & A. Tversky (Eds.), *Judgment under uncertainty: Heuristics and biases* (pp. 249–267). Cambridge: Cambridge University Press.

Eichler, A., & Vogel, M. (2010). Die (Bild-)Formel von Bayes (A visual formula of Bayes' rule). *PM: Praxis der Mathematik, 52*(32), 25–30.

Friendly, M. (1999). Extending mosaic displays: Marginal, conditional, and partial views of categorical data. *Journal of Computational and Graphical Statistics, 8*(3), 373–395.

Gigerenzer, G. (2002). *Calculated risks: How to know when numbers deceive you*. New York: Simon & Schuster.

Gigerenzer, G., & Hoffrage, U. (1995). How to improve Bayesian reasoning without instruction: Frequency formats. *Psychological Review, 102*(4), 684–704. https://doi.org/10.1037/0033-295X.102.4.684.

Johnson, E. D., & Tubau, E. (2015). Comprehension and computation in Bayesian problem solving. *Frontiers in Psychology, 6*(938). https://doi.org/10.3389/fpsyg.2015.00938.

Kahneman, D., Slovic, P., & Tversky, A. (Eds.). (1982). *Judgment under uncertainty: Heuristics and biases*. Cambridge, MA: Cambridge University Press.

Kahneman, D., & Tversky, A. (1982). Variants of uncertainty. In D. Kahneman, P. Slovic, & A. Tversky (Eds.), *Judgment under uncertainty: Heuristics and biases* (pp. 509–520). Cambridge, MA: Cambridge University Press.

Khan, A., Breslav, S., Glueck, M., & Hornbæk, K. (2015). Benefits of visualization in the Mammography Problem. *International Journal of Human-Computer Studies, 83*, 94–113. https://doi.org/10.1016/j.ijhcs.2015.07.001.

Lesage, E., Navarrete, G., & de Neys, W. (2013). Evolutionary modules and Bayesian facilitation: The role of general cognitive resources. *Thinking & Reasoning, 19*(1), 27–53. https://doi.org/10.1080/13546783.2012.713177.

Martignon, L. & Erickson, T. (2014). Proto-Bayesian reasoning of children in fourth class. In K. Makar, B. de Sousa, & R. Gould (Eds.), *Sustainability in statistics education. Proceedings of the Ninth International Conference on Teaching Statistics*. Voorburg: International Association of Statistics Education. Online: https://iase-web.org/documents/papers/rt2016/Martignon.pdf.

Oldford, R. W. (2003). *Probability, problems, and paradoxes pictured by eikosograms*. http://www.stats.uwaterloo.ca/~rwoldfor/papers/venn/eikosograms/examples/paper.pdf. Accessed May 31, 2017.

Presmeg, N. (2014). Contemplating visualization as an epistemological learning tool in mathematics. *ZDM Mathematics Education, 46*(1), 151–157. https://doi.org/10.1007/s11858-013-0561-z.

Rathouz, M. (2011). Visualizing decimal multiplication with area models: Opportunities and challenges. *IUMPST: The Journal, 2* (Pedagogy).

Rittle-Johnson, B., & Schneider, M. (2014). Developing conceptual and procedural knowledge of mathematics. In R. Cohen Kadosh, A. Dowker, B. Rittle-Johnson, & M. Schneider (Eds.), *The Oxford handbook of numerical cognition* (Vol. 1). Oxford University Press.

Sedlmeier, P., & Gigerenzer, G. (2001). Teaching Bayesian reasoning in less than two hours. *Journal of Experimental Psychology: General, 130*(3), 380–400. https://doi.org/10.1037//0096-3445.130.3.380.

Sloman, S. A., Over, D., Slovak, L., & Stibel, J. M. (2003). Frequency illusions and other fallacies. *Organizational Behavior and Human Decision Processes, 91*(2), 296–309. https://doi.org/10.1016/S0749-5978(03)00021-9.

Spiegelhalter, D., Pearson, M., & Short, I. (2011). Visualizing uncertainty about the future. *Science, 333*, 1393–1400. https://doi.org/10.1126/science.1191181.

Sturm, A., & Eichler, A. (2014). Students' beliefs about the benefit of statistical knowledge when perceiving information through daily media. In K. Makar, B. de Sousa, & R. Gould (Eds.), *Proceedings of the Ninth International Conference on Teaching Statistics*. Voorburg: International Association for Statistics Education. http://icots.info/icots/9/proceedings/pdfs/ICOTS9_7D1_STURM.pdf.

Tufte, E. R. (2013). *The visual display of quantitative information*. Cheshire, CONN.: Graphics Press.

Wassner, C. (2004). *Förderung Bayesianischen Denkens. Kognitionspsychologische Grundlagen und didaktische Analysen* (Promoting Bayesian thinking—Cognition-psychological foundations and didactical analyses). Hildesheim: Franzbecker.

Chapter 6
Probability Concepts Needed for Teaching a Repeated Sampling Approach to Inference

Hollylynne S. Lee

Abstract This paper uses theoretical and empirical perspectives to articulate what learners should understand about a repeated sampling approach to inference that emphasizes a process of randomizing data, repeating through simulation, and rejecting any model with observed data in the extreme of a distribution that does not fit the model. Key probability concepts, such as a probability model and data distributions, are identified and discussed as to why and how they can assist learners in developing richer understandings and capabilities to a repeated sampling approach to inference.

Keywords Repeated sampling · Inference · Randomization · Probability distribution · Model

6.1 Introduction

Repeated sampling approaches to inference have recently become prominent in reform-oriented statistics curricula in some countries, often using the power of computing tools for creating and running simulations of a repeated sampling process. The term repeated sampling is being used to represent any of the approaches to inference that involve repeating some sampling process, such as a Monte Carlo simulation(resampling), re-randomization (permutation tests), bootstrapping, or jacknifing (see for example, Batanero and Borovcnik 2016). Researchers have argued that a simulation that uses repeated sampling can be an important tool to help students develop a deep understanding of the abstract statistical concepts involved in these types of approaches to inferential reasoning (Maxara and Biehler 2006). In 2007, Cobb suggested that educators help students develop an understanding of this particular approach to inference through the "three R's: randomize data, repeat by simulation, and reject any model that puts your data in its tail"

H. S. Lee (✉)
NC State University, 502C Poe Hall Campus, Box 7801, Raleigh, NC 27695, USA
e-mail: Hollylynne@ncsu.edu

C. Batanero and E. J. Chernoff (eds.), *Teaching and Learning Stochastics*,
ICME-13 Monographs, https://doi.org/10.1007/978-3-319-72871-1_6

(Cobb 2007, p. 12). In this paper, I aim to unpack this "three R's" approach to illuminate the role of probability concepts in a simulation approach to inference and to offer a framework that captures the key probabilistic conceptions and capabilities that need to be addressed when teaching a repeated sampling approach to inference.

Simulations of repeated sampling have been used in several collegiate curricula efforts in the United States, and researchers have reported modest results in improvement of students' understandings of inference through this approach (e.g., Garfield et al. 2012; Tintle et al. 2012). In addition, new curriculum standards in some countries, such as the United States and New Zealand, suggest such an approach for high school students (Council of Chief State School Officers 2010; New Zealand Ministry of Education 2006).

Many curriculum developers recommend that learners experience repeated sampling methods in a physical way before using computing power (e.g., Cobb 2007; Rossman 2008). The physical experiences are intended to assist learners in conceiving of the process of sampling as a repeatable action (e.g., Watson and Chance 2012). These physical experiences serve as a way to reveal the underlying probability assumptions in a problem (e.g., is it equally likely for two events to occur? Does each person have an equal chance to being assigned to a treatment group?). However, in many curricula, the processes of the repeated sampling are often created by the instructor (or curriculum), and students are asked to use this prescribed process with physical objects or are told exactly what to input in a computer simulation (e.g., Cumming et al. 2014; Roy et al. 2014). Such prescriptions likely serve to mask both the underlying randomization that is taking place and the probability assumptions that are being made. A simulation approach, using physical and computer tools, seems to be an appropriate way to help students develop statistical inference conceptually. Thus, while the "three R's" highlight important elements of a simulation model, understanding each part of a simulation and the relationships among the parts is conceptually complicated. In fact, students who know *how* to conduct a simulation may not have a robust understanding of *why* they are conducting a simulation, *what* is being simulated, and *how* to make appropriate conclusions based on a simulation.

6.2 Theoretical Perspectives

This section provides an overview of two important theoretical perspectives used to inform a repeated sampling approach to inference: (1) conceiving and using probability as a model, and (2) prior curriculum and research that has engaged students in repeated sampling and simulations.

6.2.1 Probability as a Model

Similar to the work of Borovcnik (2014) and others, I consider learners' use of probability models as essential to conceptualizing a repeated sampling approach to inference. When conducting a simulation, one is trying to model some process so as to potentially better understand the inputs, inner workings, and outputs of the process. At the core of a simulation involving a stochastic process is randomness—the first of the "three R's". However, at every step of a simulation-based repeated sampling approach to inference there are other issues concerning probability models that learners need to understand. What follows are descriptions of the theoretical perspectives we use in our work related to probability models, the models and modelling perspective on learning, and repeated sampling representations and processes described by others.

Probability is an abstract concept and is used as an expression of likelihood of an event. There are many objects and real world events whose behaviour and outcomes cannot be completely determined ahead of time, even when there is a great deal of information about actions on the object or event in world. Whether we are faced with many or few constraints, we can often build models to express a probability of an event occurring. A relatively simple example is to consider that in a given toss of a number cube, we cannot determine the outcome of which side will land facing up; but, we can build a probability model to estimate the likelihood of each side facing up. How one engages in building such a model may differ based on the approach or perspective taken (e.g., classical, frequentist or subjective) (Borovcnik and Kapadia 2014).

Chaput et al. (2011) described three parts of a probability modelling process that includes translating observations and assumptions of contextual problem into a pseudo-concrete working model, mathematizing the model into a hypothesis-driven probability model that can be enacted, and validating a model through examining how a model fits with empirical data and interpreting the model within the context of the problem. Many have advocated that students and teachers should understand the bi-directional relationship between probability models and data, and between empirically-developed models, typically from a frequentist perspective, and theoretically-developed models, typically from a classical or a subjective perspective (e.g., Eichler and Vogel 2014; Konold and Kazak 2008; Lee et al. 2010; Lee and Lee 2009; Pfannkuch and Ziedins 2014; Pratt 2011; Stohl and Tarr 2002; Wild 2006). When using a repeated sampling approach to inference, the probability assumptions and the model-building process in a simulation should be made more explicit. Pfannkuch and Ziedins' (2014) description of probability models and their purpose provide a useful perspective on the strong role that probability models have in a repeated sampling approach to inference:

> A probability model will often be associated with the idea of a system evolving dynamically over time…a model is usually built to answer a particular question or questions about a system, sometimes just to understand its behaviour better, but often in order to optimize some measure of its performance, or alternatively, to predict performance under some alternative

> scenario. ...although they [models] are only approximations to what happens in the real world, these approximations can help us better understand the behaviour in the real world (p. 103).

In a repeated sampling approach to inference, learners are building and using models as approximations to what happens in the real world under conditions of randomness. To begin with, learners should be conceiving of observed outcome(s) from an observational study or an experiment as resulting from a process that is repeatable, and that repeating the process may result in a different outcome. This is a frequentist perspective of probability. A key question becomes, how unusual is what happened in the particular instance that we just observed given the assumed model? In other words, what is the likelihood of a particular observed outcome occurring if a process based on an assumed model is repeated many times? That is the end goal of a repeated sampling approach to inference. But to achieve this, we need to make sense of the problem we are trying to solve and consider some of the underlying assumptions, what process is being repeated, and what may be the role of randomness and probability in that process. All of these considerations require a model building process.

Given a focus on probability as a model, it makes sense to consider a models and modelling perspective on teaching and learning mathematics, as articulated by Lesh and Doerr (2003). Using this perspective, the goal for a learner is to build a model that can be generalized to other situations and productively re-used. Thus, learners should develop a robust model of using repeated sampling for making inferences that includes understanding relationships among a problem situation, physical enactments of sampling, representations of those enactments, computer representations, the underlying randomization (i.e., the probability models discussed above), the distribution of the statistics of interest, given the model, and how to interpret and use such a distribution to make a decision about likelihood of an event under some hypotheses.

6.2.2 Use of Repeated Sampling and Simulations

Much research and curriculum development in recent years has focused on understanding inference and simulation approaches. For example, Saldanha and Thompson (2002) report that when students can visualize a sampling process through a three-tier scheme, they develop a deeper understanding of the process and logic of inference in the resampling approach. This scheme is centred around "the images of repeatedly sampling from a population, recording a statistic, and tracking the accumulation of statistics as they distribute themselves along a range of possibilities" (p. 261). In their work, Saldanha and Thompson explicitly have students experience a three-level sampling process that includes: (1) randomly drawing items from a given population or a model to form a sample and record a statistic of interest, (2) repeating this process a large number of times and accumulating a collection of sample statistics, and (3) partitioning the collection of statistics to determine what proportion lies beyond a given value.

Lane-Getaz (2006) also describes the process of using a simulation to develop this logic of inference starting with a question in mind, "what if?", to investigate a problem including three tiers: population parameters, random samples with statistics for each sample, and a distribution of sample statistics where one should ask if the observed statistic, and those more extreme, is rare or whether the event could be explained by chance in case the population has a given parameter. In line with Lane-Getaz's suggestion, Garfield et al. (2012) used a generalized structure of the logic of a simulation approach to inference in their curriculum materials. Their structure includes specifying a model, using the model to generate simulated data for a single trial and then multiple trials, each time collecting a statistic of interest, and finally using the distribution of collected statistics to compare observed data with the behaviour of the model. One way to specify a model is to establish a hypothesis about a parameter(s) in the population, which is typically how formal inference procedures begin. Garfield et al. (2012) research indicates that, with repeated experiences with this model-building process, students do make considerable gains in their thinking about using simulated data from these models to make an inference.

In 2014, Saldanha and Liu described work with learners in repeated sampling tasks and made the case that students should develop a stochastic conception of an event that "entails thinking of it as an instantiation of an underlying repeatable process, whereas a non-stochastic conception entails thinking of an event as unrepeatable or never to be repeated" (p. 382). Such a stochastic conception includes seeing an event as an expression of some process that could be repeated under similar conditions that produces a collection of outcomes and "reciprocally, seeing a collection as having been generated by a stochastic process" (p. 382). This approach was also successfully used with preservice teachers by Lee et al. (2014) in their work using empirical sampling distributions to help learners compare the likelihood of real world events by examining which event occurred less often under a repeated stochastic process.

6.3 Empirical Perspectives

For the past several years, I have worked closely with colleagues in the design and implementation of a graduate course for those interested in teaching statistical thinking. Participants in the course tend to be those interested in learners ages 13–20. In this paper, experiences and data are drawn from two sections of the course offered in Spring 2014, with 27 participants total (21 female, 6 male). Participants consisted of one pre-service teacher (5th year senior), six pre-service and 11 in-service teachers in masters programs, one full-time masters student in mathematics education, seven doctoral students in Mathematics or Mathematics Education (three currently teaching in post-secondary contexts), and one doctoral student with interests in statistics education.

We used a models and modelling perspective (Lesh and Doerr 2003) to design a sequence of model development activities consisting of structurally related tasks that begin with a *model eliciting activity*, and are followed by *model exploration activities* and *model application activities* (Ärlebäck et al. 2013). Our aim was to foster the development of understandings of how probability models are critical for understanding repeated sampling approaches to inference.

In this work with teachers as learners, the learning goal is for teachers to develop a stochastic conception of events and a conceptual model that they can use to approach inference situations using a repeated sampling approach, and for them to be able to assist others in using such an approach. This conceptual model includes understanding the relationships among the problem situation, physical enactments of sampling, representations of those enactments, computer representations, and the underlying randomization (i.e., the probability models discussed above), the distribution of the statistics of interest and how to interpret and use such a distribution to make a decision. In order for teachers to develop that model (and the entailments needed for teaching that model to other learners), we hypothesized that the teachers should be able to make connections to and use an underlying probability model of repeatable actions with unpredictable outcomes. We used both observational studies and experimental designs in our instructional tasks. Our instructional design sequence and analysis of teachers' conceptualizations is not the focus of this paper, but can be read in other publications (Arnold et al. in press; Lee et al. 2015, 2016). For illustration purposes, two key activities are described here along with description of the most salient aspects that our learners struggled with.

A few lessons into the lesson sequence, learners engaged with an adapted version of the Paul the Octopus task (Lock et al. 2013) concerning predictions of winners in World Cup soccer games. In the real context in this task, the famous octopus, Paul, had achieved eight out of eight correct predictions by swimming in a tank to choose one of two boxes marked with the team's country flags to eat from. His choice of boxes was deemed as his prediction and then compared with the actual winner of each game to see if he was correct. This task was designed to elicit learners' initial models of repeated sampling, using coins in a cup as a repeatable random process, to determine the likelihood of a single proportion, namely correctly guessing the winners of eight out of eight games (100%).

The learners were able to see the choice of two boxes (representing teams) as being modelled as equiprobable outcomes and that under normal circumstances, one would expect about four correct guesses out of eight. Given a set of eight coins and a cup, the learners were asked to generate a way to simulate the actions in the real scenario that were being repeated. They were easily able to see how they could use one coin toss to simulate the octopus choosing a box to eat from. However, a rich discussion ensued concerning whether to simulate another coin toss for the actual winner and compare it to the prediction coin toss. In addition, when it was suggested to consider the coin toss as representing whether the prediction was correct or not, a further discussion ensued about whether they needed to do eight tosses with one coin, or if they could toss eight coins simultaneously. This discussion made it clear that the mapping of the real world scenario to the simulation was not at all obvious.

After several more additional tasks using both hands on and computer simulations, the following task was used to allow our learners an opportunity to express their developing conceptions of the simulation process in terms of how they would help their students to understand the process. In groups of 2–3 they had to create diagrams on a large poster to explain using a repeated sampling approach with their students to help them use a simulation (with physical objects or computer models) to investigate if an observed statistic is likely to occur, or not unlikely to occur under given assumptions. The intended audience for this representation would be the future students of learners and this representation hence served a perceived purpose of explaining the structure of their understanding of using repeated sampling for inference tasks to other learners. There was a wide variety of representations expressed in participants' diagrams. Many attended to the need to build a model based on contextual assumptions in the problem; however, many participants seemed to be "hand-waving" about the "magic" that happens in the randomize and repeat phases in a simulation approach (e.g., Fig. 6.1). For example, consider how the poster in Fig. 6.1 includes key ideas about phases in a simulation and emphasizes the use of distribution of sample statistics for decision making, but does not explicate the randomize and repeat phases. All of the posters included a distribution of sample statistics as an object that resulted from the repeated process and that was used to make a decision.

Those teachers who could only vaguely state that they needed a model to begin with often had difficulty in analysing a new context and designing an appropriate way to use a simulation with physical or computer tools. It was the teachers who were able to carefully unpack the assumptions in the context and conceive of what actions were being repeated in the context who were able to apply such understandings to build new models and design appropriate simulations in new contexts.

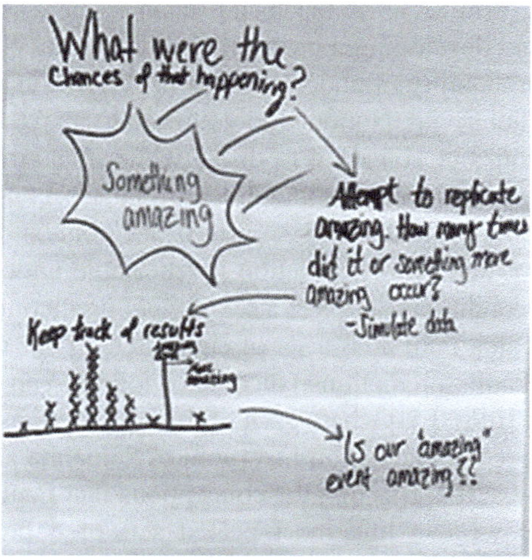

Fig. 6.1 Learners' diagram of key questions in a repeated sampling approach

6.4 Unpacking Probability Conceptions in Inference

To identify the key probabilistic conceptions for teaching a repeated sampling approach to inference, I draw on empirical data from our research with teachers and the theoretical perspectives represented in literature. If a goal is to have learners understand how and why a repeated sampling approach to inference works, more attention needs to be given to the modelling process, the explicit role of probability in inference, and use of probabilistic language. Several relevant conceptualizations seem to be important for building learners' conceptual model for a repeated sampling approach to inference. These conceptualizations are unpacked using Cobb (2007) "three R's" of randomize, repeat, and reject to organize the discussion.

6.4.1 Randomize

Most previous theoretical work refers to this first tier as being a single "population" or model level (e.g., Garfield et al. 2012; Lane-Getaz 2006). This does not really describe well what exactly is being randomized. There seems to be a two-level modelling process that should be made explicit to learners. The first level is the process of creating a local specific model of the real world context in statistical terms to make clear what a null hypothesis would be if we assume the status quo. For example, in an observational study like in the Paul the Octopus task, we need to think about the expected number of correct predictions out of eight that should occur if Paul was randomly guessing. This requires assuming Paul is equally likely to pick the winning team as he is the losing team if there is nothing out of the ordinary going on, and developing a probability model of an equiprobable uniform assignment for the probability of choosing each of the two teams. Conceiving this probability model is not trivial. Not only do you need to map the real context onto a probability distribution, but you need to do so while imagining the usefulness of this model as a way of describing the likelihood of the aspect of the situation that is repeatable (Pfannkuch and Ziedens 2014). So for the context of Paul the Octopus, one needs to conceive that every prediction is repeatable under the same (or close enough to the same) conditions, and that a set of eight predictions is also repeatable, i.e., we can conceive that Paul does a set of eight predictions, over and over again. (Note: the count of Paul's predictions (success or failure) can be described with a binomial distribution [$B(8, 0.5)$] where each of the 8 predictions has a probability of success of 0.5.) This recommendation incorporates Saldanha and Liu (2014) focus on repeatable processes. It seems that the conceptions that Paul does 8 independent predictions, and that we can imagine repeated sets of 8 predictions, would be supported by past experiences with developing probability models and from applications of a frequentist approach to probability.

If a learner is able to conceive of the aforementioned probability ideas in this first part of the randomize phase, this affords the opportunity to consider what results would be considered unusual, or what would be considered usual or likely to happen. Learners would then be able to express a usual expectation as a null hypothesis and to specify which statistic of interest from the context should be observed when each action is repeated (in this case the number of correct guesses in eight predictions). Thus, in accord with Chaput et al. (2011), the observed statistic from the original context also becomes a target for wondering about, i.e., would it be unusual to get that observation if the probability model, assumptions, and null hypothesis are true?

The second aspect of the randomize phase is conceiving of a method for creating a simulation process that embodies the repeatable actions in the original context and can be used to generate random samples. Again, the focus on the repeatable event is key and can allow learners to think critically about developing a simulation model, whether this is by hand or on a computer, that embodies the probability model for the context and provides a mechanism for repeating the entire event (e.g., Paul's eight predictions and whether each was correct or not). In addition to the key probability concepts already discussed, an understanding of independence is critical here. The simulation needs to be designed so that each sample is independent from each other. Learners need to realize that as their simulation runs, future outcomes should not be dependent on prior outcomes.

6.4.2 Repeat

Part of building a simulation method and model is conceiving of repeated sampling as a way to generate simulated statistics, as was described in the three-tier scheme by Saldanha and Thompson (2002). The method needs to include a mechanism for keeping track of the statistic of interest (e.g., proportion of success, or difference of two means if comparing results from a control and experimental group) and recording it. However, in order to meaningfully keep track of the sample statistic, a learner should be able to conceive of the entire sample space of possibilities. For example with Paul the Octopus, there could be anywhere from 0 to 8 successful predictions, mapping onto the discrete sample space of (0, 12.5, 25, 37.5, 50, 62.5, 75, 87.5, 100%) for the possible values for a statistic of interest of percent correct. If learners are doing a repeated sampling approach by hand, then they need to find a way to organize how many times each of the possible elements of the sample space occurs. However, if the repeated sampling is being performed on a computer, then aspects of the sample space may become hidden and learners may not explicitly realize the full range of all the possibilities that occur under the probability model. Understanding the range of possibilities in a sample space, and the need to keep track of how many times each of those possibilities occurs, are essential conceptualizations in the *repeat* phase and can assist learners in developing another important conception. As they are keeping track of how many times each possibility occurs, they are building a distribution of sample statistics and should conceive of this distribution as an empirical sampling distribution from which they can estimate a probability.

6.4.3 Reject

While students are building this empirical sampling distribution, variation from the expected value of the statistic of interest (e.g., 50% success in Paul's 8 predictions) should be noticed. As more repeated samples are generated, learners should begin to notice that those possible outcomes closer to the expected value seem to occur more often than those further away. If they conceive of this distribution as an approximation to a theoretical probability distribution, then they will think about those closer to the expected value (the center of the full range in the sample space) as having higher probability of occurring, and those further away as less likely to occur, with a lower probability.

Aligned with the work by Garfield et al. (2012) and others, once a distribution of sample statistics is formed (be it theoretical or empirical), learners should conceive of the inferential decision as deciding if an observed statistic from the original context (e.g., if Paul the Octopus had predicted 7 out of 8 or 87.5% successfully) and those more extreme are likely to happen under the assumptions built in the model in the randomize phase. This is the phase where learners need to evoke bi-directional reasoning between the empirical sampling distribution they have created through the repeated sampling and the original model for the null hypothesis; i.e. does the observed statistic lies on this sampling distribution close or far from the expected value of the probability model (e.g., in Paul's case that he should get about 50% of his predictions correct under an assumption of equally likely outcomes).

Learners should be able to quantify the likelihood of getting the observed statistic or those more extreme under the given hypothesis, by conceiving of the sampling distribution as an approximation to the probability distribution and partitioning the distribution using the observed statistic as a partition border. This partition can assist learners in being able to estimate a probability by considering the number of times an outcome of interest occurred (getting the observed statistics or more extreme) divided by the number of outcomes that were generated in the repeat phase. Of course this probability has been shortened to be known as the p-value if we were to compute this for all possible combinations that could occur across the sample space of outcomes under the assumptions in the probability model (not in the empirical sampling distribution).

When students work with simulations, instead of working with the theoretical sampling distribution (which is a probability distribution), they do need to recognize they are working with an approximation (the empirical sampling distribution). In fact, the empirical sampling distribution is a data distribution, as we build it with observed data (not with theoretical probabilities). What we get with the empirical sampling distribution is frequencies and not probabilities. If the number of simulations is big, we can estimate probabilities for each event in the sample space; the estimations will be more accurate as we increase the number of simulations. But an empirical sampling distribution does introduce an additional error in the inference process (as compared with working with the theoretical sampling distribution), and thus students need to be aware that p-values generated from simulations are approximations.

6.5 Implications

The conceptualizations described above seem important for learners to develop in order to have a robust way of conceiving how a repeated sampling approach using simulations can be used to engage in this particular approach to inference in a range of contexts. There was evidence of this in the learners in our graduate course who *were* developing stronger conceptions, and those who struggled may have benefitted from additional learning experiences that would help them develop such conceptualizations.

If the statistics education community is going to continue suggesting that repeated sampling approaches such as resampling, re-randomization and bootstrapping be a useful approach to inference for secondary and postsecondary statistics learners, then there also needs to be attention to how learners can either co-develop or have prior experiences with the probability concepts outlined in this paper. Of course there are other valid approaches to inference that could also be taught with little formalization. Students do not need large doses of probability theory, but I posit that the probability conceptions discussed above are a strong part of understanding repeated sampling approaches to inference, and if these concepts are not fully developed, then students' and instructors uses of this approach may not lead to a better understanding of this particular approach to inference.

References

Ärlebäck, J. B., Doerr, H. M., & O'Neil, A. H. (2013). A modelling perspective on interpreting rates of change in context. *Mathematical Thinking and Learning, 15*(4), 314–336. https://doi.org/10.1080/10986065.2013.834405.

Arnold, P., Confrey, J., Jones, S., Lee, H. S., & Pfannkuch, M. (in press). Learning trajectories in statistics education. In D. Ben-Zvi, J. Garfield, & K. Makar (Eds.), *First international handbook of statistics education*. New York: Springer.

Batanero, C., & Brovcnik, M. (2016). *Statistics and probability in high school*. Rottterdam: Sense Publishers.

Borovcnik, M. (2014). Modelling and experiments—An interactive approach towards probability and statistics. In T. Wassong, D. Frischemeier, P. R. Fischer, R. Hochmuth, & P. Bender (Eds.), *Mit Werkzeugen Mathematik und Stochastik lernen—Using tools for learning mathematics and statistics* (pp. 267–281). Fachmedien Wiesbaden: Springer.

Borovcnik, M., & Kapadia, R. (2014). A historical and philosophical perspective on probability. In E. J. Chernoff & B. Sriraman (Eds.), *Probabilistic thinking: Presenting plural perspectives* (pp. 7–34). New York: Springer.

Chaput, B., Girard, J. C., & Henry, M. (2011). Frequentist approach: Modelling and simulation in statistics and probability teaching. In C. Batanero, G. Burrill, & C. Reading (Eds.), *Teaching statistics in school mathematics-challenges for teaching and teacher education* (pp. 85–95). Dordrecht: The Netherlands: Springer.

Cobb, P. (2007). The introductory statistics course: A ptolemaic curriculum? *Technology Innovations in Statistics Education, 1*(1), 1–16. http://escholarship.org/uc/item/6hb3k0nz#page-1. Accessed June, 2017.

Council of Chief State School Officers. (2010). *Common core state standards for mathematics*. Washington D.C.: Author.

Cumming, J., Miller, C., & Pfannkuch, M. (2014). Using bootstrap dynamic visualizations in teaching. In K. Makar, B. de Sousa, & R. Gould (Eds.), *Proceedings of the Ninth International Conference on Teaching Statistics Voorburg*. The Netherlands: International Statistical Institute. http://iase-web.org/icots/9/proceedings/pdfs/ICOTS9_3D1_CUMMING.pdf. Accessed June, 2017.

Eichler, A., & Vogel, M. (2014). Three approaches for modelling situations with randomness. In E. J. Chernoff & B. Sriraman (Eds.), *Probabilistic thinking: Presenting plural perspectives* (pp. 75–99). Dordrecht, The Netherlands: Springer.

Garfield, J., delMas, R., & Zieffler, A. (2012). Developing statistical modelers and thinkers in an introductory, tertiary-level statistics course. *ZDM, Mathematics Educatioin, 44*, 883–898.

Konold, C., & Kazak, S. (2008). Reconnecting data and chance. *Technology Innovations in Statistics Education, 2*(1), 1–37.

Lane-Getaz, S. J. (2006). What is statistical thinking and how is it developed? In G. Burrill (Ed.), *Thinking and reasoning with data and chance: Sixty-eighth yearbook* (pp. 272–289). Reston, VA: National Council of Teachers of Mathematics.

Lee, H. S., Angotti, R. L., & Tarr, J. E. (2010). Making comparisons between observed data and expected outcomes: Students' informal hypothesis testing with probability simulation tools. *Statistics Education Research Journal, 9*(1), 68–96.

Lee, H. S., Doerr, H. M., Tran, D., & Lovett, J. N. (2016). The role of probability in developing learners' models of simulation approaches to inference. *Statistics Education Research Journal, 15*(2), 216–238.

Lee, H. S., & Lee, J. T. (2009). Reasoning about probabilistic phenomena: Lessons learned and applied in software design. *Technology Innovations in Statistics Education, 3*(2). http://www.escholarship.org/uc/item/1b54h9s9. Accessed June, 2015.

Lee, H. S., Starling, T. T., & Gonzalez, M. D. (2014). Using data to motivate the use of empirical sampling distributions. *Mathematics Teacher, 107*(6), 465–469.

Lee, H. S., Tran, D., Nickell, J., & Doerr, H. M. (2015). Simulation approaches for informal inferences: Models to develop understanding. In K. Krainer, & Vondorva, N. (Eds.), *Proceedings of the Ninth Congress of European Society for Research in Mathematics Education* (pp. 707–714). Prague, Czech Republic. https://hal.archives-ouvertes.fr/hal-01287093/document. Accessed June, 2017.

Lesh, R. A., & Doerr, H. M. (Eds.). (2003). *Beyond constructivism: Models and modelling perspectives on mathematics problem solving, learning, and teaching*. Mahwah, NJ: Lawrence Erlbaum.

Lock, P. F., Lock, R. H., Morgan, K. L., Lock, E. F., & Lock, D. F. (2013). *Statistics: Unlocking the power of data*. Hoboken, NJ: Wiley.

Maxara, C., & Biehler, R. (2006, July). Students' probabilistic simulation and modeling competence after a computer-intensive elementary course in statistics and probability. In A. Rossman & B. Chance (Eds.), *Proceedings of the Seventh International Conference on the Teaching of Statistics*. Salvador de Bahia (Brazil): International Statistical Institute. https://www.stat.auckland.ac.nz/~iase/publications/17/7C1_MAXA.pdf. Accessed May, 2015.

New Zealand Ministry of Education. (2006). *The New Zealand curriculum: Mathematics and statistics*. Wellington: Author. http://nzcurriculum.tki.org.nz/. Accessed May, 2015.

Pfannkuch, M., & Ziedins, I. (2014). A modelling perspective on probability. In E. J. Chernoff & B. Sriraman (Eds.), *Probabilistic thinking, Presenting plural perspectives* (pp. 101–116). Dordrecht, The Netherlands: Springer.

Pratt, D. (2011). Re-connecting probability and reasoning about data in secondary school teaching. In *Proceedings of the 58th World Statistics Conference*. Dublin: International Statistical Institute. http://2011.isiproceedings.org/papers/450478.pdf. Accessed June, 2017.

Rossman, A. (2008). Reasoning about informal statistical inference: One statistician's view. *Statistics Education Research Journal, 7*(2), 5–19.

Roy, S., Rossman, A., Chance, B., Cobb, G., Vanderstoep, J., Tintle, N., & Swanson, T. (2014). Using simulation/randomization to introduce *p*-value in week 1. In K. Makar, B. de Sousa, & R. Gould (Eds.), *Proceedings of the Ninth International Conference on Teaching Statistics*. Voorburg, The Netherlands: International Statistical Institute. http://iase-web.org/icots/9/proceedings/pdfs/ICOTS9_4A2_ROY.pdf. Accessed June, 2017.

Saldanha, L., & Liu, Y. (2014). Challenges of developing coherent probabilistic reasoning: Rethinking randomness and probability from a stochastic perspective. In E. J. Chernoff & B. Sriraman (Eds.), *Probabilistic thinking, presenting plural perspectives* (pp. 367–396). Dordrecht, The Netherlands: Springer.

Saldanha, L., & Thompson, P. W. (2002). Conceptions of sample and their relationship to statistical inference. *Educational Studies in Mathematics, 51*(3), 257–270. https://doi.org/10.1023/A:1023692604014.

Stohl, H., & Tarr, J. E. (2002). Developing the notions on inference using probability simulation tools. *Journal of Mathematical Behavior, 21*(3), 319–337. https://doi.org/10.1016/S0732-3123(02)00132-3.

Tintle, N., Topliff, K., Vanderstoep, J., Holmes, V.-L., & Swanson, T. (2012). Retention of statistical concepts in a preliminary randomization-based introductory statistics curriculum. *Statistics Education Research Journal, 11*(1), 21–40.

Watson, J., & Chance, B. (2012). Building intuitions about statistical inference based on resampling. *Australian Senior Mathematics Journal, 26*(1), 6–16.

Wild, C. (2006). The concept of distribution. *Statistics Education Research Journal, 5*(2), 10–26.

Chapter 7
Characterizing Probability Problems Posed in University Entrance Tests in Andalucia

Carmen Batanero, María del Mar López-Martín, Pedro Arteaga and María M. Gea

Abstract The compulsory university entrance tests in Andalucía (Spain), which traditionally include a probability problem, often determine whether the student is admitted to study in his or her desired university program. Consequently, it is important to ensure that the content of the test is directly related to the content included in the high school curriculum. The aim of this research was to investigate the distribution of the main variables characterizing the probability problems posed in these tests in Andalucía. Specifically, we examined all the problems posed to students in the period 2003–2014 (n = 144 problems). We considered the following variables: type of experiment and sample space considered, type of probability that should be computed, theorems or properties needed to find the solution, format of the data, and context. The results of the analysis reveal the potential difficulty of the problems, as well as the relevance given to conditional probability in the tests, in comparison to other curricular contents.

Keywords Assessment · University entrance tests · Conditional probability Compound probability

C. Batanero (✉) · M. del MarLópez-Martín · P. Arteaga · M. M. Gea
Facultad de Ciencias de la Educación, University of Granada,
Campus de Cartuja, 18071 Granada, Spain
e-mail: batanero@ugr.es

M. del MarLópez-Martín
e-mail: mariadelmarlopez@ugr.es

P. Arteaga
e-mail: parteaga@ugr.es

M. M. Gea
e-mail: mmgea@ugr.es

C. Batanero and E. J. Chernoff (eds.), *Teaching and Learning Stochastics*,
ICME-13 Monographs, https://doi.org/10.1007/978-3-319-72871-1_7

7.1 Introduction

Given their important role in many disciplines, as well as in planning and decision making, statistics and probability are today included as topics of study at all school levels (Batanero and Borovcnik 2016; Franklin et al. 2007). For example, in the Programme for International Student Assessment (PISA) framework (Organisation for Economic Co-operation and Development [OECD] 2015), it is recognised that uncertainty is at the heart of the mathematical analysis of many problem situations, and that consequently, students should understand random variation and its mathematical representation, as well as basic principles of probability, sampling, and inference.

One factor that has a strong influence on the way in which probability is taught at the high school level in Spain is the University Entrance Tests (UET), which are used to evaluate the maturity of students who intend to enter university with regards to the knowledge and skills they have developed during their high school studies. These tests are also used as screening tools in admitting students to specific universities and programs, and therefore, the scores on these tests influence the student's likelihood to be admitted to a chosen school or university. Consequently, it is important to ensure that the content of the test is directly related to the content included in the high school curriculum.

7.1.1 Probability and Statistics in High School Curricula

The need for probability education has been recognized in Spanish curricular guidelines, which include probability content in all grades starting in primary school. At the high school level, probability has been taught as part of the mathematics curriculum within the specialties of Science and Technology and Humanities and Social Sciences (Ministerio de Educación y Ciencia [MEC] 2007) with the following contents:

- *Mathematics for Sciences and Technology I*: Joint, compound, and conditional probability; binomial and normal distributions (p. 45449).
- *Mathematics for Social Sciences I*: Descriptive statistics; tables and graphs; measures of central tendency and spread; bidimensional distributions; correlation and linear regression; assigning probabilities to events; binomial and normal distributions (p. 45475).
- *Mathematics for Social Sciences II*: Prior and posterior probability; joint, conditional, and total probability; Bayes' theorem; binomial and normal distributions; practical implications of the central limit theorem; approximation of the binomial distribution by the normal distribution; the law of large numbers; sampling; sampling distributions for means and proportions; confidence intervals and hypothesis tests for means and for the difference of means (normal distribution with known standard deviation); confidence intervals and hypothesis tests for proportions in large samples (p. 45476).

Today, we are facing changes in the Spanish curricular guidelines (Ministerio de Educación, Cultura y Deporte [MECD] 2015), with only three specialities established for high school (Arts, Sciences, and Humanities and Social Sciences). Probability is addressed in the mathematics curriculum for the Sciences and Humanities specialty and the Social Sciences specialty. The topics addressed in these courses are as follows:

- *Mathematics I for Sciences* (This course is taught to students in the Science and Technology specialty): Bidimensional descriptive statistics; contingency tables; joint, conditional, and marginal distributions; dependence and independence; interpreting scatter plots; covariance and correlation; linear regression (p. 418).
- *Mathematics II for Sciences*: Events; assigning probability to events using Laplace's rule and using relative frequency; Kolmogorov axioms; applying combinatorics to compute probabilities; simple and compound experiments; conditional probability; dependence and independence; the law of total probability and Bayes' theorem; prior and posterior probabilities; likelihood of an event; random discrete variables; probability distributions; mean, variance, and standard deviation; binomial and normal distributions; standardisation; fitting a normal distribution to the binomial distribution (p. 422).
- *Mathematics Applied to Social Sciences I* (This course is taught to students in the Humanities and Social Sciences specialty): Bidimensional descriptive statistics; contingency tables; joint, marginal, and conditional distributions; dependence and independence; covariance and correlation; linear regression; events; assigning probability to events using Laplace's rule and using relative frequency; simple and compound experiments; conditional probability; dependence and independence; random discrete and continuous variables; probability distributions and density functions; mean, variance, and standard deviation; binomial and normal distributions; standardisation; fitting a normal distribution to the binomial distribution (p. 385).
- *Mathematics Applied to Social Sciences II*: Reinforcing the theory of probability; Kolmogorov axioms; assigning probability to events by Laplace's rule and using relative frequency; simple and compound experiments; conditional probability; dependence and independence; the law of total probability and Bayes' theorem; prior and posterior probabilities; likelihood of an event; sample and population; sampling methods; parametric inference; parameters and statistical summaries; point estimation; mean and standard deviation of the simple mean and proportion in large samples; confidence intervals; relationships between confidence, error, and sample size; confidence intervals for the mean in a normal distribution with known standard deviation; confidence intervals for means and proportions in large samples (p. 389).

Assessment is an important element of teaching practice that guides teachers and students towards achieving learning goals. Assessment has an especially important role in the transition from high school to university in Spain, because a student's score on the compulsory UETs often determines whether the student is admitted to study in their desired university program (and, in the future, whether the student is able to practice in a desired career). Traditionally, the tests administered to Humanities and Social Sciences students include a probability problem and an inference problem, while, paradoxically, no statistical or probability problems are included in the UET administered to Science and Technology students, despite the fact that statistics and probability are also taught to these students (as described in the above paragraphs).

Starting from the above considerations, the aim of this work is to analyse the probability problems that were posed in the UETs for Humanities and Social Science students in Andalucía, Spain in the period 2003–2014. These tests require students to solve four different problems, one of which is related to probability; therefore, a correct solution of the probability problem has a 25% weight on the final mark received by students (and their likelihood to be able to follow their chosen studies). Since the educational opportunities of students at the high school level (and the mathematical content that is taught to them by the teachers) are often determined by what is addressed on the UET, we indirectly examine the variables that are considered relevant in determining competence in problem solving in the probability education of these students. Given that the curricular guidelines include specific assessment criteria and learning standards, we also analyse the correspondence between the intended and the assessed curriculum in probability.

In the following sections, we present the research background and method, as well as the main results. We then analyse all of the problems posed in the Andalucía UETs in the period 2003–2014, discuss the results, and conclude with some recommendations to improve the teaching and assessment of probability at the high school level.

7.2 Background

7.2.1 Theoretical Framework

Our work is based on the onto-semiotic approach to mathematics education research, which assumes that mathematical knowledge has a socio-epistemic dimension, since it is linked to the activity in which the subject is involved and depends on the institutional and social context in which it is embedded. Mathematical activity is described in terms of practices or sequences of actions, regulated by institutionally established rules, and directed towards solving a problem (Drijvers et al. 2013).

In this framework, the meaning of mathematical objects is linked to the mathematical practices carried out by a person or an institution to solve specific mathematical problems. Around the mathematical practices linked to these specific problems, different rules (concepts, propositions, procedures) emerge (Godino et al. 2007). These rules are supported by mathematical language (terms and expressions, symbols, graphs, etc.), which is itself regulated by the rules. All of these objects are linked to arguments that are used to communicate the problem solutions, properties and procedures, as well as to validate and generalize them to other contexts and problems. An epistemic configuration (either institutional or personal) is the system of objects involved in the mathematical practices carried out to solve a specific problem.

We conceive different types of institutional meanings for a mathematical object (in this case, probability): (a) *reference meaning*, or the system of practices used as reference in research (for example, the formal meaning of the object in mathematical practice); (b) *intended meaning*, which is the part of the meaning that is planned for teaching (for example, the practices proposed in the curricular guidelines); (c) *implemented meaning*, which refers to what is actually taught to the students; and (d) *assessed meaning*, is the content of assessment. In our research, we try to identify the assessed meaning of probability in the UETs and compare this meaning with the intended institutional meaning for these students, as defined in the curricular guidelines.

Godino (1996) describes understanding as competence (a subject is said to understand a mathematical object when he/she uses the object in a competent way in different practices). The understanding of a mathematical object by a student is a psychological construct that cannot be directly observed; however, it can be inferred from the student's response to assessment tasks. Students' mathematical practices in solving these tasks (e.g., their solutions, strategies, arguments, symbols, etc.) help to indirectly assess their learning. It is therefore essential that assessment tests be valid, that is, that there is an adequate correspondence between the intended and assessed institutional meaning for a given content. This correspondence between the intended and assessed meaning of probability in the Andalucía UETs in the 2003–2014 period is analysed in this paper.

7.2.2 Analysis of Probability Problems in Other Documents

Our research is also based on two previous research studies that have analysed the probability problems in some Spanish textbooks, which were used to identify the variables used in the present study. Ortiz (2002) studied all of the probability problems posed in 22 high school mathematics textbooks and identified the main concepts implicit in these problems, which included: simple and compound experiments; dependence and independence; simple, conditional, and joint probability; total probability; and Bayes' theorem. In our analysis, we considered the need for students to apply each of these objects in the problems proposed in the UET.

Díaz (2004) studied the probabilistic problems included in 17 university textbooks aimed at Social Science students who had just completed high school and were in their first year of university. She classified the problems according to the following variables: (a) type of experiment (simple or compound); (b) dependence (whether the simple experiments in a compound experiment were dependent or independent); (c) property or theorem used in the solution (addition rule, computing conditional probability from simple and compound probabilities, total probability or Bayes' theorem); and (d) type of situation (synchronic, which refers to simultaneous experiments, or diachronic, which refers to consecutive experiments). Although these authors identified variables and categories that are relevant in the analysis of probability problems, they limited themselves to identifying the presence or absence of each category in the sample of textbooks they analysed.

The only analysis of the probability problems proposed in the university entrance test is reported in our previous paper (Contreras et al. 2015), where they superficially analysed the problems posed in three different years to identify the presence of the mathematical objects identified by Díaz (2004). Consequently, this paper is both a new contribution and a complement to these previous studies.

7.2.3 Students' Difficulties with Conditional Probability

As described later, a vast proportion of the probability problems posed in the UETs ask students to compute a conditional probability. Although the mathematical definitions of independence and conditional probability are simple to state, psychological research shows that their application in certain circumstances is far from easy. In such situations, people have generally been found to rely on biases and, consequently, to make wrong decisions (see Batanero and Borovcnik 2016; Borovcnik 2012; or Díaz et al. 2010 for a survey). The most common difficulties, as identified by Díaz and Batanero (2009), Díaz et al. (2012) and Díaz and de la Fuente (2007), are the following:

- *Conditioning and causation*: There are two different possible relationships when assessing a conditional probability $P(A|B)$: If B is perceived as a cause of A, the relationship is referred to as causal, while when A is perceived as a possible cause of B, then we deal with a diagnostic relationship (Tversky and Kahneman 1982). Although from the mathematical point of view both relationships are equivalent, causal data have a greater impact on our perceptions and inferences, and people tend to interpret conditional probability as a causal relationship.
- *Time-axis fallacy*: Falk (1986) suggested that students who confuse conditional and causal reasoning believe that an event cannot condition another event that occurs before it.
- *Transposed conditional fallacy*: The probability $P(A|B)$ is often confused with the probability $P(B|A)$ and even with the joint probability $P(A \cap B)$. Pollatsek et al. (1987) suggest that part of this problem is due to verbal ambiguity in expressing conditional probabilities. Data from different research (e.g., Contreras 2011; Díaz et al. 2010) also suggest that another major source of error is the confusion between conditional probabilities and joint probabilities.
- *Confusing mutually exclusive events with independent events*. Due to verbal ambiguity, many students confuse independent and mutually exclusive events.

Students also make many errors in applying Bayes' theorem, such as incorrect identification of data, wrong partition of the sample space, difficulties in producing a tree diagram, or confusion in applying the total probability theorem. To sum up, previous research (e.g., Huerta 2014) suggests that conditional probability, independence, and Bayes' theorem are difficult for a vast majority of students. The need for the students to apply all this knowledge in the UETs has not previously been analysed and is therefore an important goal in our research.

7.3 Method

The Spanish regulations and the admission procedures to Spanish public universities (Ministerio de la Presidencia [MP] 2008) require that all the students pass a maturity test (UET) that assesses their knowledge and abilities. The UET mathematics test for students in the Humanities and Social Sciences specialty consists of two options. Students can chose one of these options and are expected to solve all of the problems posed in the selected option. The content of the tests (4 problems) is similar for both options: an algebra problem, a calculus problem, a probability problem, and a statistical inference problem. The correct solution to each problem is marked out of 2.5 points, for a total of 10 points in the full test.

In this study, we examined the probability problems in the UET mathematics tests administered to Social Science students in Andalucía during the period 2003–2014. In each of these years, 6 different versions of the tests were used (depending on the semester and city). Since each test has two options, we analysed a sample of 144 problems. Given that we only considered one Spanish region, our study is exploratory, which is common in qualitative research. We therefore do not intend to extrapolate the results to other contexts. However, we believe that our findings can be used to formulate some preliminary hypotheses about the probability content of UET in other Spanish regions, which would be tested in further studies. Our results can also inform teachers about the probability knowledge that students need to pass the UET, as well as identify criteria for improving future UETs.

We performed a content analysis for each problem, starting from the solution of the problem and its semiotic analysis (Godino et al. 2007), which helped us to identify the mathematical objects involved in the solution and the categories for each variable. This technique helps to decompose a text into analysis units that can be classified into a reduced number of categories, which involve underlying variables that help performing inferences about its content (Krippendorff 2013). Using an inductive and cyclical procedure, we coded and analysed the data to produce graphs and tables that helped us to draw conclusions about the distribution of the categories in each variable, the evolution of the variables with time, and the problem's potential difficulty.

7.4 Results and Discussion

When analysing the mathematical objects needed to solve the probability problems included in the UET, we noticed that the problems are quite complex and involve a high level of difficulty for students. In the sections below, we summarise the main results concerning each variable analysed.

7.4.1 Type of Experiment

Each probability problem presents a random experiment that must be identified by the student in order to understand the situation. Following the suggestions by Ortiz (2002), we first classified the experiment in each problem as a simple or compound random experiment. In the former case, the experiment cannot be decomposed, whereas in the latter case, the experiment is defined by a composition of two or more simple random experiments. For example, flipping a fair coin is considered to be a simple experiment, whereas flipping two or more coins is considered to be a compound experiment because it involves several simple experiments, each consisting of flipping only one coin.

Thus, the experiment in the problem posed in Example 1 (included in the Appendix) is a simple experiment because there is no reference to compound probability and it is possible to compute the probability of union and conditional probability in a simple experiment. Moreover, the experiment underlying Example 2 (see Appendix) is a compound experiment, because it is composed of two different simple experiments: E1, which involves selecting a person at random and observing the gender (man or woman), and E2, asking the person who has been selected about the way he or she travels to work (whether the person travels by car or not). Compound experiments introduce higher complexity in the problem because it is necessary to consider whether the experiments are dependent or independent in order to compute the desired probabilities.

In Fig. 7.1 we classify UET probability problems according to the number of experiments involved. As in Ortiz (2002), problems included simple experiments (20%) and compound experiments (80%). Given that most problems involved compound experiments, in general students were required to analyse two experiments (76%) or three experiments (4%) that are combined.

Figure 7.1 reveals that there is great variability over time: in some years (e.g., 2012), almost all of the experiments in the problems were compound experiments, while the distribution in other years was more heterogeneous (e.g., 2013). However, in all years the number of compound experiments outnumbered simple experiments. Given that problems involving compound experiments are more complex, this finding points to the difficulty of the proposed problems.

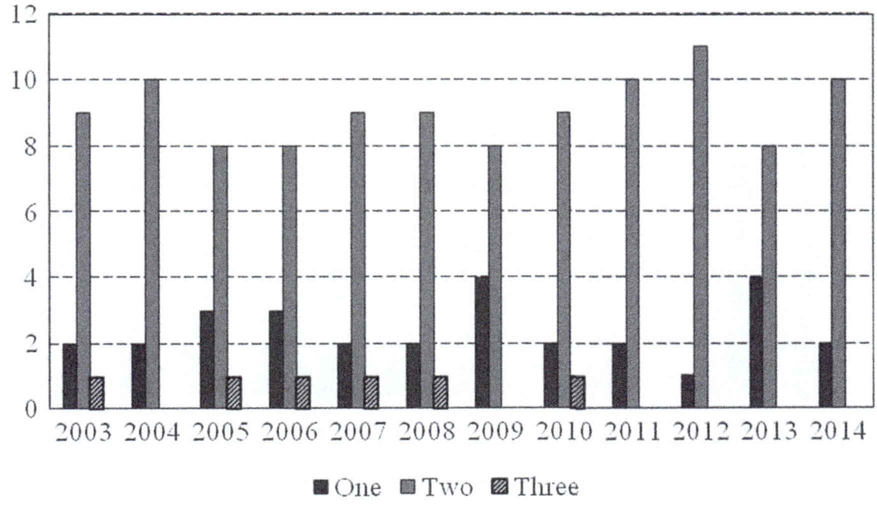

Fig. 7.1 Number of experiments involved in the problem

7.4.2 Dependence or Independence

In the problems analysed we can find dependence among the events of a single experiment or among different experiments. Since most problems were based on compound experiments, we analysed whether or not in these cases the simple experiments comprising the compound experiment were independent of each other.

In most cases, the problem statement does not indicate dependence or independence between experiments. Rather, this feature of the problem must be deduced by the student, either by recognising a context he or she is familiar with and recalling whether the experiments are dependent or independent from previous experience, or by checking whether the definition of independence applies to the problem context. We found two categories for this variable:

(a) All of the experiments in the problem are independent of each other. In Example 3 (see Appendix), the students must deduce independence from previous experience because no data about the probabilities are given. In particular, students must assume that the letter selected by each person will not depend on the letter chosen by the other person.
(b) There are some dependent experiments involved in the problem. In general, this must be deduced from the data in the problem statement. Thus, in Example 4, dependence is deduced by the fact that the probability that a book is written in Spanish or English changes depending on the book subject (mathematics or physics). Dependence is not so evident in Example 2, since we need to compute a conditional probability to deduce it.

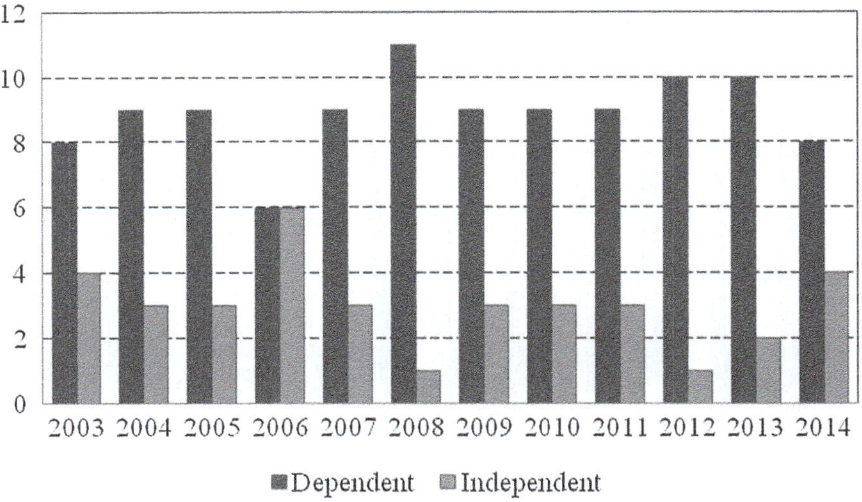

Fig. 7.2 Dependent or independent experiments

Results for this variable are presented in Fig. 7.2. 75% of the compound experiments involved dependent experiments, which, according to Díaz (2004), are more complex for students. Only 25% of compound experiments involved independent experiments. There is some variation over time, but only in 2006 was there an equal number of independent and dependent experiments. In the remaining years, most problems involved dependent experiments.

7.4.3 Sample Space

Following Ortiz (2002) in his study of textbook problems, we discriminated the following types of sample space: (a) two equiprobable events; (b) two non-equiprobable events; (c) three or more equiprobable events; (d) three or more non-equiprobable events; and (e) imprecise, where the sample space is abstract and there is no way to know the number of events or their probabilities. Ortiz suggests that the difficulty of the problem increases from categories (a) to (e).

An example of a sample space with two equiprobable events is the simple experiment in Example 5 (throwing a fair coin three times), with the two equiprobable events being heads and tails. The sample space in Example 3 includes five equiprobable events (writing a vowel at random). In Example 7, the two events in the sample space are non-equiprobable. An example of a sample space with more than two non-equiprobable events is the compound experiment in Example 6.

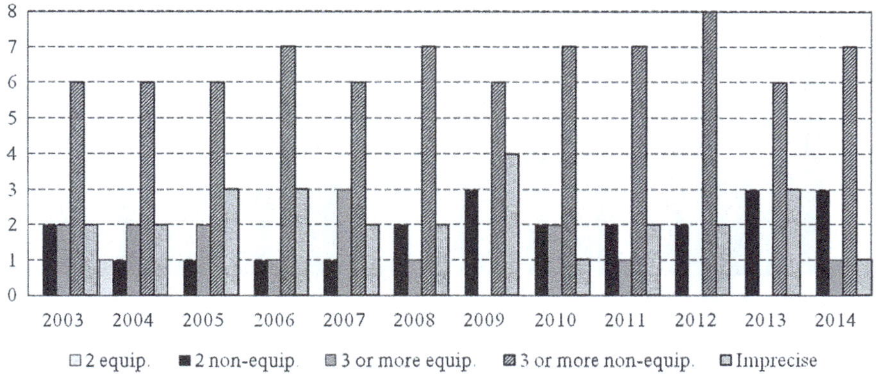

Fig. 7.3 Type of sample space according to number and type of element

Results for this variable are presented in Fig. 7.3. Only two problems involved a sample space with only two equiprobable experiments; 16% of the problems included more than two equiprobable events; 10% of the problems involved two non-equiprobable events; 54% of the problems included more than two non-equiprobable events; and 19% of the problems had an imprecise sample space consisting of abstract events (no probability given). These findings do not take into account the intuitive belief of many students in the equiprobability of results in a random experiment (Batanero et al. 1996). Consequently, with regard to sample space, the difficulty of the problems is moderate to high (Ortiz 2002).

7.4.4 *Type of Probability That Needs to Be Computed*

Regardless of the type of experiment or sample space presented in the problem, the student may be requested to compute simple, compound, or conditional probabilities—usually, two or three different probabilities in each problem. For instance, in Example 4, the student is asked to compute a compound probability in part (a) and a conditional probability in part (b). In addition, the problem statement requires students to identify a simple and a conditional probability.

Results for this variable are presented in Fig. 7.4, which shows that most problems require more than one probability to be computed (since there are only 12 problems per year). In summary, 78% of the problems involved conditional probability, which is difficult for students (Pollatsek et al. 1987); 66% requested students to compute a simple probability, and 58% asked students to compute a compound probability. As these findings reveal, conditional probability was given greater emphasis in these tests, which is reasonable, given the relevance of conditional probability in the understanding of many statistical concepts, such as correlation and regression, confidence intervals, and hypothesis tests (Batanero and Borovcnik 2016; Falk 1986). Furthermore, conditional probability is relevant in many situations where one must make decisions whose consequences affect one's personal or professional life.

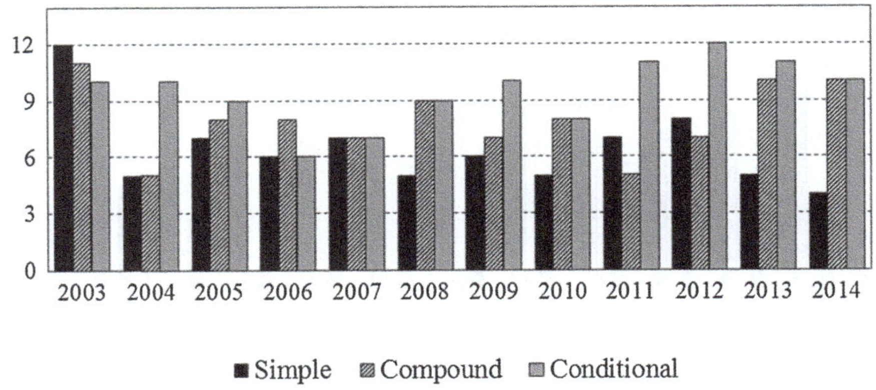

Fig. 7.4 Probability that should be computed

When studying the distribution of this variable throughout time, we observe that in many years, there is a strong presence of the three types of probability. Apart from 2007, conditional probability appeared in eight or more of the 12 problems posed in the tests in every year (ten or more in seven different years). On the contrary, simple probability appeared in six or less problems in six different years, while in 2003 all of the problems asked for computation of a simple probability. Compound probability was requested in 2003 in 11 of the 12 problems, and between six and 10 problems in the remaining years.

Our analyses also revealed a tendency to use only one problem to assess the understanding of and discrimination between two or three different types of probabilities, a discrimination that has been proven to be difficult in previous research (e.g., Díaz 2004).

7.4.5 Properties or Theorems That Should Be Applied

We solved all the problems posed and identified the theorems or properties that students should apply in order to compute the requested probabilities. Seventy-two of the 144 items analysed involved the application of either the law of total probability or Bayes' theorem. These theorems are difficult for students because their application requires a previous identification of the different partitions in the sample space, as well as computation of the probabilities of the events that comprise these partitions. A failure in this identification produces a wrong solution, as shown by Díaz (2004). The difficulty is further increased in the given problems because the student is not directly asked to apply these theorems, whose relevance must therefore be recognised from the problem statement.

Example 8 requires application of both the law of total probability and Bayes' theorem. To solve the problem, we first recognised from the problem statement that the sample space is divided into two events: Event B = "*travelling by bus*" and Event C = "*travelling by car*," which complement each other and constitute a partition of the sample space. A second partition of the same sample space is described by the events L = "*arriving late*" and its complement T = "*being on time*." We also identified the following probabilities from the problem statement: $P(B) = 0.8$, $P(L|B) = 0.20$, and $P(T|C) = 0.10$. The student should interpret the event "*arriving on time and travelling by bus*" as the intersection between T and B. Given that these events are dependent, the student should use the decomposition of conditional probability to compute the compound probability: $P(T \cap B) = P(T|B) \cdot P(B) = 0.80 \cdot 0.80 = 0.64$.

Part b of the problem may confuse the student, who may interpret it as a simple probability. However, the student is asked to compute the probability of an event in the second partition of the sample space, and should therefore use the total probability formula:

$$P(L) = P(B) \cdot P(L|B) + P(C) \cdot P(L|C) = 0.8 \cdot 0.2 + 0.2 \cdot 0.9 = 0.34.$$

Finally, in part c the student is asked to compute the probability that the student in the problem statement travelled by bus if we know that he was late. This probability is obtained using Bayes' formula:

$$P(B|L) = \frac{P(B) \cdot P(L|B)}{P(B) \cdot P(L|B) + P(C) \cdot P(L|C)}$$

The denominator was obtained in part b, and therefore, $P(B|L) = 0.47$.

In Fig. 7.5, we observe that for most problems in the time period analysed, both the law of total probability and Bayes' formula should be applied. It is not a coincidence that in five years, there is the same number of items that request each of these theorems, since Bayes' theorem usually requires application of the law of total probability. The problem becomes more complex when Bayes' theorem is requested with no previous computation of total probability (e.g., in 2004 problems); again, this is an indication that the authors of the tests tried to increase the level of difficulty of the problems.

Most problems required that the student apply a decomposition of probability; usually, the formula of conditional probability should be deduced from compound and simple probability (69% of problems). In about 43% of the problems the probability of the contrary event, the rule of total probability and the Bayes' theorem were needed to solve the problem; and in 40% of the problems the student had to apply the union of compatible events.

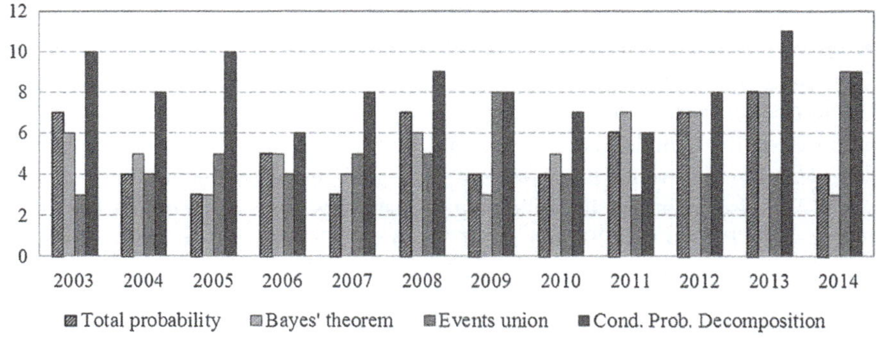

Fig. 7.5 Properties and theorems that should be applied

7.4.6 *Format of Data*

Gigerenzer (1994) remarked that the difficulty of conditional probability problems diminish considerably when the data are given as absolute frequencies. The author suggested that both simple and conditional probabilities are a source of miscommunication, because this usually leaves open the class of events to which the probability refers. On the contrary, frequency data always specify a reference class and can help avoid these misunderstandings. Moreover other researchers (e.g., Hoffrage et al. 2002) have shown that a change of format to absolute frequencies reduces the complexity of conditional and Bayes' problems.

In spite of this research, very few problems posed in the UETs used this format when providing the data, as in Example 6 (only 33% of problems). 25% of problems, such as Example 1 presented the data using a probability format and 43% of problems used percentage format (see Example 5). A few problems (e.g., Example 2) used a mixture of format in the data. In Fig. 7.6, we observe a great variation over time with no clear tendency in the format of the data presented.

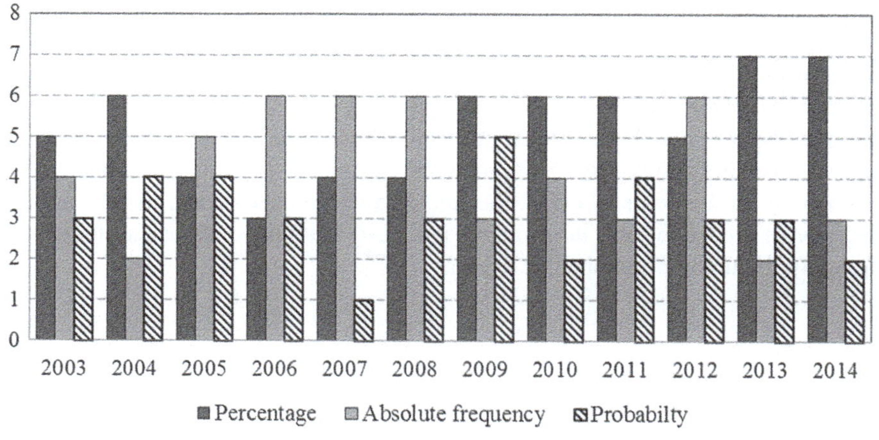

Fig. 7.6 Format of data

7.4.7 Context of the Problem

Finally, we analysed the problem contexts, which were classified according to the categories proposed in the PISA tests (OECD 2015), which have contributed to the current renewed interest in context-based education. The main goal of the context is to show students the link between the concepts and the real-world situations from which these concepts emerge, so that students make sense of their learning and increase their interest towards the learning of mathematics by recognizing its application to different situations. Wild and Pfannkuch (1999) also suggested that the ultimate goal of statistical investigation is learning in context, and that any statistical investigation is carried out to improve our knowledge about the context; this knowledge may then be used in decision making or simply for conceptual understanding. In line with the interest in context-based education, Spanish curricular regulations recommend that teachers develop "the competence to formulate, use and interpret mathematics in different contexts" in their students (MECD 2015, p. 11) by solving real-life problems.

When analysing this variable, we used the following categories:

- *Personal context*: These problems focus on activities that may be carried out by the student, by members, by his or her family or peers (e.g., cooking, shopping, playing games, taking care of one's health, personal transportation, playing sports, traveling, personal scheduling, and managing personal finances).
- *Professional context*: Problems classified in the professional context category are focused on the world of work. They include situations related to measuring, costs or benefits, salary, quality control, design/architecture, and other job-related problems.
- *Social context*: Problems classified in this category focus on the student's community (whether local, national, or global). They may involve voting systems, public transport, government, public policies, demographics, advertising, national statistics, and economics. We also include games of chance in this category.
- *Scientific context*: Citizens are confronted with a wide variety of situations involving science and technology, personal health, natural resources, the environment, and the frontiers of science and technology. Problems classified in the scientific category relate to the application of mathematics to the natural world and to issues and topics related to science and technology. Particular contexts might include weather or climate, ecology, medicine, space exploration, genetics, measurement, and the world of mathematics.
- *No context*: These are abstract mathematical situations that do not include any application to real life (see Example 1). The consequence of reliance on such problems may be routine learning, where students apply formulas without any deep understanding of the underlying concepts. These types of problems contradict current recommendations for the teaching of statistics (e.g., Batanero and Borovcnik 2016) because they conceal the multiple applications of statistics, which has shaped modern research methodology and is now an integral part of empirical research.

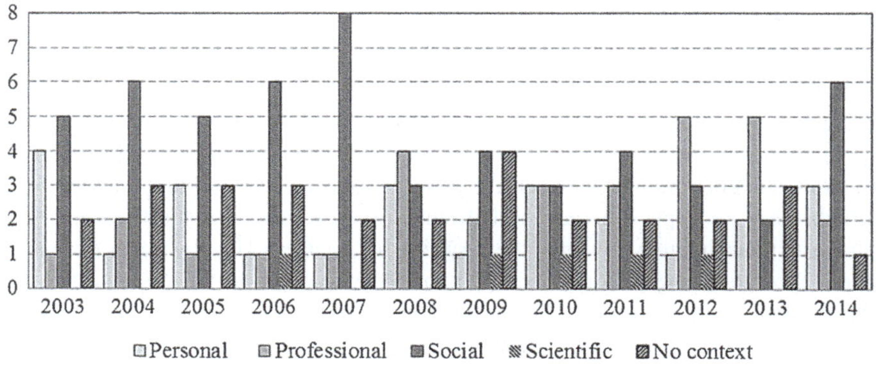

Fig. 7.7 Context of the problems

In Fig. 7.7, we observe the distribution of contexts over the years in the study period. Overall, the social context (38% of the problems) was predominant, in part due to the fact that games of chance were also included in this context. Scientific contexts were very scarce (3%); although some problems described diseases and treatments, we classified these in the social context because scientific concepts were unlikely to be identified by students in these problems and because the problem background referred to a community that was likely similar to the student's community.

Professional context (21%) and personal context (18%) were nearly balanced in representation. We also found that a large proportion of problems (20%) did not refer to a particular context, which is contrary to current recommendations for statistics education. These problems ask the student to transform or simplify algebraic expressions, or to provide justification for abstract probability rules. The goal of these problems is to verify whether students have learned the relevant formulas and whether they can correctly use relevant symbols.

There is high variability of context through time (Fig. 7.7), although social contexts predominate. In certain years, however, professional contexts, where students can appreciate the application of probability to different professional activities, were more frequent.

7.5 Implications for Teaching and Assessment

All of the UETs analysed include a probability problem (one of the four problems posed to the students) related to compound or joint probability, and, therefore, these concepts are given great relevance in these tests. However, as described in the introduction, in Spain, the high school curriculum for Social Sciences and Humanities students includes other probability and statistical content, such as descriptive statistics, correlation and regression, contingency tables, binomial and normal distributions, the central limit theorem, and the normal approximation to the binomial distribution or the law of large numbers, that are hardly taken into account. Although the tests also include other problems related to inference (statistical tests or confidence intervals), it is clear that the greater emphasis in the evaluation of the statistical knowledge of these students is given to joint and conditional probability.

In addition to statistics and probability, the curriculum for Social Sciences and Humanities students includes two additional content areas (calculus and algebra). While statistics and probability constitutes a third of the curricular content (and, usually, only a few pages in textbooks), half of the content of the tests is devoted to probability and statistics. In the case of Science and Technology students, probability and statistics is only taught in the first year of high school, along with three other content areas (algebra, calculus, and geometry); however, no probability or statistics problems are included in the UETs for these students. These results suggest the existence of a mismatch between the high school curriculum and the UET content, which should be revised by the educational authorities in charge of designing the tests.

Probability knowledge and reasoning is needed in everyday and professional situations, and the assessment instruments used reflect the type of content students are expected to learn. In problems lacking context, the multiple applications of probability to different sciences are concealed from students, and probability reasoning is reduced to algebraic reasoning. It is important to substitute these types of decontextualized problems with other situations where students can link mathematics to its applications and make sense of the concepts they apply in solving the problem.

Another conclusion that we have drawn from our analyses is that the probability problems on the UETs pose a high level of difficulty, due to the complex sample space, emphasis on dependent situations, multiple properties and theorems that should be applied, and the format in which the data are provided. The scores obtained by the students on these tests often determine whether they can enter their chosen career; consequently, our results should be taken into account by the test designers in order to build more reasonable assessment tests in the near future that better reflect the curricular content. In addition to including statistics and probability problems in the tests aimed at Science and Technology students, the probability problem posed should take into account the results of educational research. For example, it is important to reduce problem difficulty by providing data in frequency format when available, as recommended by Gigerenzer (1994), and to improve the problem wording so as to avoid confusion between joint and conditional probabilities. Students should also be encouraged to use representations such as the tree diagram or the unit squared (Böcherer-Linder et al. 2015) to describe their reasoning process while solving the problem, and should be scored not only for the correct numerical solution of the problem but also for the quality of their probabilistic reasoning.

Finally, teachers should also take into account the variables analysed in this paper when teaching probability to their students and prepare them to discriminate between dependent and independent experiments, conditioning and causation, as well as joint and conditional probability. It is also important that students become aware of their reasoning biases, such as the time-axis fallacy and the transposed conditional, at the same time that they learn how data can be translated between different representations and how different visualisation tools may help them solve probability problems (Borovcnik 2012).

Acknowledgements Research Projects EDU2016-74848-P (AEI, FEDER), and EDU2013-41141-P (MINECO), and Research Group FQM126 (Junta de Andalucía).

Appendix: Examples of Problems

Example 1 Consider two events A and B:

(a) Describe, using set operations, the following events: (1) None of the two events happen; (2) At least one of the two events happen; (3) B happens and A does not happen.
(b) Suppose it is known that $P(A) = 0.5$, $P(B) = 0.5$, and $P(A|B) = 0.3$. Find $P(A \cup B)$.

Example 2 55% of the Spanish population are women, 23% of which use travel to work by car. The probability that a person (either man or woman) travels by car to work is 0.52.

(a) We select a man at random, what is the probability that this man travels to work by car?
(b) We select a person at random and this person travels to work by car. Compute the probability that this person is a woman.

Example 3 Blanca and Alfredo write a vowel at random on different pieces of paper.

(a) Determine the sample space in this experiment.
(b) Find the probability that they write different vowels.

Example 4 There are only physics and mathematics books in a library and the books are written either in English or in Spanish. 70% of the books are physics books, 80% of the books are written in Spanish, and 10% of the books are mathematics books written in English.

(a) What percentage of books are physics books written in Spanish?

(b) We select a physics book at random, what is the probability that the book is written in Spanish?

Example 5 Consider the random experiment consisting of flipping a fair coin three times.

(a) List the associated sample space and compute the probabilities of the elementary events.
(b) Let A be the event: "obtain at least one head," and let B be the event: "obtain a head in only one of three throws." Calculate P(A) and P(B). Are A and B independent?

Example 6 Two urns A and B contain coloured balls. Urn A contains 4 white balls and 3 red balls. Urn B contains 5 white balls, 2 red balls, and 1 black ball. A die is rolled. If the result is 1, 2, 3 or 4, a ball is randomly selected from urn A; if the result is 5 or 6, a ball is randomly selected from urn B.

(a) Find the probability of getting a red ball.
(b) Find the probability of getting a black ball.
(c) If a white ball was obtained, find the probability that a 5 or 6 was rolled on the die.

Example 7

(a) Let A and B be two events from the same sample space. Suppose it is known that $P(A) = 0.5$, $P(B) = 0.4$, and $P(A \cup B) = 0.8$. Determine $P(A|B)$.
(b) C and D are two independent events from the same sample space. $P(C) = 0.3$ and $P(D) = 0.8$. Determine $P(C \cup D)$.

Example 8 A student travels to school by bus on 80% of school days and travels by car the remaining days. When he travels by bus, he is late 20% of the time, but when he travels by car he is on time only 10% of the time. A day is selected at random.

(a) Find the probability of arriving on time and travelling by bus.
(b) Compute the probability of arriving late.
(c) If the student arrived on time, compute the probability that he had travelled by bus.

References

Batanero, C., & Borovcnik, M. (2016). *Statistics and probability in high school*. Rotterdam: Sense Publishers.

Böcherer-Linder, K., Eichler, A., & Vogel, M. (2015). Understanding conditional probability through visualization. In H. Oliveira, A. Henriques, A. P. Canavarro, C. Monteiro, C. Carvalho,

J. P. Ponte, R. T. Ferreira, & S. Colaço (Eds.), *Proceedings of the International Conference Turning data into knowledge: New opportunities for statistics education* (pp. 14–23). Lisbon, Portugal: Instituto de Educação da Universidade de Lisboa.

Batanero, C., Serrano, L., & Garfield, J. B. (1996). Heuristics and biases in secondary students' reasoning about probability. In L. Puig, & A. Gutiérrez (Eds.), *Proceedings of the 20th Conference of the International Group for the Psychology of Mathematics Education* (Vol. 2, pp. 43-50). Valencia, Spain: PME group.

Borovcnik, M. (2012). Multiple perspectives on the concept of conditional probability. *Avances de Investigación en Educación Matemática, 2,* 5–27.

Contreras, J. M. (2011). *Evaluación de conocimientos y recursos didácticos en la formación de profesores sobre probabilidad condicional* (Assessing knowledge and didactic resources in training teachers to teach conditional probability). PhD dissertation. University of Granada, Spain.

Contreras, J. M., López-Martín, M. M., Arteaga, P., & Carretero, M. (2015). Probability content in the entrance to university tests in Andalucía. In H. Oliveira, A. Henriques, A, Canavarro, C. Monteiro, C. Carvalho, J. P. Ponte, R. Ferreira, & S. Colaço (Eds.), *Proceedings of the International Conference. Turning data into knowledge: New opportunities for statistics education* (pp. 24–33). Lisbon: Instituto de Educaçao da Universidade de Lisboa.

Díaz, C. (2004). *Elaboración de un instrumento de evaluación del razonamiento condicional* (Building an instrument to assess conditional probability reasoning). Master's Thesis. University of Granada, Spain.

Díaz, C., & Batanero, C. (2009). Students' formal knowledge and biases in conditional probability reasoning. Do they improve with instruction? *International Electronic Journal of Mathematics Education, 4*(2), 131–162.

Díaz, C., & de la Fuente, I. (2007). Assessing students' difficulties with conditional probability and Bayesian reasoning. *International Electronic Journal of Mathematics Education, 2*(3), 128–148.

Díaz, C., Batanero, C., & Contreras, J. M. (2010). Teaching independence and conditional probability. *Boletín de Estadística e Investigación Operativa, 26*(2), 149–162.

Díaz, C., Contreras, J. M., Batanero, C., & Roa, R. (2012). Evaluación de sesgos en el razonamiento sobre probabilidad condicional en futuros profesores de educación secundaria (Assessing reasoning biases in conditional probability of prospective secondary school teachers). *Bolema, 26*(22), 1207–1226.

Drijvers, P. G., Godino, J. D., Godino, F. V., & Trouche, L. (2013). One episode, two lenses. A reflective analysis of student learning with computer algebra from instrumental and onto-semiotic perspectives. *Educational Studies in Mathematics, 82,* 23–49.

Falk, R. (1986). Conditional probabilities: Insights and difficulties. In R. Davidson & J. Swift (Eds.), *Proceedings of the Second International Conference on Teaching Statistics* (pp. 292–297). Victoria, Canada: International Statistical Institute.

Franklin, C., Kader, G., Mewborn, D., Moreno, J., Peck, R., Perry, M., & Scheaffer, R. (2007). *Guidelines for assessment and instruction in statistics education (GAISE) report: A Pre-K–12 curriculum framework.* Alexandria, VA: American Statistical Association. http://www.amstat.org/Education/gaise/. Accessed 14 April 2017.

Gigerenzer, G. (1994). Why the distinction between single-event probabilities and frequencies is important for psychology (and vice-versa). In G. Wright & P. Ayton (Eds.), *Subjective probability* (pp. 129–161). Chichester: Wiley.

Godino, J. D. (1996). Mathematical concepts, their meanings and understanding. In L. Puig & A. Gutiérrez (Eds.), *Proceedings of the Twentieth Conference on the Psychology of Mathematics Education* (Vol. 2, pp. 417–424). University of Valencia: PME Group.

Godino, J. D., Batanero, C., & Font, V. (2007). The onto-semiotic approach to research in mathematics education. *ZDM. The International Journal on Mathematics Education, 39*(1–2), 127-135.

Hoffrage, U., Gigerenzer, G., Krauss, S., & Martignon, L. (2002). Representation facilitates reasoning: What natural frequencies are and what they are not. *Cognition, 84*(3), 343–352.

Huerta, M. P. (2014). Researching conditional probability problem solving. In E. J. Chernoff & B. Sriraman (Eds.), *Probabilistic thinking. Presenting multiple perspectives* (pp. 613–639). New York: Springer.

Krippendorff, K. (2013). *Content analysis: An introduction to its methodology*. London: Sage.

Ministerio de Educación y Ciencia. (2007). *Real Decreto 1467/2007, de 2 de noviembre, por el que se establece la estructura del Bachillerato y se fijan sus enseñanzas mínimas* (Royal Decree establishing the structure and minimum content of the high school curriculum). Madrid: Author.

Ministerio de Educación, Cultura y Deporte. (2015). *Real Decreto 1105/2014, de 26 de diciembre, por el que se establece el currículo básico de la Educación Secundaria Obligatoria y del Bachillerato* (Royal Decree establishing the basic curriculum of high school). Madrid: Autor.

Ministerio de la Presidencia. (2008). *Real Decreto 1892/2008, de 14 de noviembre, por el que se regula las condiciones para el acceso a las enseñanzas universitarias oficiales de grado y los procedimientos de admisión a las universidades públicas españolas* (Royal Decree establishing the conditions and procedures of entrance to public universities). Madrid: Author.

Organisation for Economic Co-operation and Development. (2015). *PISA 2015. Assessment and analytical framework*. Paris: Author.

Ortiz, J. J. (2002). *Significado de los conceptos probabilísticos elementales en los textos de Bachillerato* (Meaning of elementary probabilistic concepts in high school textbooks). PhD dissertation. University of Granada, Spain.

Pollatsek, A., Well, A. D., Konold, C., & Hardiman, P. (1987). Understanding conditional probabilities. *Organization, Behavior and Human Decision Processes, 40,* 255–269.

Tversky, A., & Kahneman, D. (1982). Judgements of and by representativeness. In D. Kahneman, P. Slovic, & A. Tversky (Eds.), *Judgment under uncertainty: Heuristics and biases* (pp. 84–98). New York: Cambridge University Press.

Wild, C. J., & Pfannkuch, M. (1999). Statistical thinking in empirical enquiry. *International Statistical Review, 67*(3), 223–265.

Chapter 8
Random Walks in the Didactics of Probability: Enactive Metaphoric Learning Sprouts

Jorge Soto-Andrade, Daniela Díaz-Rojas and Pamela Reyes-Santander

Abstract We argue that random walks are a foremost avatar of the key notion of randomness and that their enactive and metaphoric exploring by the learners may facilitate the access to probabilistic thinking of non-mathematically inclined students, from primary school onwards. We propose to this end so called "learning sprouts" for randomness based on random walks that grow from primary school to post graduate education. We discuss the underlying theoretical background, involving metaphorisation, enaction and (a)didactical situations, and illustrate our theoretical developments with some examples of learning sprout design and related didactic "thought situations", which could be meaningful for a broad spectrum of learners, including primary and secondary school students as well as university humanistic and scientific students and teacher students.

Keywords Random walks · Enactive · Metaphor · Learning sprouts

8.1 Introduction

Randomness is a key notion in mathematics as well as in mathematics education. In fact probability and statistics are making their way nowadays into elementary school curricula almost everywhere the world. Nevertheless the way they have been traditionally taught might not foster the development of probabilistic or statistic

J. Soto-Andrade (✉)
Departamento de Matemáticas Facultad de Ciencias, Universidad de Chile,
Casilla 653, Santiago, Chile
e-mail: sotoandrade@uchile.cl

D. Díaz-Rojas
University of Oxford, Oxford, UK
e-mail: daniela.diazrojas@wolfson.ox.ac.uk

P. Reyes-Santander
Universität Bielefeld, Bielefeld, Germany
e-mail: pamela.reyes-santander@uni-bielefeld.de

C. Batanero and E. J. Chernoff (eds.), *Teaching and Learning Stochastics*,
ICME-13 Monographs, https://doi.org/10.1007/978-3-319-72871-1_8

insight among the learners exposed to them (Borovcnik 1992, 2013; Borovcnik and Kapadia 1991; Chernoff and Sriraman 2014; Fischbein and Schnarch 1977).

In this theoretical paper, inspired by a stance in mathematics education that emphasises metaphorising and enaction (Gallagher and Lindgren 2015; Lakoff and Núñez 2000; Proulx and Simmt 2013; Sfard 2009; Soto-Andrade 2014, 2015), we intend to develop an enactive and metaphoric approach to the didactics of probability, aiming at rendering probabilistic insight more accessible to most learners. More precisely, our main motivating questions are:

- Could metaphorising and enacting help in making probabilistic insight and thinking accessible to most students? In which ways?
- How do these two fundamental processes interact and intertwine in the context of understanding randomness? Particularly, in creative problem solving and posing in probability?

Among various possible paths, we choose here to "enter" probability through random walks, a foremost and ubiquitous embodied avatar of randomness, as we claim. We are especially interested in random walks because we hypothesise that when enactively and metaphorically approached, they may help non-mathematically inclined learners to grasp probabilistic ideas, insights and contents,[1] which would remain otherwise out of reach and cryptic for most of them.

To support our claim regarding random walks, let us point out first that random walks are transversal; they cross boundaries, arising in both the "natural" and the "cultural" realm. In the natural realm we find Brownian motion of pollen micro particles (Powles 1978) and also Brownian motion in metallic alloys, foreseen by Einstein in 1905 (Preuss 2002). It might be argued however that Brownian motion was anticipated by Lucretius in 60 BC, who observed dust particles "skirmishing" under sunlight and thought this to be caused by "motions of matter latent and unseen at the bottom" (Powles 1978). Other examples of random walks in the natural realm are mosquito flights (Pearson 1905), foraging patterns in human hunter–gatherers (Raichlen et al. 2014), and erratic fluctuations of stock markets (Bachelier 1900).

In the cultural realm, we find the construction of random hexagrams when consulting Yi Jing, the ancient Chinese oracle (Wilhelm and Baynes 1967), that may be seen as a 6-step symmetric random walk in the binary tree, as drawn by Chinese mathematician Shao Yong (Marshall 2015). Another example is provided by Saint Francis of Assisi's friars walking across medieval Italy's road network to preach the Gospel, trying to be just instruments of God's will by choosing randomly at each crossroad with the help of the following clever method, devised by Saint Francis himself: At every road junction, he had a friar to whirl nonstop in spite of dizziness and nausea, until he collapsed and fell. Then the whole company would choose the road closest to the direction shown by the friar's head (Soto-Andrade 2013).

[1]This, from primary school all the way up to postgraduate school, didactically embodied in what we call "learning sprouts".

The term *random walk* was first used by Pearson (1905) in his query to the journal Nature about the probability distribution of the distance from the origin of a random flying mosquito (a vector of malaria), after a given lapse of time. We claim that random walks are indeed a visual embodiment of randomness, because we literally *see* randomness in a random walk, much better than in the different outcomes arising, for example, when throwing a dice. Random walks are easily enacted and simulated, starting at primary school (López-Fuentealba 2014) and generating "learning sprouts", (see Sect. 8.3), that "grow" up to graduateteaching. We can approach them in manifold ways: statistically, metaphorically, or probabilistically. They provide "universal models" and metaphors for varied phenomena, learning itself included. Indeed, simple random walks may be explored "bare handed", with no previous statistic or probabilistic tools or concepts. Students may tackle the paradigmatic "impossible question": *Where is the walker, after a given number of steps?* equipped with sheer common sense. Most important, they realize quickly that there are several levels of answers to this sort of question, beginning with: "*Nobody knows!*" up to a precise quantification of the degree of likelihood of each different possible location where the walker might be.

After setting up our theoretical background, we describe our "learning sprouts", sketch the design of a learning sprout (in fact, a bundle of them) for randomness, based on random walks, and present and comment a couple of illustrative examples that could be implemented at different educational levels. These learning sprouts seem to be novel in the literature and have not made their way yet to the curriculum as far as we know. So although this paper is framed as a theoretical paper, we comment some didactical phenomena, besides "thought experiments", that may throw some light on processes involved in the presence or absence of metaphorisation and enaction in the learning of probability, particularly in problem solving activities related to randomness. Our examples may also afford a falsification of the current assumption that mathematical problem solving is a purely mental activity where corporeality and concrete handling are completely irrelevant. Although suggested by theoretical considerations, we believe that our examples and thought experiments could provide inspiration and stimulus as templates for future empirical research involving didactical design development in a variety of contexts.

8.2 Theoretical Framework

In this section we recall the role of metaphorising in cognitive science and mathematics education as well as the fundamental aspects of the enactivist perspective and the theory of (a)didactical situations in mathematics education. We finally describe some useful and recurrent metaphors in the study of random variables and random walks.

8.2.1 Metaphors in Cognitive Science and Mathematics Education

Widespread agreement has been reached in cognitive science that metaphorising (looking at something and seeing something else) serves as the often unacknowledged foundation for human thought (Gibbs 2008) since our ordinary conceptual system, in terms of which we both think and act, is fundamentally metaphorical in nature (Johnson and Lakoff 2003). Particularly, from an epistemological viewpoint we may claim, metaphorically, that theories are just "unfoldings" of metaphors (Soto-Andrade 2014), i.e. theories are generated by grounding metaphors. Recall for instance the "tree of life" metaphor in Darwin's theory of evolution, the "multiplication is concatenation" metaphor in category theory in mathematics (Soto-Andrade 2014) or the nowadays familiar "probability measure" metaphor in probability theory. In what follows we will find often convenient to use the metaphorical approach as a meta-theory to tersely describe other theoretical frameworks we will use.

In mathematics education it has been progressively recognized during the last decade (English 1997; Lakoff and Núñez 2000; Sfard 2009; Soto-Andrade 2007 and many others) that metaphors are not just rhetorical devices, but powerful cognitive tools, that help us in building or grasping new concepts, as well as in solving problems in an efficient and friendly way (for a recent survey see Soto-Andrade 2014). More precisely, we make use of conceptual metaphors (Lakoff and Núñez 2000), seen as mappings from a "source domain" into a "target domain" that carry the inferential structure of the source into the one of the target, and enable us to understand the latter, usually more abstract and opaque, in terms of the former, more down-to-earth and transparent.

Although the terms "metaphor" and "representation" are often used as synonyms, we distinguish here between them as in Soto-Andrade (2014). Like Sfard (2009) we see metaphor as bringing the target concept into being rather than just shedding a new light on an already existing notion, as representation does. For instance, we bring the concept of probability into existence when, while studying a symmetric random walk on the integers, we see the walker splitting into two equal halves instead of jumping equally likely right or left (Soto-Andrade 2013). Since new concepts arise from a crossbreeding of several metaphors rather than from a single one, multiple metaphors, as well as the ability to move between them, may be necessary for the learner to make sense of a new concept (Sfard 2009). Teaching with multiple metaphors, as an antidote to unwanted entailments of one single metaphor, has been recommended (Chiu 2000; Sfard 2009).

Conceptual metaphors in mathematics education turn out to be quite old. Indeed, they are operationally equivalent to the Grundvorstellungen (basic ideas or notions) of the German school of didactics of mathematics, whose origin may be traced back to Pestalozzi in 1803, although their use is more prescriptive in the German tradition than in ours (vom Hofe 1995; Soto-Andrade and Reyes-Santander 2011). Recall that already Euler (1802) metaphorised negative numbers as debts.

8.2.2 *Enactive Metaphors and the Enactivist Perspective in Mathematics Education*

Metaphors most relevant in mathematics education are *enactive* (Varela et al. 1991), instead of just being "sitting metaphors" in the sense of Gallagher and Lindgren (2015). On the other hand, when Varela introduced the enactivistic approach in cognitive science, he metaphorized enaction as the laying down of a path in walking (Varela 1987), echoing Machado:

> "Caminante, son tus huellas el camino, y nada más; caminante, no hay camino, se hace camino al andar" [Wanderer, your footsteps are the path, nothing else; there is no path, you lay down a path in walking]. (Machado 1988, p. 142)

In Varela's own words:

> The world is not something that is given to us but something we engage in by moving, touching, breathing, and eating. This is what I call cognition as enaction since enaction connotes this bringing forth by concrete handling. (Varela 1999, p. 8)

A related albeit less radical use of the term enaction in mathematics education may be traced back to Bruner (1966) who, inspired by Dewey's "learning by doing" (Dewey 1910/1997), introduced *enactive*, *iconic* and *symbolic* modes of representation.

A well-known example of an enactive metaphor for a mathematical notion is "polygons are closed paths" (Chiu 2000), that enables students to immediately "see" that the sum of a polygon's exterior angles corresponds to a complete revolution when they enact the polygon themselves "lying down a polygon in walking" (Díaz-Rojas and Soto-Andrade 2015). In the probabilistic realm, hydraulic and pedestrian metaphors for random walks (see Sect. 8.2.4) are enactive, in the sense that they can be acted out, generating even choreographies for the latter (see Sect. 8.3.2).

An enactivistic approach in the learning of mathematics incorporates gestures, movements and, more broadly, corporeality, particularly in problem solving (Libedinsky and Soto-Andrade 2015). During the last decade such a perspective in mathematics education has arisen (Proulx and Simmt 2013; Reid 2014), inspired by Varela's enactive theory of mind (Varela 1987, 1999; Varela et al. 1991) as well as Maturana and Varela's work on autopoiesis and cognition (Maturana and Varela 1980, 1987; Maturana 1987, 2002), although Maturana himself never uses the words "enactive" or "enaction". For recent surveys see Goodchild (2014) and Reid and Mgombelo (2015). Very recent advances (Gallagher and Lindgren 2015) parallel our exploration of enactive metaphorising (Díaz-Rojas and Soto-Andrade 2015) in mathematics education, providing evidence that metaphoric and radically enactive and embodied approaches to cognition may be successful in understanding and fostering the learning of mathematics and that sensorimotor enaction in the field of promoted action gives birth to mathematical concepts (Abrahamson and Trninic 2015).

8.2.3 *Didactic Situations*

The theory of didactic situations (Brousseau and Warfield 2014) may be described as an unfolding of the emergence metaphor for mathematical content: mathematical concepts or procedures we intend to teach should *emerge* in a suitable challenging situation the learner is enmeshed in, as the only means to "save his life" (real learning is not fostered when mathematical concepts are parachuted from Olympus, we might say). This type of situation is called a *didactic situation*, because of the avowed didactical intent of the teacher who set it up. It becomes an *(a)didactic* situation when the teacher steps back to let the learners interact (enactively) on their own with the situation, with no hope of fathoming beforehand the teacher's didactical design or the mathematical content she is aiming at. Metaphors play an important role in (a)didactical situations, best described with the help of a "voltaic metaphor": Key idiosyncratic metaphors are likely to emerge as sparking voltaic arcs, in and among the learners, suggesting a way to cope with the challenge, when enough "didactical tension" is built up in a didactical situation for them. This requires of course setting up a suitable didactical situation and sustaining the necessary "didactical tension", which is not an easy task.

8.2.4 *Metaphors in the Study of Random Variables and Random Walks*

We present below some recurrent metaphors related to random variables and random walks, which—we hypothesise—could provide a friendly access to non-mathematically inclined learners.

8.2.4.1 Metaphors for Random Variables

A pictorial metaphor: Snapshots and the portrait of a random variable. Given a random variable X defined by a random experiment E, we may repeat E a certain number of times and get the associated relative frequency distribution for (the values of) X, that we might call *a snapshot* of X.[2] On the other hand, we might call the probability distribution of X, *the portrait* of X. A snapshot of X being associated to a given number of repetitions of E, we would expect that our humanistic students (first year university students majoring in social sciences and humanities, many of them with mediatic interests), once they are suggested this photographic metaphor,

[2]From a statistical viewpoint, this new distribution describes a new variable called statistics variable, which is different from the random variable; the distribution of the statistics variable is the frequency distribution, while the distribution of the random variable is the probability distribution.

tend to spontaneously associate this number of repetitions with the number of pixels, or the resolution, of the snapshot. We have indeed noticed this reaction, that in our view shows that they have assimilated the metaphor: They say that three tails and one head, i.e. the relative frequency distribution 3/4 for 0 and 1/4 for 1, when flipping a coin four times, is a "low resolution snapshot" of the random variable X = *number of heads obtained when flipping one coin*, but 54 tails and 46 heads, when flipping a coin 100 times (i.e. the relative frequency distribution 54/100 for 0 and 46/100 for 1) is a "higher resolution snapshot" of X. Comments like the following are then likely to emerge: "Among several snapshots of X those with higher resolution are more "reliable" and more resemblant of the portrait". "The portrait of X (two bars of length 1/2 at 0 and 1), on the contrary, is unique and idealized, as portraits of kings or dictators usually are, in real life".

Another more frequent metaphor is visualising relative frequencies and probabilities as masses (or weights). This mechanical metaphor looks at a distribution of a statistics variable (snapshot) or the distribution (portrait) of a random variable and sees a mass distribution of total mass 1, standing, for example, on a titanium ruler (of negligible weight). We may then draw on our physical experience and intuition regarding centres of mass or balance points, and rotational inertia (also called moment of inertia) to metaphorise as follows.

- The expectation of a random variable is the balance point or barycentre of its portrait (seen as a mass distribution).
- The standard deviation of a random variable is the (lever) arm of a "dumbbell" of total mass 1 which has the same rotational inertia as its portrait.

Here our "dumbbell" will have a weight of 1/2 at each end. Its lever arm is the distance from each weight to the midpoint of the dumbbell, which is the rotation pivot. We rotate the portrait, seen as a mass distribution, around its barycentre. The arm of the dumbbell is to be adjusted to mimic the rotational inertia of the portrait. Notice that in each case, static (expectation) or dynamic (standard deviation), we replace the given system of masses (the portrait) by the simplest mass system that is (statically or rotationally) "equivalent" to it: mass 1 at the barycentre or a (1/2, 1/2) dumbbell with midpoint at the barycentre of the portrait.

8.2.4.2 Metaphors for Random Walks

We sketch below a list of helpful metaphors, most frequently used by learners to explore and figure out random walks on graphs, at various educational levels.

Solomonic metaphor. Consider a random walk. This metaphor sees the random walker deterministically *splitting* into pieces instead of *walking randomly* according to the given transition probabilities, as King Solomon threatened to do with the disputed baby. For instance, looking at a frog jumping equally likely right or left on a row of stones in a pond, the Solomonic metaphor sees the frog splitting into two halves that go simultaneously right and left, and so on. This friendly metaphor turns the random walk into a *deterministic* fission process whose description boils down to a calculus of fractions, and enables the learners to construct the notion of probability.

Hydraulic metaphor. This metaphor is a variant of the previous one, where we look at the random walk and we see instead water draining down through a tree (or more generally a network) of ducts, bifurcating according to the given transition probabilities (see Sect. 8.3 for examples of its use). This metaphor suggests that an analogical model made out of a network of ducts and stopcocks could be constructed by the learners to figure out the frog's random walk and other more general random walks.

Pedestrian metaphor. We may prompt learners who dislike calculating with fractions or halving frogs, to look for other (idiosyncratic) metaphors that would avoid this, until the last minute at least. In this way, the idea usually emerges of unleashing an army of frogs from the starting stone, and splitting it into halves at each stage. Fittingly, 16 frogs for a 4-step symmetric walk, or 1 mega (1024) of frogs for a 10-step symmetric walk and so on. With the help of this metaphor, students just need to count how many frogs are crouching at each stone after the given number of steps and divide by the total number of frogs, to quantify likelihood. More generally this metaphor looks at a random walk on a graph and sees a company of pedestrians splitting into smaller groups as they progress along a road network. Notice that this provides a natural pedestrian approach to Pascal's triangle.

8.3 Learning Sprouts

In our view, learning sprouts can become didactic instruments to be implemented and tested by researchers, educators and teachers in various educational contexts and settings. This motivates our presenting below their conceptual description and some illustrative examples of learning sprout design leading eventually to didactic "thought experiments", close to a priori *analyses* in didactical engineering (Artigue 2009).

8.3.1 Conceptual Description

Our learning sprouts are macro (a)didactical sequences (Brousseau and Warfield 2014), strongly metaphoric and enactive in nature, growing across several educational levels (from primary to tertiary), like learning paths or trajectories unfolding on a "didactic topography". They grow from a "seed", i.e. a fundamental mathematical idea or notion, in our case, *randomness*; their tissue is provided by some avatar or embodiment of this fundamental notion, in our case, *random walks*. Their starting point, often at the elementary level, is usually a "baby version" or "toy model" of the fundamental idea. Actually, they are constructed by the learners themselves, who enactively "lie down paths in walking", as in Machado (1988, p. 142). We usually speak of *a* learning sprout associated to random walks, but in fact this sprout is manifold, rather a sprout bundle, since in different classrooms, environments, and moments, it may unfold differently. Learning sprouts have a significant random component besides a deterministic one, so they are not just "stairways to be climbed stepwise" (the prevalent metaphor in traditional teaching, as in NAEYC/NCMT 2002).

To situate our learning sprouts in the realm of current didactic techniques, we may contrast them with the closely related trajectories or paths for fundamental mathematical ideas, proposed by Wittmann (2012), which also develop across different educational levels. Wittmann's specific learning paths however are usually conceptualized as linear sequences of steps to be followed in a predetermined order, akin to the "stairways metaphor" underlying proposals like those in NAEYC/NCMT (2002). Moreover, the enactive, cooperative and metaphoric aspects are usually neither considered in this conceptualization, nor in the work of Wittmann. We emphasise the open and random aspect of our sprouts, emerging from the situated relationship between students and teachers, since from our perspective real and meaningful learning is more likely attained through random walking in a didactic landscape than through following predetermined trajectories.

8.3.2 Learning Sprout Design in Probability

We illustrate in this section learning sprout design related to randomness, based on random walks. To this end we sketch first the generation of a didactical and mathematical "topography" or landscape, where these learning sprouts may grow. As we will see below, these learning sprouts enable the learners to metaphorically construct the notion of probability and to encounter and study down to earth examples of random variables, which are in fact the main characters in probabilistic problems and situations. Moreover they will be able to deal with conditional probabilities and Bayesian problems from the very beginning, in an elementary and friendly way.

As mentioned before, a paradigmatic and natural example of random walk is *Brownian motion*, which can be thought of as a random walk with continuous time or as a limit of usual random walks. From this natural albeit complex example of random walk, a learning sprout can be designed, starting with a "Baby version" of Brownian motion, where we replace its *2D* continuous plane stage by the discrete plane $Z^2 = Z \times Z$ (*Z*: the integers), where the walker chooses equally likely any cardinal direction at each node (abridged 2D BB motion). By zooming out the trajectories, the learners may realise intuitively that this BB motion "tends" to the adult one, when the mesh of the grid ($\mathbf{Z}^2$) tends to 0. So classical Brownian motion suggests a starting seed for a learning sprout for random walks, to wit BB motion in 2D, or even in 1D. Since in our view, contextualisation is very relevant for non-mathematically inclined students, we may tell them the following story:

Brownie's walk. A puppy, called Brownie, escapes randomly from home, when she smells the shampoo her master intends to give her. At each street corner, confused and stressed by the traffic's noise and smells, escaping barely from being overrun, she chooses equally likely any of the four cardinal direction and runs nonstop a whole block until the next corner. Exhausted after, say, four blocks, she collapses at some corner. Her master would like to know where to look for Brownie and also to estimate how far she will end up from home[3]

If we are interested in primary mathematics education, we can think of even simpler versions of BB motion, in *1D* for example, like the symmetric random walk of a frog jumping on a row or a regular polygon of stones in a pond (Soto-Andrade 2006). More precisely, a learning sprout for random walks, could start—at primary school, 4th or 5th grade—with a frog jumping symmetrically on a row of nine stones in a pond, simulated with the help of a coin, or a dice (easier to toss). The fundamental but awkward question emerges:

Where will the frog be after one, two, three, four jumps, if it starts from the central stone?

The learners may *enact* the frog themselves or simulate it, flipping a coin to choose right or left. They can bet on some stone, where the frog would land after two or three jumps, for instance. A collective simulation may be carried out, where every student in the class would simulate a frog and then a snapshot would be taken of the final total frog distribution on the stones. Children may realise that to use a pedestrian metaphor, unleashing an army of 2^n frogs for n jumps is convenient. Also, thanks to a betting game, most of them will learn the hard way that not all stones are equally likely (8-year-olds seem to realise this more easily than 6-year-olds). Students in 4th or 5th grade can represent data obtained through simulation and, from there, propose an "ideal" distribution of frogs.

These simple symmetric random walks can even choreographed. Indeed, the idea may arise, for the 4-step frog's symmetric random walk on a row of stones, say, to set up a "choreography", where 16 students try to devise an efficient way to simultaneously enact each one a different trajectory of the frog. The interesting challenge (also for teachers) is here whether this needs to be coordinated hierarchically by a "Magister Ludi" or whether the group of students can carry out the choreography in a non-hierarchical collaborative way.[4] Our learning sprout grows further when we realise that our frog can jump randomly in many other universes (for grades 6th to 8th, and later in grades 9th to 12th), such as:

- Bounded linear universes with boundary, like a finite row of rocks, where the two end rocks (the boundary) are in fact alligator heads… More formally, a random walk with absorbing barriers.

[33]The implementation of this 2D BB motion has been discussed in detail in Soto-Andrade (2013).

[4]Implementation of this learning sprout in a 30 student 5th grade class, has been reported in López-Fuentealba (2014) and Soto-Andrade (2015).

- Bounded linear universes, with no boundary, like regular polygons, where the symmetric frog (i.e. the "heads or tails frog") leads students to discover parity dependent behaviour, the inscribed n-gons in an $2n$-gon and oscillatory dynamics.
- Our frog may turn into an ant that crawls randomly along the edges of bounded discrete universes with no boundary, like a wire tetrahedron, cube, or other regular polyhedra. Here again our humble ant leads us to discover the two inscribed tetrahedra in a cube, for instance. From grade 7th onwards, students may even explore random walks of an ant in regular polytopes, like the hypercube.
- More generally, from secondary school to university, we may enter the realm of random walks on graphs, regular and non-regular. For symmetric walks on non-regular graphs (like a square with just one diagonal), or asymmetric walks on regular graphs, predicting intuitively or heuristically the final equilibrium state, if any, may be a motivating and challenging task.

Motivated by our last example, of asymmetric random walks in regular graphs, our learning sprout may grow in the direction of economic systems dynamics, for instance. Even the simplest case of a frog random jumping asymmetrically between just two rocks A and B may appear as a metaphor (or a model, if you wish) of an unmerciful market struggle between two yoghurt producers A and B, which month after month seduce each other's consumers according to a fixed pattern. For instance: A entices 1/2 of B's consumers while B seduces only ¼ of A's consumers. Looking at the dynamics of this consumer market, we would see metaphorically a frog randomly jumping between rocks A and B in such a way that when it sits sunbathing on A it will jump to B one out of four times while when crouching on B it will jump to A one out of two times. The fraction of the market controlled by A at a given time, becomes the probability of finding the frog at rock A at this time, and the same for B. Of course an initial condition is also required, e.g. the market being initially equally shared between A and B. In that case, the initial state of the frog was quantum: it squatted with equal probability 1/2 on each rock.

We could say that the random walk is a metaphor for the market struggle, or the other way around, according as to whether the involved learner is more acquainted with jumping frogs or with market struggles. Moreover in all these examples, Solomonic, pedestrian, hydraulic or other metaphors may emerge, eventually after some prompting, necessary to counter the effects of the current didactical contract (Brousseau et al. 2014) that usually does not allow for metaphorising or enacting in the mathematics classroom.

At the university level and already at grades 11th and 12th, we can work systematically with the limit states of these random walks, especially in universes with boundary, e.g. random walks with absorbing or reflecting barriers. Here hydraulic or Solomonic metaphors are especially useful. Also the notion of the (deterministic) *barycentric walk* associated to a random walk may emerge (i.e. the motion of the centre of mass of the system of masses that metaphorises the random walk). Interesting infinite series as well as their sums appear here, as well as fractal properties of the walk, the learners can take advantage of (see specific examples in

Sect. 8.3.3). Notice that the arithmetic and algebraic skills involved here are simpler than those involved in the unending lists of senseless algebra exercises our students are forced to solve in most of our secondary schools.

We remark that manifold relations with other fields appear at a more advanced level: Fourier analysis of random walks, harmonic functions, heat equation, recursive series, fractals, process boundaries (like the 2-adic integers, metaphorised as the *boundary* of the infinite rooted binary tree, that may be identified with the set of "end points" of infinite geodesic symmetric random walks starting from the root, arising from indefinitely tossing a coin).

Last but not least, learners can discover that sundry stochastic or deterministic processes and situations, like the ruin problem, repeated drawings without replacement from an urn, unfinished tournaments, false positive problems, Bayesian inference problems, market struggles, etc., can be metaphorised (or modelled) by random walks, at a second glance. We will turn to giving some specific examples of these below.

8.3.3 Didactic Learning Sprouts Based on Random Walks

In this section, we briefly describe some illustrative examples of learning sprouts for randomness, based on random walks, suggested by our theoretical perspective. Our examples, resembling the "thought experiments" imagined by theoretical physicists, or *a priori* analyses in didactical engineering (Artigue 2009), describe possible and plausible reactions of the learners to given challenging situations. To our knowledge, these examples seem to be rather new in the literature, although we have found an interesting convergence of the work of Gigerenzer and collaborators on the use of natural frequencies in Bayesian problems (Zhu and Gigerenzer 2006; Meder and Gigerenzer 2014) with our pedestrian metaphoric approach to these problems seen as random walks (see below, the example of Rayen's fall) and also close connections of previous work of Borovcnik and collaborators (Borovcnik 1992, 2013; Borovcnik and Kapadia 1991) with our own enactivistic approaches to probability. For a discussion of other examples and empirical evidence related to our enactivistic metaphoric approach to probability and statistics, see Soto-Andrade (2013, 2015) and Díaz-Rojas and Soto-Andrade (2015).

8.3.3.1 Bayesian Problems as Random Walks

Awesome Bayes theorem becomes dispensable if problems involving "probabilities of the causes" are metaphorised as (typically 2-step) random walks on a tree, or better, on a grid. Figuring out the walk with the help of a pedestrian metaphor (Soto-Andrade 2013), some students might even able to solve the problem mentally. We will consider the following example:

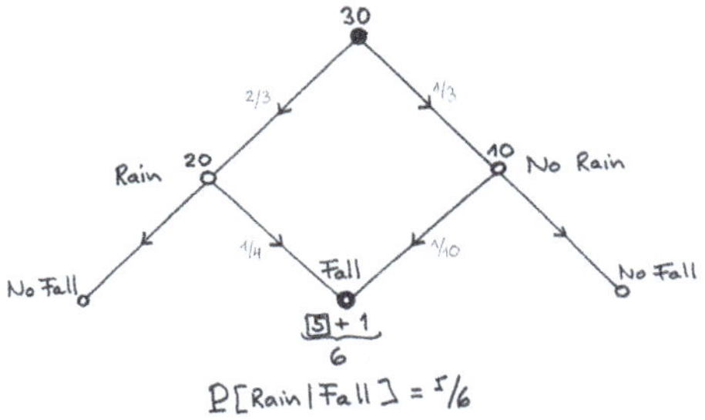

Fig. 8.1 Rayen's fall

> *Rayen's fall.* Rayen lives in the south of Chile and rides her bike along a windy downhill road to school every weekday in winter, when it rains two days out of three. On wet road she falls from her bike one out of four times, on dry road, only one out of ten, on the average. If you learned that Rayen fell from her bike today, how likely is that this was a rainy day?

Figure 8.1 shows a "pedestrian resolution" of this Bayesian problem. In fact, students might refer indifferently to 30 days or to 30 pedestrians, in this case noticing that the problem boils down to count how many of the six walkers arriving at destination "Fall" came through the "Rain" gate.

From our viewpoint we foresee that non-mathematically inclined students could be at least as efficient in solving this type of Bayesian problems as mathematically trained students, if they take advantage of such a friendly and intuitive metaphoric random walk approach. In this way, those using a pedestrian metaphor to figure out this random walk, would converge with the natural frequencies approach promoted by Gigerenzer (Zhu and Gigerenzer 2006; Meder and Gigerenzer 2014).

8.3.3.2 The Frog's Random Walk

Below we describe a second example that can be worked out with different metaphors.

> A frog called Filomena jumps randomly and symmetrically, starting at rock 3, on a row of 6 rocks (tagged 0 to 5) between two alligators, called Anibal and Artemio, camouflaged as rocks 0 and 5. Of course, Filomena is instantly swallowed if it ever lands on the head of either alligator. What can you say about Filomena's fate?

More formally, we have a symmetric random walk on the integers, starting at rock 3 with absorbing barriers at rocks 0 and 5. Students may take here advantage of a hydraulic metaphor (1 litre liquid draining down a suitable vertical grid, at discrete instants of time, represented downwards in the vertical axis) to easily calculate the relevant probabilities of Filomena being still alive, squatting on some rock, or having been eaten, after n jumps, for small n, the latter probabilities becoming just the total amounts of liquid trapped at sinks 0 and 5, at step n (Fig. 8.2).

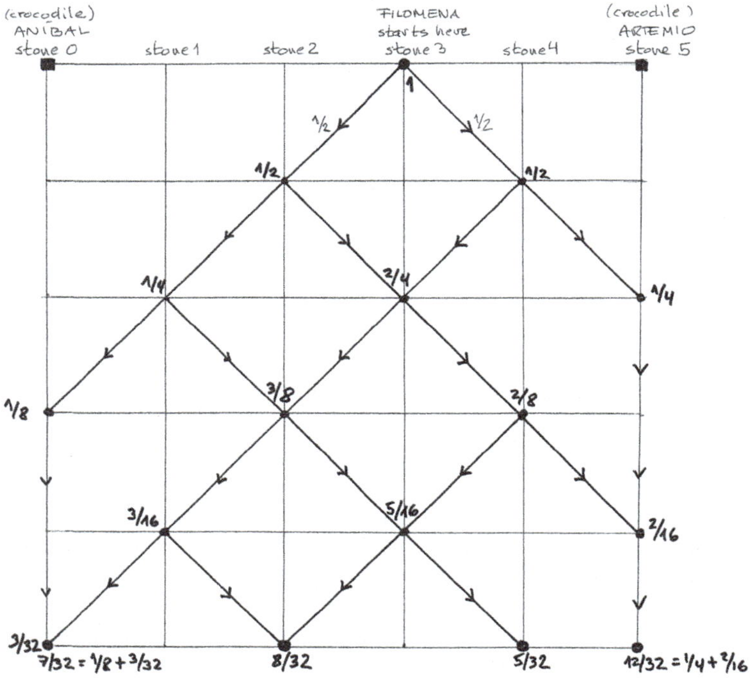

Fig. 8.2 The frog's random walk

Reactions of students and teacher students to this problem may be varied. We would expect non-mathematically inclined students to just estimate probabilities of Filomena being eaten by each alligator in the long run. Students with more mathematical training (e.g. prospective secondary school mathematics teachers) would diligently compute stepwise the amounts of liquid trapped in each sink, to find in the long run two infinite series, one for each sink. They may realise that the general term of these series is the n-th Fibonacci number F_n divided by 2^{n+1}, with n odd for the right sink (i.e. $1/2^2$, $2/2^4$, $5/2^6$, …) and with n even for the left one (i.e. $1/2^3$, $3/2^5$, $8/2^7$, …). With some toil, they may discover that these two series add up to 1 (because the amount of liquid circulating in between the sinks tends to 0 in the long run) and that their values are 3/5 and 2/5 respectively. More advanced students may go even further, using either a Solomonic or a hydraulic metaphor, to solve the problem in different and more transparent ways.

The first strategy takes advantage of the "fractality" of the drain to obtain the *averaging property* (*harmonicity*) for the amount of liquid $L_5(m)$ finally trapped at sink 5, as a *function* of the location m of the initial litre to drain, which we now let vary! So they may notice that $L_5(m) = (1/2)[L_5(m-1) + L_5(m+1)]$ and $L_5(0) = 0$, $L_5(5) = 1$. They can realise then that the harmonic function L_5 must look graphically as a line, connecting points (0,0) and (5,1), so its value at 3 must be 3/5.

A different strategy may emerge however, suggested by the Solomonic metaphor. Enactively, the students see a unit mass starting at location 3 and splitting repeatedly into halves. At each stage they visualise then a system of particles (fractions of the unit mass) and may have the idea to look at its *centre of mass* or *barycentre*. The symmetry of the splitting entails that this centre of mass, will remain motionless at rock 3. We have nevertheless two "black holes", we might say, at rocks 0 and 5, that in the long run will "swallow" all the mass. Since the centre of mass will remain forever at rock 3, the masses finally trapped at 0 and 5 must necessarily be in the ratio of 2 to 3, so they will be 2/5 and 3/5. Metaphorising may enable here students with lesser mathematical training to keep the pace of more advanced students. Working in mixed groups students may also generalise the problem to black holes at rocks m and n, including the case n equal to infinity, i.e. just one black hole at finite distance. They may also become motivated to explore systematically the (deterministic) motion of the barycentre of the system of masses that metaphorises Solomonically any random walk (which just describes the expected location of the walker after n steps).

In our view it is important to notice that the frog's random walk between two alligators is also a metaphor for the classical "ruin problem" where two players A and B, having a certain amount of coins each, flip repeatedly a fair coin to decide who wins. After each flip the winner receives a coin (all coins having the same value) from the loser. No credit is available, so the game ends when any player runs out of coins (is "ruined"). In our case, A would have an initial "fortune" of 3 coins, and B would have only 2 coins. A's ruin means that Filomena was eaten by Anibal. The questions arise:

How likely is that in the long run A or B becomes ruined? How likely is that the game goes on forever?

In fact, students faced with the ruin problem might alternatively move to the frog's walk, the Solomonic metaphor or the liquid draining process.

This rather open ended example of a learning sprout, shows then a possible chain of metaphors that may emerge among the students, where the ruin problem, a heads and tails game, becomes a more concrete random walk of a frog, that in turn may be metaphorised as a liquid draining or a particle splitting. Notice that the notion of barycentre, and its (deterministic) motion, emerges naturally when looking at the particle splitting and affords a transparent way of figuring out the long term fate of the system.

Among other questions, it is here interesting to investigate the reaction of students having being proposed both the frog's random walk and the ruin problem (with the same data) in a test. We hypothesise that students that have being exposed to traditional teaching, leaving no room for metaphorising, would have more trouble in recognising both problems as the "same" problem. They might even solve it twice, maybe one right and the other wrong!

8.4 Summary and Some Open Questions

In this paper, metaphorising has played a manifold role: as a means to making sense or constructing mathematical concepts, like probability, but also as a way of characterising different theoretical approaches in didactics of mathematics and also in describing our instructional approach to randomness. Motivated by the need to render probabilistic thinking and insight accessible to non-mathematically inclined learners, we have first described random walks as a foremost *avatar*—metaphorically speaking—of the key notion of randomness. We noticed then that random walks, because of their ubiquitous and embodied nature, lend themselves naturally to a metaphoric and enactive approach that can be meaningful for many learners, and can be carried out across different educational levels. We sketched examples of helpful metaphors to study random walks as well as several naturally arising "learning sprouts" for randomness, based on random walks, whose open ended design we analysed with some detail. We have suggested in particular how this approach could demystify probabilistic results unfriendly for most learners, like Bayes Theorem, that becomes now just a matter of noticing which way pedestrians followed to reach a common destination (see Sect. 8.3.3.1). We have given other illustrative examples where this approach would enable learners with no previous knowledge of probability to solve probabilistic problems, and eventually construct probability ideas along the way (see Sect. 8.3.3). We have seen how felicitous metaphorising may open up new ways to approach the same problem, some of them much friendlier for many students, besides helping them to notice unexpected connections with other orders of ideas and to glean surprising analytical results (see Sect. 8.3.3.2).

Regarding the intertwining of metaphorising and enaction, in our examples often, but not always, emerging metaphors have an important embodied and enactive component. Last but not least, we may notice that random walks besides providing universal models for many probabilistic and deterministic problems also work as a metaphor for our learning sprouts themselves, in the sense that these are not deterministic learning paths to be dutifully followed, but open ended trajectories whose unfolding depends randomly on the emergent class dynamics and many other (hidden) variables.

In retrospect, progress was made in designing and implementing learning sprouts for randomness, based on random walks, which could be used with a broad spectrum of learners, all the way from primary school to advanced graduate math courses. Many open ends and questions remain or emerge however, that may be investigated empirically, such as:

- Exploring under which conditions learners feel more at ease with the original random walk or some of its various metaphors, like the Solomonic, pedestrian or hydraulic metaphors. For instance, for several years we have consistently noticed that humanistic university students tend to prefer hydraulic to pedestrian metaphors, contrary to Zhu and Gigerenzer (2006) prediction in favour of natural frequencies, and in spite of not feeling comfortable with fractions (essentially avoided by pedestrian metaphors). Apparently these students find the hydraulic metaphor conceptually more transparent. It may also happen that this metaphor helps them to re-signify fractions and their calculus (particularly multiplication of fractions and infinite series with fractional terms).

- Investigating the way in which different types of learners react to our learning sprouts.
- The weight of the didactical contract regarding metaphorising and enacting deserves further systematic exploration, as well as the time and conditions required for the emergence of metaphors.
- Comparing enactive and non-enactive approaches to the same (a)didactical situations, in particular comparing "sitting" and enactive metaphors, in the sense of Gallagher and Lindgren (2015), with various cohorts of learners.
- Undertaking a longitudinal study involving the subsequent professional practice —related to the teaching of probability—of prospective and in service teachers. In fact, in our experience with some in service primary school teachers, they have already designed activities for their classrooms inspired by learning sprouts that we have discussed in professional development activities with them.

Acknowledgements PIA-CONICYT Basal Funds for Centres of Excellence Project FB 0003 and University of Chile Domeyko Fund (Interactive Learning Networks Project).

References

Abrahamson, D., & Trninic, D. (2015). Bringing forth mathematical concepts: Signifying sensorimotor enactment in fields of promoted action. *ZDM Mathematics Education, 47*(2), 295–306.

Artigue, M. (2009). Didactical design in mathematics education. In C. Winsløw (Ed.), *Nordic Research in Mathematics Education. Proceedings of NORMA 08* (pp. 7–16). Rotterdam: Sense Publishers.

Bachelier, L. (1900). Théorie de la spéculation. *Annales de l'Ecole Normale Supérieure, 17*, 21–86.

Borovcnik, M. (1992). Stochastik im Wechselspiel von Intuitionen und Mathematik (Stochastics in the interplay of intuitions and mathematics). Mannheim: Bibliographisches Institut.

Borovcnik, M. (2013). Bedingte Wahrscheinlichkeit—Ein Schlüssel zur Stochastik (Conditional probability—a key to stochastics). *Schriftenreihe zur Didaktik der Mathematik der Österreichischen Mathematischen Gesellschaft, 46*, 1–18.

Borovcnik, M., & Kapadia, R. (Eds.) (1991). *Chance encounters: Probability in education.* Berlin: Springer.

Brousseau, G., Sarrazy, B., & Novotna, J. (2014). Didactic contract in mathematics education, In S. Lerman (Ed.), *Encyclopedia of mathematics education* (pp. 153–159). Berlin: Springer.

Brousseau, G., & Warfield, V. (2014) Didactic situations in mathematics education. In S. Lerman S. (Ed.) *Encyclopedia of mathematics education* (pp. 163–170), Berlin: Springer.

Bruner, J. (1966). *Toward a theory of instruction.* Harvard, MA: Belknap Press.

Chernoff, E. J., & Sriraman, B. (Eds.) (2014). *Probabilistic thinking: Presenting plural perspectives.* Berlin/Heidelberg: Springer Science.

Chiu, M. M. (2000). Metaphorical reasoning: Origins, uses, development and interactions in mathematics. *Educational Journal, 28*(1), 13–46.

Dewey, J. (1910/1997). *How we think*. Mineola, NY: Dover.

Díaz-Rojas, D., & Soto-Andrade, J. (2015). Enactive metaphoric approaches to randomness. In K. Krainer, N. Vondrová (Eds.). *Proceedings of the 9th Congress of European Research in Mathematics Education* (pp. 629–636). Prague: ERME.

English, L. (Ed.) (1997). *Mathematical reasoning: Analogies, metaphors, and images*. London: Lawrence Erlbaum.

Euler, L. (1802). *Vollstaendige Anleitung zur Algebra (1. Teil)*. (Comprehensive guide to algebra, Part 1). St. Petersburg: Kaiserliche Akademie der Wissenschaften.

Fischbein, E., & Schnarch, D. (1977). The evolution with age of probabilistic, intuitively based misconceptions. *Journal for Research in Mathematics Education, 28*(1), 96–105.

Gallagher, S., & Lindgren, R. (2015). Enactive metaphors: Learning through full body engagement. *Educational Psychology Review, 27*(1), 391–404. https://doi.org/10.1007/s10648-015-9327-1.

Gibbs, R. W. (Ed.) (2008). *The Cambridge handbook of metaphor and thought*. Cambridge, UK: Cambridge University Press.

Goodchild, S. (2014). Enactivist theories. In S. Lerman (Ed.). *Encyclopedia of mathematics education* (pp. 209–214). Berlin: Springer.

Johnson, M., & Lakoff, G. (2003). *Metaphors we live by*. New York, NY: The University of Chicago Press.

Lakoff, G., & Núñez, R. (2000). *Where mathematics comes from*. New York, NY: Basic Books.

Libedinsky, N., & Soto-Andrade, J. (2015). On the role of corporeality, affect and metaphoring in Problem solving. In P. Felmer, J. Kilpatrick, & E. Pehkonen (Eds.), *Posing and solving mathematical problems: Advances and new perspectives* (pp. 53–67). Berlin: Springer.

López-Fuentealba, K. (2014). *El paseo de la rana—Probabilidades para 5º básico* (The frog's walk -Probability for 5th graders). Santiago: University of Chile.

Machado, A. (1988). *Selected poems*. Cambridge, MA: Harvard University Press.

Marshall, S. (2015). *Reading hexagrams off the Xiantian diagram*. www.biroco.com/yijing/xiantian.htm. Accessed 27 June 2017.

Maturana, H. (1987). Everything said is said by an observer. In W. Thompson (Ed.), *Gaia: A way of knowing* (pp. 65–82). Hudson, NY: Lindisfarne Press.

Maturana, H. (2002). Autopoiesis, structural coupling and cognition: A history of these and other notions in the biology of cognition. *Cybernetics & Human Knowing, 9*(3/4), 5–34.

Maturana, H., & Varela, F. (1980). *Autopoiesis and cognition: The realization of the living*. Dordrecht: Reidel.

Maturana, H., & Varela, F. (1987). *The tree of knowledge: The biological roots of human understanding*. Boston: New Science Library/Shambhala.

Meder, B., & Gigerenzer, G. (2014). Statistical thinking: No one left behind. In E. Chernoff & B. Sriraman (Eds.), *Probabilistic Thinking* (pp. 127–148). Dordrecht: Springer.

NAEYC/NCMT. (2002). *Learning paths and teaching strategies in early mathematics*, USA: National Association for the Education of Young Children. https://oldweb.naeyc.org/about/positions/pdf/mathchart.pdf. Accessed 27 June 2017.

Pearson, K. (1905). The problem of the random walk. *Nature, 72*(294), 318–342.

Powles, J. G. (1978). Brownian motion—June 1827. *Physics Education, 13*, 310–312.

Preuss, P. (2002). *Brownian motion in metal. Science Beat*. http://www.lbl.gov/Science-Articles/Archive/MSD-Brownian-motion.html. Accessed 27 June 2017.

Proulx, J., & Simmt, E. (2013). Enactivism in mathematics education: moving toward a re-conceptualization of learning and knowledge. *Education Sciences & Society, 4*(1), 59–79.

Raichlen, D. A., Wood, B. M., Gordon, A. D., Mabulla, A. Z. P., Marlowe, F. W., & Pontzer, H. (2014). Evidence of Lévy walk foraging patterns in human hunter–gatherers. *Proceedings of the National Academy of Science, 111*(2), 728–733.

Reid, D. (2014). The coherence of enactivism and mathematics education research: A case study. *AVANT 5*(2), 137–172. https://doi.org/10.12849/50202014.0109.0007.

Reid, D. A., & Mgombelo J. (2015). Key concepts in enactivist theory and methodology. *ZDM Mathematics Education 47*(2), 171–183.

Sfard, A. (2009). Metaphors in education. In H. Daniels, H. Lauder, & J. Porter (Eds.), *Educational theories, cultures and learning: A critical perspective* (pp. 39–50). New York, NY: Routledge.

Soto-Andrade, J. (2006). Un monde dans un grain de sable: Métaphores et analogies dans l'apprentissage des mathématiques (A world in a grain of sand: Metaphors and analogies in the learning of mathematics). *Annales de Didactique et de Sciences Cognitives, 11*, 123–147.

Soto-Andrade, J. (2007). Metaphors and cognitive styles in the teaching-learning of mathematics. In D. Pitta-Pantazi & J. Philippou (Eds.), *Proceedings of the 5th Conference of the European Research Society for Mathematics Education* (pp. 191–200). http://ermeweb.free.fr/CERME5b/WG1.pdf. Accessed 27 June 2017.

Soto-Andrade, J. (2013). Metaphoric random walks: A royal road to stochastic thinking. In B. Ubuz, C. Haser, & M. A. Mariotti (Eds.), *Proceedings of the 8th Conference of the European Research Society for Mathematics Education* (pp. 890–900). Antalya: ERME. http://www.mathematik.uni-dortmund.de/~erme/doc/CERME8/CERME8_2013_Proceedings.pdf. Accessed 9 August 2017.

Soto-Andrade, J. (2014). Metaphors in mathematics education. In S. Lerman (Ed.) *Encyclopedia of Mathematics Education* (pp. 447–453). Berlin: Springer.

Soto-Andrade, J. (2015). Une voie royale vers la pensée stochastique: les marches aléatoires comme pousses d'apprentissage (A royal road to stochastic thinking: random walks as learning sprouts). *Statistique et Enseignement, 6*(2), 3–24.

Soto-Andrade, J., & Reyes-Santander, P. (2011). Conceptual metaphors and "Grundvorstellungen": A case of convergence? In M. Pytlak, T. Rowland, & E. Swoboda (Eds.), *Proceedings of the 7th Conference of the European Research Society for Mathematics Education* (pp. 735–744). Rzészow: University of Rzészow, ERME.

Varela, F. J. (1987). Lying down a path in walking. In W. I. Thompson (Ed.), *Gaia: A way of knowing* (pp. 48–64). Hudson, NY: Lindisfarne Press.

Varela, F. J. (1999) *Ethical know-how: Action, wisdom and cognition*. Stanford, CA: Stanford University Press.

Varela, F. J., Thompson, E., & Rosch, E. (1991). *The embodied mind: Cognitive science and human experience*. Cambridge, MA: The MIT Press.

vom Hofe, R. (1995). *Grundvorstellungen mathematischer Inhalte*. (Fundamental notions for mathematical contents). Heidelberg: Spektrum Akademischer Verlag.

Wilhelm, R., & Baynes, C. F. (1950/1967). *The I Ching or book of changes* (3rd ed.). Bollingen Series XIX. Princeton NJ: Princeton University Press.

Wittmann, E. (2012). Teaching and learning mathematics along fundamental mathematical ideas from kindergarten to the matura. In R. S. Zavod (Ed.) *1st International Conference on Learning and Teaching Mathematics* (pp. 13–25). Ljubljana: El. Knjiga.

Zhu, L., & Gigerenzer, G. (2006). Children can solve Bayesian problems: The role of representation in mental computation. *Cognition, 98*(3), 287–308.

Chapter 9
The Role of Statistics Anxiety in Learning Probability

Caterina Primi, Maria Anna Donati and Francesca Chiesi

Abstract The current study aimed at investigating the effect of statistics anxiety on learning probability, while taking into account mathematics anxiety and basic probabilistic reasoning, in order to provide insight into the direct and indirect relationships among affective and cognitive factors. Data were collected on a sample of undergraduate students attending an introductory statistics course. Through a step-wise regression model and a mediation model we showed that statistics anxiety mediates the relationship between probabilistic reasoning ability and achievement outcomes. Educational implications are discussed.

Keywords Statistics anxiety · Math anxiety · Probability · Learning
Statistics education

9.1 Introduction

Statistics anxiety is described as an anxiety encountered when taking statistics in any form, and which appears to involve a complex array of emotional reactions (Onwuegbuzie and Wilson 2003). More specifically, Zeidner (1991) defined statistics anxiety as feelings of extensive worry, intrusive thoughts, mental disorganization, tension, and physiological arousal when being exposed to statistics content or problems and instructional situations, or evaluative contexts that deal with statistics.

C. Primi (✉) · M. A. Donati · F. Chiesi
Department NEUROFARBA, Section of Psychology, University of Florence, Italy, Via di S. Salvi 12—Padiglione 26, 50135 Florence, Italy
e-mail: primi@unifi.it

M. A. Donati
e-mail: mariaanna.donati@unifi.it

F. Chiesi
e-mail: francesca.chiesi@unifi.it

C. Batanero and E. J. Chernoff (eds.), *Teaching and Learning Stochastics*, ICME-13 Monographs, https://doi.org/10.1007/978-3-319-72871-1_9

Statistics anxiety is a pervasive problem in many fields of study because a large proportion of students identify statistics courses as the most anxiety-inducing courses in their curriculum (Zeidner 1991) and experience problems over the course of their statistics education. For example, they use inadequate learning behaviours and invest less effort and time for learning (Macher et al. 2011); at the same time, they tend to delay or postpone learning, completing assignments, or preparing for an examination (Onwuegbuzie 2004). As a consequence, statistics anxiety has a negative direct or indirect impact on academic performance (e.g., Chiesi and Primi 2010; Macher et al. 2011, 2013) because students are so troubled in dealing with statistics exams that they delay the completion of the degree program, or in some cases they fail to complete it (Onwuegbuzie 2003; Walsh and Ugumba-Agwunobi 2002). For all these reasons, experiences of statistics anxiety have an influence beyond a single examination situation and may have long-lasting effects (Matcher et al. 2013; for a review, Ruggeri et al. 2008).

Several studies investigated statistical anxiety antecedents (for a review see Onwuegbuzie and Wilson 2003) concluding that statistics anxiety is a multi-dimensional construct that it is related to three types of antecedents: situational antecedents, such as previous math experience and skills (e.g., Baloglu 2003; Hong and Karstensson 2002); dispositional antecedents, such as math self-concept or self-esteem (e.g., Zeidner 1991), self-efficacy (e.g., Onwuegbuzie 2003), attitude toward statistics (e.g., Chiesi and Primi 2010); and personal factors, such as age (e.g., Fitzgerald et al. 1996), and gender (e.g., Baloglu 2003; Zeidner 1991). Among them, particular attention has been payed to the relationship between statistics anxiety and mathematics anxiety, which is commonly defined as an adverse emotional reaction to mathematics or the prospect of doing mathematics (Hembree 1990) and a state of nervousness and discomfort brought upon by the presentation of mathematical problems (Ashcraft and Moore 2009). Although several studies look at statistics anxiety and mathematics anxiety as similar constructs (e.g., Murdock 1982), other studies view them as two different constructs (Benson and Bandalos 1989; Cruise et al. 1985; Onwuegbuzie et al. 1997; Zeidner 1991) documenting that students who report high math anxiety do not necessarily report high statistics anxiety. Indeed, a moderate positive association between mathematics anxiety and statistics anxiety was found (Baloglu 2001; Birenbaum and Eylath 1994) indicating that less than 50% of the variance in statistics anxiety can be explained referring to mathematics anxiety. Statistics anxiety may be conceptualized as a different construct from mathematics anxiety because, according to Cruise et al. (1985), statistics involves different cognitive processes and requires more than manipulation of mathematical symbols. Thus, even though statistics employs basic mathematical concepts, it is more closely related to verbal reasoning than mathematical reasoning and it was found that, unlike mathematics anxiety, statistics anxiety was significantly correlated with inductive reasoning ability (Baloglu 2004; Birenbaum and Eylath 1994).

As stated above, statistics anxiety has frequently been shown to affect students' academic outcomes. Nonetheless, results concerning the specific association between statistics anxiety and statistics achievement attested in prevalence that this relationship is moderate, e.g., the highest correlation between anxiety and performance did not exceed $r = -.40$ (for a review see Macher et al. 2015). These findings suggest to look not only at the correlations between statistics anxiety and achievement but to investigate the interrelations among other related variables. For example, Chiesi and Primi (2010) conducted a study involving undergraduate students, wherein they aimed to model the effects of attitude toward statistics, mathematical ability, and statistics anxiety on students' achievement. The authors found that mathematical ability had a direct negative effect on anxiety and a direct positive effect on achievement. They also found a direct negative effect of anxiety on end-of-course attitude, which, in turn, had a direct positive effect on achievement. These results indicate that lower anxiety is associated with better end-of-course attitudes toward statistics, and with better achievement outcomes.

Following this premise, in the current study we aimed at investigating the effect of statistics anxiety on achievement, taking into account other affective and cognitive factors, such as mathematics anxiety, and mathematics knowledge. The explored relationships can be detailed as follow.

First of all, since statistics anxiety and mathematics anxiety can be defined as a separate constructs, the use of a specific measure of anxiety as direct predictor of statistics performance was chosen and, in line with previous researches (e.g., Finney and Schraw 2003; Furnham and Chamorro-Premuzic 2004; Vigil-Colet et al. 2008), we expected to find a negative relationship between achievement and statistic anxiety. Nonetheless, since statistics anxiety and mathematics anxiety are separate but related constructs, we included in our study both these types of content-specific anxiety in order to better ascertain the role of mathematics anxiety on statistics anxiety and performance.

Secondly, since several studies indicated cognitive variables (like basic mathematical abilities) as predictors of the performance in statistics, we took into account students' individual cognitive ability. In particular, the achievement was measured referring to probability, which is one of the main topics of an introductory statistics course and an important objective of statistics classes. Indeed, teaching students to reason about probabilities means to provide them with tools to think about uncertain outcomes and to make decisions on the basis of probabilistic information (see Chernoff and Sriraman 2014, for a review). As a consequence, basic probabilistic reasoning grounded on the mathematical ability to deal with ratios, proportions, and percentages was the cognitive variable included in our study and expected to be related to achievement. Additionally, because mathematical background is considered one of the predictors of statistics anxiety, we also expected a negative relation between this reasoning ability and statistics anxiety.

Finally, since Macher et al. (2015) declared that there is a lack of studies that investigated the immediate effects of statistics anxiety in an examination together with its antecedents and consequences, to further understand the mechanisms underlying relationships among all these variables, we tested a mediation model. Mediation implies a situation where the effect of the independent variable (X) on the dependent variable (Y) can be explained using a third mediator variable (M) which is caused by the independent variable and is itself a cause for the dependent variable. By modelling an intermediate variable, the overall effect between X and Y can be decomposed into component parts called the direct effect of X on Y and the indirect effect of X on Y through M (i.e. the mediated effect). Thus, we aimed at testing a mediation model in which basic probabilistic reasoning ability, was associated with achievement through statistics anxiety. We expected that lower ability was associated with higher anxiety and, as consequence, with poorer achievement.

9.2 Method

This section, organized in subsections, provides the method used in the research study including participants, measures and procedure. Additionally offers a description of the course. Finally presents the data analysis used in this study. In sum, this section offers enough information to allow other researchers to replicate this study.

9.2.1 *Participants*

Participants were 172 psychology students (mean age = 21.5 years, *SD* = 4.9) attending the University of Florence in Italy and enrolled in an undergraduate introductory statistics course. They were first year students who did not have previous experience with the discipline at the university level but they might have encountered the discipline before in school-related contexts or in their out-of-school lives. Most of the participants were women (70%). This proportion reflects the gender distribution of the population of psychology students in Italy. All students participated on a voluntary basis after they were given information about the general aim of the research.

9.2.2 *Description of the Course*

The course covered the usual introductory topics of descriptive and inferential statistics and their application in psychological research. Referring specifically to probability, basic concepts of probability theory and calculus (including simple, conditional, and conjunct probabilities) were presented as well as the more common sample distributions and probability distributions. The course was scheduled to take place over 10 weeks, and takes 6 h per week (for a total amount of 60 h). During each class some theoretical issues were introduced followed by exercises done by either paper-and-pencil procedure (about one hour per week) and computer package *R-commander* (about one hour per week). Students followed the course in a large group whereas they were allowed to work in small groups when doing exercises. The performance of students was assessed through two exam tests (midterm and end of the course). Students who obtained completely sufficient performances were not required to sit a final examination.

9.2.3 Measures

During the course participants were administered the following scales; The *Probabilistic Reasoning Scale* (PRS; Primi et al. 2017) was developed to measure basic probabilistic reasoning skills applying Item Response Theory (IRT) that is an appropriate approach to construct scales aimed to accurately measure specific levels of the assessed ability (Embreston and Reise 2000). Some items of the scale measure not only the ability to normatively reason in probabilistic terms but they also indicate susceptibility to specific heuristics and biases. Indeed, each item presented a multiple choice question (one correct among three alternatives) and in the case of some items, among the incorrect options, one corresponded to a specific bias (e.g. the gambler's fallacy or ratio bias). The scale was proved to be reliable and valid. Indeed, IRT offers a better assessment of the measurement precision of scales than CTT. Instead of providing a single value (e.g., coefficient alpha) for reliability, through the amount of information, IRT methods can quantify the information value of both individual items and the overall test, and this information can be evaluated at any level of the latent trait (Hambleton et al. 1991). Additionally, the nomological validity of the scale was attested showing that the PRS scores correlated positively with numerical skills, math-related attitudes, statistics achievement, IQ, reasoning skills, and negatively with risky choices (Primi et al. 2017).

In detail, the PRS consists of 16 multiple-choice questions (Example 1: "*A ball was drawn from a bag containing 10 red, 30 white, 20 blue, and 15 yellow balls. What is the probability that it is neither red nor blue?* (a) 30/75; (b) 10/75; (c) 45/75". Example 2: "*Smokers are about 35% of the population. At the airport in the waiting room there are 200 passengers. How many smokers are there*? (a) 70; (b) 35; (c) 20". Example 3: "*A marble bag contains 10 blue and 20 green marbles. After you drew 5 marbles (the marble drawn was always put back into the bag), a sequence of 5 green marbles was obtained. What is the most likely outcome if a marble is drawn a sixth time*? (a) a green marble; (b) a blue marble; (c) blue and green are equally likely"). A single composite score was computed based on the sum of correct responses. In the present sample Cronbach's alpha was .72.

The *Abbreviated Math Anxiety Scale* (AMAS; Hopko et al. 2003; Italian version: Primi et al. 2014) is one of the most used tools to measure math anxiety in college and high school students (for a review, see Eden et al. 2013). It was chosen because it was adapted for use in the Italian educational context providing evidence for the validity and reliability of the scale when administered to Italian college students (Primi et al. 2014). In detail, the AMAS measures math anxiety experienced by students in learning and test situations. Participants have to respond on the basis of how anxious they would feel during the events specified, using a 5-point response scale (ranging from "strongly agree" to "strongly disagree"). High scores on the scale indicate high math anxiety. A single composite score was obtained, based on participants' ratings of each statement. In the present sample Cronbach's alpha was .83.

The *Statistical Anxiety Scale* (SAS; Vigil-Colet et al. 2008; Italian version: Chiesi et al. 2011) is a self-reported measure of anxiety experienced when dealing with statistics. The scale was chosen because we confirmed its validity and reliability when administered to Italian college students (Chiesi et al. 2011). It consists of 24 items with a five-point rating scale. The SAS has a three-factor latent structure: *Examination anxiety* (8 items, e.g., "Studying for examination in a statistics course"), *Asking for help anxiety* (8 items, e.g., "Asking the teacher how to use a probability table"), and *Interpretation anxiety* (8 items, e.g., "Trying to understand a mathematical demonstration"). High scores on the scale indicate high statistics anxiety. A single composite score was obtained, based on participants' ratings of each statement. In the present sample Cronbach's alpha was .93.

The *Achievement in probability* was assessed with a test that was a part of the midterm examination and consisted of four exercises with data presented in a contingency table. Students were requested to read data relying on probabilistic reasoning. A total score (range 0–4) was computed based on the sum of correct responses.

9.2.4 Procedure

Participants completed the measures individually in a self-administered format in the classroom. Each task was briefly introduced, and instructions for completion were given. The answers were collected in a paper-and-pencil format. All participants completed the PRS and AMAS during the first week of the introductory statistics course. The SAS was completed during the fifth week of the course and the *Achievement in probability* was assessed at the midpoint (about the sixth week) of the course.

9.2.5 Data Analyses

To investigate our hypothesis about the relationships between basic probabilistic reasoning ability, statistics anxiety, and achievement in probability, we computed Pearson correlations among these variables.

To test the effect of statistics anxiety on achievement, a step-wise multivariate regression was conducted. In the stepwise procedure we proceeded forward adding one variable at a time. In the first step, basic probabilistic reasoning ability was included as unique predictor, in the second step we added mathematics anxiety, and statistics anxiety in the third step. At each step, the analysis allows to determine the contribution of the predictor variable entered later in the equation. Statistically, this contribution is determined through the significant change in R-squared (ΔR^2).

To further enhance the understanding of mechanisms underlying the relationships among these variables, a mediation model was tested. Specifically, statistics anxiety was modelled as the intermediate variable (M). Thus, the overall effect between basic probabilistic reasoning ability (X) and achievement (Y) can be decomposed into component parts, called the direct effect of X on Y and the indirect effect of X on Y through M, i.e., we analyzed if basic probabilistic reasoning ability influenced achievement directly (path c′ in Fig. 9.1, where dotted line denotes the direct effect when the mediator is not included. a, b, c, and c′ are unstandardized ordinary least squares (OLS) regression coefficients.) and indirectly (path ab in Fig. 9.1) through statistics anxiety. The indirect effect was estimated from probabilistic reasoning ability to statistics anxiety (path a in Fig. 9.1) and from statistics anxiety to probabilistic reasoning ability controlling for mathematics skills (path b in Fig. 9.1). Then, the direct and indirect effects added to yield the total effect (path c in Fig. 9.1).

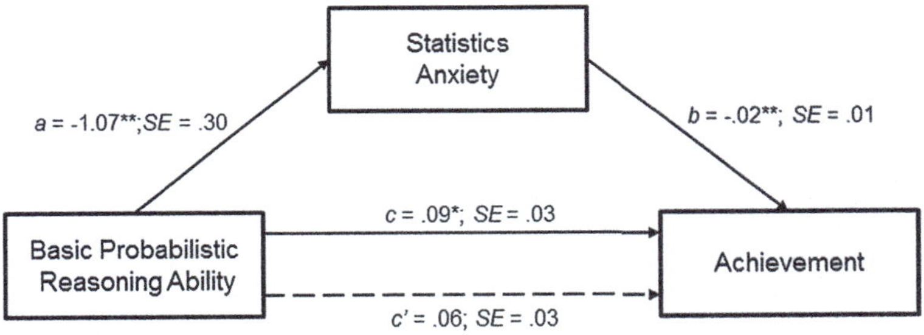

Fig. 9.1 Path coefficients for the mediation analysis. *$p < .05$; **$p < .01$

To test the mediation model we used the INDIRECT macro for SPSS (Hayes 2013). The INDIRECT macro tested the hypothesized model using a bootstrapping procedure (with 5000 bootstrap samples) to estimate the 95% confidence interval for the indirect (mediated) effect (for more details, see Preacher and Hayes 2008). Bootstrapping is a resampling strategy for estimation and hypothesis testing. With the bootstrapping method, the sample is conceptualized as a pseudo-population that represents the broader population from which the sample was derived, and the sampling distribution of any statistic can be generated by calculating the statistic of interest in multiple resamples from the dataset. The bootstrapping procedure has been suggested as representing the most trustworthy test for assessing the effects of mediation models, overcoming issues associated with inaccurate p-values that result from violations of parametric assumptions (Hayes and Scharkow 2013). Indeed, the bootstrapping procedure is advantageous because it does not impose the assumption of normality of the sampling distribution of indirect effects, and it maintains high power while maintaining adequate control over Type I error rate (Hayes 2009; MacKinnon et al. 2002, 2004; Preacher and Hayes 2008). The bootstrap test is statistically significant (at .05) if both confident limits have the same sign (e.g., both positive or both negative). This indicates that zero is not a likely value, and therefore, that the null hypothesis of the indirect effect being zero should be rejected.

9.3 Results

The correlation among all the variables in the study are reported in Table 9.1. As expected, achievement was significantly and negatively correlated with statistics anxiety and positively with basic probabilistic reasoning ability. A negative but not significant correlation was found with mathematics anxiety. Concerning the relationship between statistics anxiety and mathematics anxiety, we found a significant positive correlation.

Table 9.1 Correlations, means, and standard deviations for variables in the study

Pearson's correlations	1	2	3	4
1. Achievement	–			
2. Probabilistic reasoning	.22*	–		
3. Mathematics anxiety	−.14	−.24**	–	
4. Statistics anxiety	−.25**	−.17*	.60***	–
Mean (standard deviation)	2.84 (1.06)	12.09 (2.71)	26.58 (7.59)	71.41 (14.89)

*$p < .05$; **$p < .001$; ***$p < .001$

The step-wise multivariate regression results are reported in Table 9.2. Basic probabilistic reasoning ability was a significant predictor ($F(1,129) = 6.65, p < .05$) that accounted for the 5% of the variability in achievement. Adding mathematics anxiety, results showed that the model was significant ($F(2,128) = 3.85$, $p < .05$). Nonetheless, mathematics anxiety had not a significant predictive power on achievement. In the next step, statistics anxiety was added to the model and the introduction of this variable explained an additional 5% ($\Delta R^2 = .05$) of the variance. This increase in the amount of variance explained was statistically significant ($p < .01$). In this model, both basic probabilistic reasoning ability ($\beta = .20, p < .05$) and statistics anxiety ($\beta = -.28$; $p < .01$) had a significant predictive power on achievement ($p < .001$) whereas contrary to initial expectations, mathematics anxiety was not a significant predictor.

Table 9.2 Step-wise regression analysis with achievement as dependent variable

Predictors	β	t	p	R^2	R^2 change	F change	p
Model 1							
PRS	.22	2.58	<.05	.05	–	–	–
Model 2							
PRS	.20	2.27	<.05	.06	.01	1.05	*n.s.*
AMAS	−.09	−1.02	31		–	–	–
Model 3							
PRS	.20	2.31	<.001	.11	.05	7.52	<.01
AMAS	.07	.703	.48		–	–	–
SAS	−.28	−2.74	<.01				

PRS probabilistic reasoning scale; *AMAS* abbreviated math anxiety scale; *SAS* statistical anxiety scale

In the light of these results, the mediation model was estimated to derive the total, direct, and indirect effects of probabilistic reasoning ability on achievement through statistics anxiety (Fig. 9.1). Results showed a significant positive indirect effect of basic probabilistic reasoning ability on achievement through statistics anxiety (point estimate = 0.3, 95% CI = [0.0085, 0.559]). More specifically, greater basic probabilistic reasoning ability appears to be negatively related to statistics anxiety, which in turn is negatively related to achievement.

9.4 Discussion and Conclusion

In order to understand the mechanisms underlying the role of statistics anxiety in learning probability, the aim of this study was to test the relationship between statistics anxiety and achievement in probability controlling for the effect of basic probabilistic reasoning ability and mathematics anxiety. In line with the hypotheses, our study provided evidence that statistics anxiety mediates the relationship between probabilistic reasoning ability and achievement. More specifically, greater basic probabilistic reasoning ability appears to be negatively related to statistics anxiety, which in turn is negatively related to achievement. Based on our findings, the relationship between ability and achievement can be explained by taking into account the mediating role of statistics anxiety.

From a qualitative point of view, we can argue that students with low ability and high anxiety tend to adopt a learning approach characterized by rote-learning, poor understanding, and lack of engagement in the learning process (the so called *surface* approach), which results in low achievement. As such, the current findings suggest the need to adopt intervention strategies aiming at encouraging an approach that is more likely to result in better retention and application of knowledge (the so called *deep* approach). In discipline different from statistics, surface and deep approaches have been related to students' perceptions of the learning environment (e.g., suitability of workload and assessment, teaching) but also to the students' characteristics such as skills, motivation, and anxiety (Baeten et al. 2010). Thus, beside using quantitative measures of students' learning approach when studying statistics (Chiesi et al. 2016), future research should focus on these aspects to promote high quality learning outcomes in statistics.

First of all, students with difficulties in basic probabilistic reasoning should be supported from the start of their courses with specific training activities, e.g., Primi and Chiesi (2013) provided some evidences that students can both consolidate their abilities and acquire more confidence by incrementing their mathematical prerequisites (i.e., the ability to deal with proportions and percentages) necessary to reason in probabilistic terms. Moreover, it is important to provide students with the resources to cope with statistics anxiety according to its effect on achievement. Unfortunately, less attention in the research has been devoted to reduce statistics anxiety in students (Williams 2010). Dillon (1982) demonstrated that students' anxiety can be decreased by encouraging them to talk about their fears, and then suggesting ways that they can cope with their anxiety. Schacht and Stewart (1990) explored the use of humorous cartoons in statistics classes and they found that this type of humour not only reduced the students' anxiety, but also improved their learning. Others (Wilson and Onwuegbuzie 2001) found the effect of instructor interpersonal style such conveying a positive attitude, encouragement, reassurances of the students' ability, and acknowledgement of students' anxiety. Additionally, Dolinsky (2001) suggested creating a collaborative environment in which active learning strategies were used as the primary method to teach statistics. Finally, Pan and Tang's (2004) study shows that the combining application-oriented teaching methods with instructors' attentiveness to students' anxiety is a significantly effective way to reduce students' anxiety in learning statistics.

Future studies are required to confirm and extend the current findings. For instance, it would be desirable to verify the adequacy of this mediation model in different sample of students, e.g., students attending different degrees or students who do not attend regularly the course, in order to provide evidence of the generalizability of the present results. Indeed, the current findings refer to psychology students, i.e. with students who are not primarily interested in statistics but have statistics as a compulsory course in their degrees. Additionally, this mediation model should be tested taking into account other variables, e.g., general math abilities or general test anxiety, in order to provide further insights about the interplay among cognitive and non-cognitive factors acting on probability and statistics achievement. Finally, having a more homogeneous sample for gender will allow testing differences related to this personal characteristic. Indeed, gender is certainly one of the most widely investigated antecedents of statistics anxiety but mixed results have been reported (Macher et al. 2011).

In conclusion, the findings from the present study provide insight into the relationship between students' statistics anxiety and learning probability as one of the achievement objective in statistics education.

References

Ashcraft, M. H., & Moore, A. M. (2009). Mathematics anxiety and the affective drop in performance. *Journal of Psychoeducational Assessment, 27*(3), 197–205.

Baloglu, M. (2001). *An application of structural equation modeling techniques in the prediction of statistics anxiety among college students* (Unpublished Ph.D. dissertation). Texas A&M University Commerce, USA.

Baloglu, M. (2003). Individual differences in statistics anxiety among college students. *Personality and Individual Differences, 34*(5), 855–865.

Baloglu, M. (2004). Statistics anxiety and mathematics anxiety: Some interesting differences. *Educational Research Quarterly, 27*(3), 38–48.

Baeten, M., Kyndt, E., Struyven, K., & Dochy, F. (2010). Using student-centered learning environments to stimulate deep approaches to learning: Factors encouraging or discouraging their effectiveness. *Educational Research Review, 5*(3), 243–260.

Benson, J., & Bandalos, D. L. (1989). Structural model of statistical test anxiety in adults. In R. L. Schwarzer, H. M. van der Ploeg, & C. D. Spielberger (Eds.), *Advances in test anxiety research* (Vol. 6, pp. 137–154). Amsterdam, The Netherlands: Swets & Zeitlinger.

Birenbaum, M., & Eylath, S. (1994). Who is afraid of statistics? Correlates of statistics anxiety among students of educational sciences. *Educational Research, 36*(1), 93–98. https://doi.org/10.1080/0013188940360110.

Chernoff, E. J., & Sriraman, B. (Eds). (2014). *Probabilistic thinking: Presenting plural perspectives.* Dordrecht, The Netherlands: Springer.

Chiesi, F., & Primi, C. (2010). Cognitive and non-cognitive factors related to students' statistics achievement. *Statistics Education Research Journal, 9*(1), 6–26.

Chiesi, F., Primi, C., & Carmona, J. (2011). Measuring statistics anxiety: Cross-country validity of the Statistical anxiety scale (SAS). *Journal of Psychoeducational Assessment, 29,* 559–569. https://doi.org/10.1177/0734282911404985.

Chiesi, F., Primi, C., Bilgin, A. A., Lopez, M. V., del Carmen Fabrizio, M., Gozlu, S., & Tuan, N. M. (2016). Measuring university students' approaches to learning statistics: An invariance study. *Journal of Psychoeducational Assessment, 34*(3), 256–268.

Cruise, R. J., Cash, R. W., & Bolton, D. L. (1985, August). Development and validation of an instrument to measure statistical anxiety. In *American Statistical Association Proceedings of the Section on Statistics Education* (Vol. 4, pp. 92–97).

Dillon, K. M. (1982). Statiscophobia. *Teaching of Psychology, 9,* 117.

Dolinsky, B. (2001). An active learning approach to teaching statistics. *Teaching of Psychology, 28,* 55–56.

Eden, C., Heine, A., & Jacobs, A. M. (2013). Mathematics anxiety and its development in the course of formal schooling: A review. *Psychology, 4*(6), 27.

Embretson, S. E., & Reise, S. P. (2000). *Item response theory for psychologists.* Mahwah, NJ: Lawrence Erlbaum.

Finney, S. J., & Schraw, G. (2003). Self-efficacy beliefs in college statistics courses. *Contemporary Educational Psychology, 28*(2), 161–186.

Fitzgerald, S. M., Jurs, S., & Hudson, L. M. (1996). A model predicting statistics achievement among graduate students. *College Student Journal, 30,* 361–366.

Furnham, A., & Chamorro-Premuzic, T. (2004). Personality and intelligence as predictors of statistics examination grades. *Personality and Individual Differences, 37*(5), 943–955.

Hayes, A. F. (2009). Beyond Baron and Kenny: Statistical mediation analysis in the new millennium. *Communication Monographs, 76*(4), 408–420.

Hayes, A. F. (2013). Introduction to mediation, moderation, and conditional process analysis. New York: The Guilford Press.

Hayes, A. F., & Scharkow, M. (2013). The relative trustworthiness of inferential tests of the indirect effect in statistical mediation analysis: Does method really matter? *Psychological Science, 24*(10), 1918–1927. https://doi.org/10.1177/0956797613480187.

Hambleton, R. K., Swaminathan, H., & Rogers, H. J. (1991). *Fundamentals of item response theory*. London: Sage.

Hembree, R. (1990). The nature, effects, and relief of mathematics anxiety. *Journal for Research in Mathematics Education, 21*(1), 33–46.

Hong, E. & Karstensson, L. (2002). Antecedents of State Test Anxiety. *Contemporary Educational Psychology, 27*(2), 348–367. https://doi.org/10.1006/ceps.2001.1095.

Hopko, D. R., Mahadevan, R., Bare, R. L., & Hunt, M. K. (2003). The abbreviated math anxiety scale (AMAS) construction, validity, and reliability. *Assessment, 10*(2), 178–182.

Macher, D., Paechter, M., Papousek, I., & Ruggeri, K. (2011). Statistics anxiety, trait anxiety, learning behavior, and performance. *European Journal of Psychology of Education, 27*(4), 483–498. https://doi.org/10.1007/s10212-011-0090-5.

Macher, D., Paechter, M., Papousek, I., Ruggeri, K., Freudenthaler, H. H., & Arendasy, M. (2013). Statistics anxiety, state anxiety during an examination, and academic achievement. *British Journal of Educational Psychology, 83*(4), 535–549.

Macher, D., Papousek, I., Ruggeri, K., & Paechter, M. (2015). Statistics anxiety and performance: Blessings in disguise. *Frontiers in Psychology, 6*. Online: http://journal.frontiersin.org/article/10.3389/fpsyg.2015.01116/full.

MacKinnon, D. P., Lockwood, C. M., Hoffman, J. M., West, S. G., & Sheets, V. (2002). A comparison of methods to test mediation and other intervening variable effects. *Psychological Methods, 7*(1), 83–104. https://doi.org/10.1037/1082-989X.7.1.83.

MacKinnon, D. P., Lockwood, C. M., & Williams, J. (2004). Confidence limits for indirect effect: Distribution of the product and resampling methods. *Multivariate Behavioral Research, 39*(1), 99–128. https://doi.org/10.1207/s15327906mbr3901_4.

Murdock, M. (1982). Teaching to overcome statistics anxiety. *Proceedings of the Statistical Education Section of the American Statistical Association*, (pp. 92–93). Alexandria, VA: American Statistical Association.

Onwuegbuzie, A. J. (2003). Modeling statistics achievement among graduate students. *Educational and Psychological Measurement, 63*(6), 1020–1038.

Onwuegbuzie, A. J. (2004). Academic procrastination and statistics anxiety. *Assessment & Evaluation in Higher Education, 29*(1), 3–19.

Onwuegbuzie, A., & Wilson, V. (2003). Statistics anxiety: Nature, etiology, antecedents, effects, and treatments—A comprehensive review of the literature. *Teaching in Higher Education, 8* (2), 195–2009. https://doi.org/10.1080/1356251032000052447.

Onwuegbuzie, A. J., DaRos, D., & Ryan, J. M. (1997). The components of statistics anxiety: A phenomenological study. *Focus on Learning Problems in Mathematics, 19*(4), 11–35.

Pan, W., & Tang, M. (2004). Examining the effectiveness of innovative instructional methods on reducing statistics anxiety for graduate students in the social sciences. *Journal of Instructional Psychology, 31*(2), 149–159.

Preacher, K. J., & Hayes, A. F. (2008). Assessing mediation in communication research. In *The Sage sourcebook of advanced data analysis methods for communication research* (pp. 13–54).

Primi, C., Busdraghi, C., Tomasetto, C., Morsanyi, K., & Chiesi, F. (2014). Measuring math anxiety in Italian college and high school students: Validity, reliability and gender invariance of the Abbreviated Math Anxiety Scale (AMAS). *Learning and Individual Differences, 34*, 51–56.

Primi, C. & Chiesi, F. (2013). Evaluating the efficacy of a training for improving probability and statistics learning in introductory statistics courses. In B. Ubuz, Ç. Haser, & M. A. Mariotti (Eds.), *Proceedings of the Eighth Congress of European Research in Mathematics Education* (pp. 850–859). Antalya, Turkey: ERME.

Primi, C., Morsanyi, K., Donati, M. A., Galli, S., & Chiesi, F. (2017). Measuring probabilistic reasoning: The construction of a new scale applying item response theory. *Journal of Behavioral Decision Making*. https://doi.org/10.1002/bdm.2011.

Ruggeri, K., Diaz, C., Kelley, K., Papousek, I., Dempster, M., & Hanna, D. (2008). International issues in education. *Psychology Teaching Review, 14*(2), 65–74.

Schacht, S., & B. J. Stewart. (1990). What's funny about statistics? A technique for reducing student anxiety. *Teaching Sociology, 18*(1), 52–56.

Vigil-Colet, A., Lorenzo, U., & Condon, L. (2008). Development and validation of the Statistical Anxiety Scale. *Psicothema, 2*(1), 174–180.

Walsh, J. J., & Ugumba-Agwunobi, G. (2002). Individual differences in statistics anxiety: The roles of perfectionism, procrastination and trait anxiety. *Personality and Individual Differences, 33*(2), 239–251.

Williams, A. S. (2010). Statistics anxiety and instructor immediacy. *Journal of Statistics Education, 18*(2), 1–18.

Wilson, V. A., & A. J. Onwuegbuzie. (2001, November). Increasing and decreasing anxiety: A study of doctoral students in educational research courses. Paper presented at the *30th Annual Meeting of the Mid-South Educational Research Association*, Little Rock, AR: Eric. Online: http://files.eric.ed.gov/fulltext/ED459214.pdf.

Zeidner, M. (1991). Statistics and mathematics anxiety in social science students: Some interesting parallels. *British Journal of Educational Psychology, 61*(3), 319–328. https://doi.org/10.1111/j.2044-8279.1991.tb00989.

Part II
Students' Reasoning and Learning

Chapter 10
What 9- and 10-Year Old Pupils Already Know and What They Can Learn About Randomness

Peter Bryant, Terezinha Nunes, Deborah Evans, Rossana Barros, Laura Gottardis and Emanouela Terlektsi

Abstract Randomness lies at the heart of several topics that are taught at school; yet, little research has been done on how to teach children about randomness and randomising. This chapter presents an intervention study in which 75 school children, aged 9 and 10 years, were randomly assigned to three groups: (1) a group given a 5-session intervention on randomness, (2) an active control group given a 5-session course on directed numbers (3) an unseen, control group. Before the intervention started, the children were given a pre-test of their understanding of randomness and, after it, an immediate post-test and, 10 weeks later, a delayed post-test. Before the teaching began the children had a stronger understanding of randomness than most previous research suggests. The intervention did improve the intervention group's understanding of aspects of randomising and helped them to justify their answers in randomness problems.

Keywords Understanding randomness · Learning about randomness
Random sequences · Determined sequences · Recency effects

P. Bryant (✉) · T. Nunes · D. Evans · R. Barros · L. Gottardis · E. Terlektsi
Department of Education, University of Oxford, 15 Norham Gardens,
Oxford OX2 6PY, UK
e-mail: peter.bryant@education.ox.ac.uk

T. Nunes
e-mail: terezinha.nunes@education.ox.ac.uk

D. Evans
e-mail: deborah.evans@education.ox.ac.uk

R. Barros
e-mail: rossana.barros@education.ox.ac.uk

L. Gottardis
e-mail: laura.gottardis@education.ox.ac.uk

E. Terlektsi
e-mail: emanouela.terlektsi@brookes.ac.uk

C. Batanero and E. J. Chernoff (eds.), *Teaching and Learning Stochastics*,
ICME-13 Monographs, https://doi.org/10.1007/978-3-319-72871-1_10

10.1 Introduction

There is a tradition of research on schoolchildren's understanding and knowledge of the nature of random sequences and random events, but not on how they acquire this understanding. Yet, even though randomness plays an important part in many topics taught at school, such as evolution, history, sub-particle physics and statistics, little research has been done on how pupils learn and whether they can be taught about randomness and randomising.

The ramifications of randomness are complicated and often difficult for adults as well as for children to grasp (Batanero et al. 2005; Batanero and Serrano 1999; Falk and Konold 1997) and so it would be wrong to expect that primary school children could be taught about the entire topic in all its complexity and sophistication.

For example, in a wide ranging review of the subject Batanero (2016) argues convincingly that there are five different meanings of the term randomness, each of which involves its own particular kinds of problems and its own methods for solving them. Even experienced adults often find it difficult to understand a topic with so many distinctions and so many implications. They frequently fail, for example, to discriminate between random and determined sequences (Falk and Konold 1997) and their predictions about future events are, all too often, plagued by the well-known gambler's fallacy (Bryant and Nunes 2012; Falk 2016).

There are, nevertheless, some fundamental aspects of randomness that even 9- and 10-year old pupils at the end of primary school might grasp since these are part of their everyday experiences, and it is the main point of this chapter that it is possible even for primary school children to begin to learn about randomness in an intellectually respectable way with the help of these everyday experiences.

One such aspect is the connection between randomising and unpredictability (Batanero 2016). If you throw dice or pick cards from a thoroughly shuffled pack, or toss a coin, or if you play any kind of lottery, you cannot predict what will happen: each possible outcome of your action is as likely to ensue as any of the rest. Since many pupils will be familiar with the use of these forms of randomising in board games and in other sporting and competitive contexts (e.g. deciding which soccer team kicks off or who serves first in a tennis match), perhaps they will understand how randomness is a way of ensuring fairness of opportunity. Of course, these conventional methods of ensuring fairness are not perfect at producing random sequences and children's experiences with them could only give them a partial insight into the concept of randomness and would not provide them with a complete knowledge of randomness in its full mathematical sense. Nevertheless well-known fairness routines might be a useful first step of the many steps that children must take in learning about the nature of random, unpredictable events in a sophisticated way.

Another basic aspect of randomness, which may well become familiar to many pupils, is the independence of successive outcomes in a random sequence (Bennet 1998). If I want to throw a 6 and I toss the die 10 times with no success, the chances of my 11th throw producing a 6 are exactly the same as they were on my first throw: my 10 successive failures do not affect, in one way or the other, my chance of success the next time. Children who play board games will probably have had experience of runs of this kind, and it is worth knowing whether they understood that the result of the next throw does not depend on what happened with previous throws.

These two aspects of randomness, probably because of their widespread presence in everyday life, have attracted most of the attention of researchers interested in what pupils understand about randomness and about the processes of randomising. Research on the link between randomness and unpredictability has led to different conclusions in different studies. In a classic study, Piaget and Inhelder (1975) showed pupils, whose ages ranged from 4 to 12 years, a tilted tray with red and white beads at the bottom end: all the red beads were on one side and all the white ones on the other side of the tray. They asked each child what would happen to the beads if the tray were tilted the opposite way and the beads as a result rolled to the opposite end. The younger pupils insisted that the beads would still be grouped by their colours—the red to one side and the white ones to the other—after the tray was tilted once and even after it was tilted a second and a third time. Older pupils (most of them 7 years old or older) recognised that the arrangement of the two kinds of beads would be likely to change with each tilt and that the two colours would probably mingle increasingly and randomly each time. These investigators concluded that at first pupils do not understand how randomising leads to unpredictability.

Later, Kuzmak and Gelman (1986) reached a different conclusion. They presented children of 3- to 7-years with two different mechanisms, both of which dispensed balls of various colours one-by-one. In one mechanism the balls were lined up in an orderly way in a single tube, so that it was easy to see the colour of the ball that would come out next, whereas the other mechanism consisted of a complicated tangle of moving tubes designed to eject the balls in a random sequence and to make it impossible to predict the next ball out. The children were asked whether they could work out the colour of the next ball to be dispensed by each apparatus. The majority of those who were 4 or more years-old correctly said that they could make the prediction with the apparatus that was arranged in an orderly fashion but not with the disorderly apparatus, although that majority was very slim with the 4-year-olds. The experimenters argued that this showed that Piaget and Inhelder's hypothesis was wrong and that even four-year old pupils can distinguish between random and non-random sequences. Their conclusion was probably too hasty. The children could have made these judgements just on the basis of one arrangement—the tangled one—being a great deal more complicated than the other. They could easily have judged that it was harder to be sure what would happen with the complicated mechanism than with the relatively simple one, and thus their judgments may have had nothing to do with randomness.

The results of both studies can be criticised on the same count: they both conjured up forms of randomisation that were unfamiliar to the pupils. There is certainly room for experiments with more familiar kinds of randomisation. Shuffling cards and throwing dice are two obvious candidates.

Pupils' understanding of the other basic aspect of randomness, the independence of successive events in a random sequence, has also been studied intensively (e.g., by Green 1989 and Truran 1994). Chiesi and Primi (2009) reported persistent misconceptions among Italian 8- and 10-year-olds and among adults as well. Their measure of this understanding was a task in which the participants were told (with the help of a PowerPoint demonstration) about someone playing with a bag in which there were 15 green and 15 blue marbles. (In other trials in this experiment the numbers of the different coloured balls in the bag were unequal, but here we will describe only the results of the trials in which they were equal.) The character drew a marble from the bag four times and then immediately replaced the marble in the bag each of the four times. On all four times, he drew a green marble. The participants were asked what would happen on the fifth draw. Would he be more likely to draw a blue marble or to draw another green marble, or was each colour just as likely as the other?

The results were striking. *None* of the 8-year-old pupils gave the correct answer to the question, that the probability of the next marble being one colour was equal to the probability of it being the other colour: most of the 8-year olds judged that the next marble would be of the same colour as the first four marbles, an error commonly called 'positive recency', but a minority said that the fifth draw would be of the other colour, which is usually described either as a 'negative recency' error or as the 'gambler's fallacy' (Falk 2016; Fischbein and Schnarch 1997). The performance of the 10-year-olds was better, but still very low: they gave the correct answer only 40% of the time, and made roughly an equal number of positive and negative recency mistakes. Even more surprising were the scores of the adults, all of whom were college students (mean age 25 years). They were no better than the 10-year olds at producing the correct answer: roughly 40% answered the question correctly and another 40% made the negative recency mistake.

A well known study by Fischbein and Gazit (1984) with pupils aged between 10 and 12 years also showed considerable difficulty in understanding the independence of different events in a randomised sequence. The pupils were told about someone who had entered a weekly lottery over a period of two months and had never yet won. This character said that he intended to go on trying, because the fact that he had consistently lost so far made him think that he was more likely than before to win in the next few times.

Only 18% of the pupils in each of these groups recognised this example of negative recency as a mistaken belief. Some of the 10-year-olds had been given a course on probability: the others had not. However those who had taken the course were no better at answering this question than the others. These results suggest that many pupils of this age do not grasp the principle of the independence of successive events in a random sequence, and also that their misunderstanding of randomness may be impervious to teaching.

These studies converge in indicating that most pupils around the age of 10 years do not have a coherent idea of the independence of successive events in a random sequence, but they raise two questions. The first is how to settle an apparent disagreement. The 10 year olds in the Chiesi and Primi study made an equal amount of positive and negative recency errors, whereas Fischbein and Gazit report that over 80% of the pupils of this age in their study failed to judge a negative recency answer as a mistake. This inconsistency needs to be settled. The second unanswered question concerns teaching pupils about the independence of successive events in a random sequence. Fischbein and Gazit's apparent demonstration that teaching does not affect pupils' understanding of this independence was based on a comparison between pupils who had been given a course on probability and others who had not attended this course. This comparison is an inadequate measure of the effects of teaching pupils about probability because the pupils were not given a pre-test of their understanding of randomness before the probability course. The two groups may therefore have started from different levels, and the pupils in the taught group may actually have made more progress than those in the untaught group. We need a properly designed intervention study to promote pupils' understanding of randomness. As far as we know, such a study has never been done.

10.2 The Intervention Study

We now report an intervention study which compared pupils, who were taught randomness, to two other groups of pupils; one of these was given extra teaching about directed (positive and negative) numbers whereas the other received no extra teaching at all.

10.2.1 The Aim and Design of the Study

All three groups were given a pre-test before the interventions started, and two post-tests after the interventions were over, of their understanding of both the aspects of randomness discussed above. The aim of the comparison between groups was to evaluate the effectiveness of the randomness programme by comparing progress made by pupils in the randomness group and by the pupils in the other two groups.

Pupils from three Primary schools (n = 75) participated in the project. They were either 9-years (63.2%) or 10-years (36.8%) old when the project began and were in their penultimate primary school year. All three schools catered for pupils from a wide range of SES (socio-economic status). Within each school the pupils were randomly assigned to one of three groups:

(a) A randomness group (n = 24) taught about randomness.
(b) A directed numbers group (n = 22) taught about directed numbers.
(c) An unseen control group (n = 29) not involved in any intervention.

The random assignment into groups was done by randomising the order of a set of numbers in a computer generated random sequence: each number had been assigned to a particular pupil and the groups were formed by putting those that came out in the first third of the order of numbers in one group, those whose numbers came out in the middle third of the order of numbers in another group and those whose numbers were in the last third of the order of numbers in the third group.

All the pupils were given a pre-test just before the two interventions started, an immediate post-test after the interventions ended just before the Summer break, and a delayed post-test after the Summer break, when the pupils started their final year in primary school. These tests contained items on randomness and also on directed numbers. The tests were administered in the pupils' schools.

The pupils in the Randomness Group and in the Directed Numbers Group were given five weekly intervention sessions, each lasting for an hour. During each session the pupils worked in pairs and each pair was given a set of tasks, most of which involved either familiar material, like dice and playing cards, or games and puzzles presented on a laptop. After each activity, the pairs of pupils discussed their solutions in these tasks with the rest of the group.

10.2.2 The Pre- and the Post-tests

We were unable to find any standardised measures of understanding the concept of randomness; so we devised an assessment based on tasks in the literature. The pre-test and immediate post-test assessments contained nine questions on randomness. Six of these dealt with random versus determined events: in some of these six items the material was randomised and the resulting sequence or spatial arrangement was unpredictable: in others the material was arranged in an orderly pattern and was therefore entirely predictable and not random.

One randomised item was about the effect of shuffling a pack of cards. A person has eight cards of two colours. The four on the top are green and the bottom four are purple. He shuffles them a few times. What is most likely to happen? The pupils are given three choices: the top four are still green, the top four are purple, or they are mixed. A second randomised item was about throwing counters, each with one black and one white side, into the air, and which will be the uppermost colour of these counters when they land. The children were asked to choose the most likely of three possible outcomes (all black, all white, a mixture of white and black).

The third randomised item, based on the Piaget and Inhelder (1975) task was about a tray holding marbles of two different colours which were initially all at one end of the tray and arranged so that those of one colour were on one side and those of the other colour on the other side of that end of the tray. Tipping the tray would cause the marbles to roll to the other end of the tray and the three choice answers were about the most probable effect of this tipping on the spatial arrangement of the marbles: that the different colours would stay on the same side of the tray as before, that the marbles of each colour would stay together but swap sides, that the marbles would no longer be grouped by colour and the two colours would mingle with each other.

Two of the determined items dealt with counters that were black on one side and white on the other. In one the pupils were asked about the effect on a set of counters with their white side up if they were all turned over.

Another very similar determined item with counters began in the same way but this time the pupils were asked what would happen if half the counters were turned over. The third determined item was about the effect of tipping a tray with a partition in the middle on the spatial arrangements of two groups of marbles on their spatial arrangement when all the marbles rolled to the other end: the pupils were given the same three choice answers as in the randomised tray item.

The remaining three items on randomness dealt with the independence of successive events in a random sequence. These were a replication of Chiesi and Primi (2009) task, in which participants are told about someone who draws a ball from a sack containing balls of two colours four times and immediately replaces it each time before drawing another. The colour of the ball that the person draws the first four times is the same. The participants were asked whether in the fifth draw the person would be more likely to pick a ball (1) of the same colour as in the previous four draws, or (2) of the other colour, or (3) whether there was an equal chance of picking either colour this time. In one of these items the number of balls in the two colours was the same, and in the two other items it was different. The pupils were also asked to justify their choices in two of the three independence items.

The delayed post-test contained less randomness items than the pre- and immediate post-tests (because items related to randomised/determined events had shown ceiling effects in the pre- and immediate post-tests). This post-test contained an extra item on drawing two cards from a shuffled pack, in which the card drawn first was not replaced.

In all three tests, the pupils were given booklets that presented every problem pictorially (black and white) and in writing. The questions, also in the booklets, were multiple choice ones and the pupils were asked to indicate their choice of the correct answer to each problem. The researcher conducting the testing presented a PowerPoint version of each item (in colour) on a screen and read out the questions. The pupils were encouraged to ask questions if they did not understand any of the problems. Two or three other researchers were present at these group sessions to alert the presenter if any child was failing to keep up, to answer any questions by pupils who had not understood an item, and to discourage pupils from looking at the answers of a neighbouring child.

10.2.3 The Intervention Sessions

The pupils in both intervention groups were given a wide range of problems and activities. All the pupils were actively involved in solving each problem in each session. They discussed their solutions in pairs, and subsequently as a whole group. The researcher used these group discussions to ask further questions and give feedback. Concepts were introduced through practical activities presented in familiar contexts e.g. through card and counter games, playground rhymes, school meal choices. The emphasis was on reasoning and not on calculations. The pupils' reasoning was supported either by the use of manipulative objects such as counters or representational materials, or by the use of drawings and diagrams.

Discussion was a key element in the intervention sessions. All pupils, whether they had reached the right or wrong answer, were given regular opportunities to discuss their reasoning with their partner or with a small group of pupils.

10.2.4 Randomness Intervention

In the first randomness intervention session the activities centred on the effects of randomisation and on the difference between random and non-random sequences. The pairs of pupils were given, at different times, a shuffled, and therefore randomised, pack of Happy Families cards (a game played in the U.K.) and an unshuffled pack in which the cards were grouped into families; the order of each member of the family (father, mother etc.) was the same for each family in the latter pack. The pupils were asked to make predictions about which card was at the top and what would be the next card, and the next after that and so on. Then they played a round of Happy Families with both kinds of pack. They were asked to discuss and to come to a conclusion about the ease of predicting the next card to be drawn from the top of the pack in the shuffled and unshuffled packs. In the second part of this session, the pairs of pupils were given two counters and then one child secretly took either one or two counters in their hands. The other one's task was to predict whether the number of counters that held by the first was an odd or an even number. They did this several times. The task is apparently a randomised one, but often one child adopts a consistent strategy (e.g. alternating 1 and 2), which the other child discovers. At the end of the series of predictions, the pupils discussed whether either had detected this sort of departure from randomness in their game. The session ended with a discussion among the group as a whole about the fairness and also the degree of randomness of different ways of starting and playing games.

The main purpose of the second session was to extend the pupils' understanding of randomised and unpredictable sequences. They played some purpose-designed computer games in which they attempt to guess the order in which three pictures will appear. In some games, it is possible to detect a pattern; in others, the sequences are random. The pairs of pupils were asked to detect rules where there were any, and later they discussed the differences between the two kinds of sequence. The researcher warned them that apparent patterns could occur by chance in the non-rule-governed sequences, but these patterns did not hold over a larger sequence of trials.

The third session was devoted to two highly familiar ways of producing random sequences, throwing dice and tossing coins. The session began with the students throwing dice repeatedly and recording the results of each throw. Some pupils had a normal, and therefore a "fair", die: the others had a loaded die. The pupils were not told about this difference beforehand, but needed little time to detect it, and then they were asked to discuss the difference in predictability between the randomised and non-randomised sequences.

The fourth session began with the teacher asking the pupils to distinguish events that are quite unlikely but could happen from those that are quite impossible. This is a distinction which pupils normally find hard to make (Piaget and Inhelder 1975; Shtulman 2009; Shtulman and Carey 2007) but which seems significant for understanding probability. The pupils were presented with pairs of hypothetical events or actions, such as 'making an umbrella out of glass' and 'making an umbrella out of air' and were asked to discuss whether each of these was possible or quite impossible. The session ended with a modified bingo game, in which digits were drawn separately to form a two-digit number and were not replaced. The pupils were asked to discuss the predictability of the next number to be called, considering that some numbers became impossible.

The final session began with an activity that involved drawing marbles of two different colours from a bag. The important distinction in this activity was whether each marble that was drawn was then replaced or not. If the drawn marbles were replaced, the status quo was restored each time and the probability of picking each of the colours was as it was when the activity started, but when the marbles were not replaced the relative probability of the colour of the next marble changed with each draw. The aim of this activity was to draw the pupils' attention, through discussion, to the fact that the composition of the contents of the bag is all that matters. The session ended with the pupils throwing dice and making predictions for the result of each throw. In one game the pupils threw a single die and in another they threw two dice at once, in which case their prediction was about the sum of the pair of dice. The point of comparing these games was that the range of possibilities was different in the two cases, and also that the possibilities were equiprobable when one die was thrown, but not equiprobable when two dice were thrown (e.g. you are much more likely to throw a total of 7 than a total of 2 or of 12). The pupils would subsequently be introduced to problems about throwing two dice in order to work on understanding the sample space; at this stage in the programme, the focus was simply on the fact that only some outcomes were possible for the sum and that some totals were more likely than others.

10.2.5 The Directed Numbers Intervention

The Directed Numbers Group also went through five intervention sessions, which took place at the same times and in the same schools as the randomness intervention sessions. The directed number intervention sessions concentrated on the difference between direct and inverse additive reasoning problems. Direct problems are those in which there is a concordant relation between the language in the problem and the operation used in its solution: for example, the problem contains the word "more" or "gained" and the operation to be used is addition. In inverse problems, problems with the words "more" or "gained" would be solved by a subtraction, as one addend was missing. The instructor's aim was to teach the inverse relation between addition and subtraction in the context of natural and of directed numbers.

10.3 Results

The results are presented in two sections. The first deals with the baseline scores in the pre-test. The aim is to establish how well pupils in this age group understand randomness and randomising. The second section compares the progress made by the pupils in the three groups from the pre-test to the two post-tests. The purpose of this section is to see if instruction about randomness can lead to improvements in young pupils' understanding of the nature of randomness and its use in everyday life.

10.3.1 The Pre-test Results

Since the test was given before the teaching programmes began, we will present the pre-test scores of the sample as a whole in this section.

Correct choices: The scores for the number of correct choices in the three randomised and unpredictable items and in the three determined and predictable outcome items were generally high. Eighty-two per cent of the choices in these 6 items were correct: there were 89.7% correct choices in the randomised items and 74.3% in the determined items. These scores suggest that pupils of this age do understand the consequences of randomising quite well.

The pupils' scores for the three items that dealt with the independence of successive events in a random sequence were also surprisingly good. Table 10.1 shows that 74.7% of the pupils made the correct choice in the item in which the numbers of the two colours in the sack were equal. Recall that only 40% of the 10 year olds in the Chiesi and Primi study made the correct choice in this item: recall also that the college students in the study made as many errors in this kind of item as the 10-year old pupils.

Their scores in our study in the two items in which the number of balls in the two colours was unequal were also high: 67.5% correct choices when the first 4 draws were of the minority colour and 76.0% correct choices when the first four draws were of the majority colour. These high scores present a very different picture from the one given in the Chiesi and Primi study.

Table 10.1 The percentage of pupils making correct choices or making different kinds of mistaken choices in the three pre-test items on the independence of successive outcomes in a random sequence

	Correct choices	Positive recency	Negative recency	Equiprobable choice
First item—equal numbers of each colour	74.7	17.3	8.0	
Second item—unequal numbers of the two colours—first 4 draws of the minority colour	67.5	17.3		15.2
Third item—unequal numbers of the two colours—first 4 draws of the majority colour	76.0		10.7	13.3

Table 10.1 also shows more positive recency than negative recency mistakes in the item in which the number of marbles of each of the two colours was equal. However, only 7 of the 13 pupils who made the positive recency mistake in the first item went on to make the same mistake in the second of the three items, and only 3 of 6 pupils who made the negative recency mistake in the first item also made this mistake in the third item. Thus neither positive nor negative recency was a consistently adopted strategy. To summarise, the pupils in our study did much better than pupils in the same age group in the original Chiesi and Primi study and there was little evidence of individual pupils consistently making either positive or negative recency mistakes.

10.3.2 Justifications of the Independence Items: The Number of Relevant Aspects Mentioned

In two of the independence of successive events in a random sequence items, the pupils were also asked to justify their choices in writing. The basis for coding these justifications was the assumption that five possible relevant aspects to each of the items could be mentioned in the pupils' justifications. These were:

- Uncertainty of which colour would be drawn. "you don't know what will happen it is just luck".
- Randomness: the random arrangement of the balls in the sack. "because the marbles are mixed up".
- Initial relation between two quantities (equal/unequal). "It is because there is the same number of colours", "because there are more green marbles than red".
- Replacement: the fact that replacing the drawn ball after each draw kept the relation between the two colours constant. "Because she put the marbles back in the bag".
- Independence of successive events: what happened in the first four draws has no effect on the outcome of the fifth draw. "It may have been just luck 4 purple".

Two raters independently scored 20 of the participants' justifications for each item. The inter-rater reliability (Cronbach's alpha) based on intra-class correlations was 0.98 for the first item and 0.97 for the second. Because these reliability scores were high, only one rater scored the remaining justifications.

Most of the pupils were able to mention at least one aspect relevant to randomness in the justifications for their answers to the items on the independence of successive events. In the item in which the numbers of the two colours in the sack were equal, 60.8% of the pupils offered one or more relevant justification, and in the item with unequal numbers for the two colours, the equivalent figure was 71.6% of the total sample. Most of the pupils who did produce a relevant justification mentioned only one relevant aspect (24 pupils in the first of these items and 43 in the second).

By far the commonest relevant aspect mentioned was the relative number of the two colours in the sack: this was given by 35.5% of the sample with the first of the two questions and by 67.1% with the second question. In contrast, the independence of successive events (or the irrelevance of previous events) was only given as a justification by 5.2% of the sample in the first of the items and by 4.1% in the second.

Thus, most of the pupils were able to justify their answers to some extent, but their justifications tended to be terse and were often limited to the relative number of balls of the two colours in the sack.

In summary, the pre-test demonstrated that many UK pupils in the 9–10 year age group have a clear understanding of the effects of randomisation, and that their understanding of the independence of successive events in a random sequence was much better than previous research has suggested. However, their ability to produce coherent and relevant reasons for their choices was rather limited.

10.3.3 The Immediate and the Delayed Post-tests Results

A comparison of the three groups across the pre-test and the two post-tests allowed us to see if the pupils taught about randomness made more progress in the post-tests than those taught directed numbers or those given no extra teaching during the study.

Randomised and determined items scores. Table 10.2 presents the number of correct choices made in the three randomised and the three determined outcome items in the pre-test and the immediate post-test. It shows a small, but consistent improvement from pre-test to post-test in the Randomness Group and also in the Directed Numbers Group, but not in the Unseen Control group. The small improvement in the two intervention groups may have been partly due to the pre-test scores being quite high and, in fact, near to ceiling.

Table 10.2 The mean number of correct choices (with SD in brackets) by pupils in the three groups in the randomised and determined items in the pre-test and the immediate post-test

Groups	Pre-test scores (out of 3)		Immediate post-test scores (out of 3)	
	Randomised items	Determined items	Randomised items	Determined items
Randomness	2.63 (0.77)	2.38 (0.77)	2.96 (0.20)	2.63 (0.58)
Directed numbers	2.68 (0.72)	2.18 (0.85)	2.86 (0.47)	2.55 (0.67)
Unseen control	2.76 (0.44)	2.14 (0.88)	2.60 (0.89)	2.10 (1.13)

A univariate analysis of co-variance of the children's total post-test scores in these items, in which their pre-test scores for the same six items were the co-variate, produced a significant difference between the three groups ($F(2.71) = 3.36$, $p = .04$, partial $\eta^2 = 0.09$). Simple contrasts of the adjusted post-test scores, based on the ANCOVA, showed that the Randomness Intervention group's post-test scores were significantly better than those of the Unseen Control group ($p = .019$), whereas the difference in the scores for the Directed Numbers Intervention group and the Control group fell just short of significance ($p = .055$). The difference between the Randomness and the Directed Numbers groups in the immediate post-test scores was not significant.

In the delayed post-test we introduced a new item on shuffling which was the non-replacement item in which the character did not replace the first card that he drew from the pack, thus changing the sample space from one draw to the next. The children in the Randomness group did the best in this item with a success rate of 87.5% correct in this item, which suggests that instruction does help to alert children to the importance of replacement or non-replacement in such problems. However the children in the other two groups did quite well too: (Directed Numbers Group 68.2%: Unseen Control Group 78.6%). The difference between the proportion of correct responses in the three groups fell below significance in a χ^2 test.

Understanding the independence of successive events in a randomised sequence. The children were given the same three items on the independence of successive events in a randomised sequence in the pre-test and the two post- tests. Table 10.3 gives the mean number of total correct choices for these items. The scores of all three groups improved from the pre-test to the two post-tests. The Randomness group's post-test scores were higher than those of the other two groups in both post-tests. This difference, however, was quite small.

Table 10.3 Mean number of correct choices (maximum possible score 3) in the items on the independence of successive outcomes in a random sequence (SD in brackets) in the pre-test, the immediate post-test and the delayed post-test

Group	Pre-test	Immediate post-test	Delayed post-test
Randomness	2.38 (0.82)	2.88 (0.45)	2.79 (0.42)
Directed numbers	2.05 (0.99)	2.45 (0.80)	2.41 (0.80)
Unseen control	2.14 (1.06)	2.54 (0.84)	2.50 (0.92)

An analysis of covariance of the immediate and delayed post-test scores, in which the covariate was the children's scores in the pre-test, produced no significant differences between groups ($F(2.68) = 2.19$, n.s.). So, the analysis provided no support for the hypothesis that teaching children about randomness helps them to understand the independence of successive events in a randomised sequence.

Justifications for two items on the independence of successive events in all three tests. Table 10.4 gives the mean justification scores for the two items on the independence of successive events in each of the three tests. The scores of all three groups improved from pre-test to immediate post-test, but the biggest improvement was in the Randomness group's justifications in the immediate post-test. The scores of all three groups then declined in the delayed post-test. The superiority of the Randomness group's immediate post-test scores suggests that they were helped by being taught about randomness: their failure to maintain this difference in the delayed post-test suggests that this teaching effect may be short-lived. The immediate post-test scores supported the prediction that the randomness intervention would improve children's justifications, but the delayed post-test results did nothing to confirm this group difference.

An analysis of covariance of these scores in the immediate and delayed post-test, with the equivalent scores in the pre-test as the co-variate, produced a significant difference between Groups ($F(2.67) = 4.87$, $p = .004$, partial $\eta^2 = 0.152$). The adjusted post-test scores of the Randomness Group were significantly higher in Simple Contrasts based on the analysis of co-variance than those of the Directed Numbers ($p = .005$) and of the Unseen Control group ($p = .003$). These two results establish that the scores of the children in the Randomness Group improved in the post-tests more than those of the children in the other two groups.

There were no other significant differences in this analysis but there were some surprises. One surprise was that, although there seemed to be a clear decline across groups from the immediate to the delayed post-test, this did not lead to a significant difference between the two post-tests ($F(1.67) = 0.278$, $p = .600$, partial $\eta^2 = 0.004$) in the analysis. The other surprise was that the randomness group, as we have seen, made more progress than the other two groups in the immediate post-test but not in the delayed post-test, and yet this pattern of results, to our surprise, did not lead to a significant interaction between groups and post-tests in the analysis ($F(2.67) = 2.043$, $p = .158$, partial $\eta^2 = 0.032$).

Table 10.4 The mean number of relevant aspects mentioned in the pupils' justifications in the pre- and post-tests (SD in brackets)

Group	Pre-test justification scores	Immediate post-test justification scores	Delayed post-test justification scores
Randomness	1.83 (1.02)	2.71 (1.12)	2.04 (0.81)
Directed numbers	1.68 (1.21)	2.05 (0.95)	1.50 (0.67)
Unseen control	1.24 (0.98)	1.71 (0.85)	1.68 (0.86)

10.3.4 Summary of the Results

Pre-test results. The pre-test scores showed a high level of understanding by 9–10 year olds of the process of randomising, in the form of shuffling, throwing counters and marbles on a tilted surface, and of the difference between random and determined arrangements. The pupils' choices in the three pre-test items on the independence of successive events were also high and they were much better than those reported in previous work on this topic. The high numbers of correct choices, and to some extent the justifications that the pupils gave for them, established a clear understanding of the independence of successive events in a random sequence in the majority of the pupils in the sample.

Intervention results (pre-test and post-tests). All three groups showed an improvement in their understanding of the consequences of randomising from pre-test to the immediate post-test. This improvement was reliably greater in the pupils who had been taught about randomness than in the unseen control group.

The improvement by the pupils taught about directed numbers was also greater than that of the unseen control group. The latter difference was almost significant ($p = .055$).

This specific result was at the same time disappointing and interesting. It was disappointing because we had predicted that teaching children directly about randomness would help them more than teaching them about directed numbers, but the two intervention groups improved to much the same extent. It is interesting because this result suggests that teaching children about directed numbers might also have an effect on their understanding of randomness, although this is something to be explored in further research.

The scores of the three groups for correct choices in the items that tested their understanding of the independence of successive events improved in the post-tests in all three groups to a roughly equal extent. Therefore the pattern of the post-test scores for correct choices in these items showed no direct effect of the intervention.

The justifications that the pupils gave for their choices in items on the independence of successive events did show an improvement from pre-test to the immediate post-test. The pupils in the Randomness group improved from pre-test to immediate post-test more than those in the other two groups. The difference between the groups can be taken as evidence for the effectiveness of instruction about randomness, at any rate in the short term.

10.4 Discussion and General Conclusions

The aim of this section is to discuss the implications of the results of the study. We first present a framework for the discussion, then describe children's initial knowledge and finish with an analysis of children's learning.

10.4.1 Framework for the Discussion

The study had two main aims. One was to explore pupils' understanding of randomness and randomising at the time of the pre-test when they had been given no formal instruction on these topics. The other was to find out whether teaching about randomness, in the form of our teaching programme, would improve pupils' understanding of the nature and effects of randomising and of the independence of successive events in a random sequence. We shall discuss these two questions in turn.

10.4.2 What the Pupils Already Knew About Randomness

The pupils' scores in the pre-test items on randomising were higher than expected, given previous research on this question. The three forms of randomising in these items were shuffling cards, throwing a set of counters, each with a black and white side, in the air, and tilting a tray so that the marbles at one end, which were initially arranged by their colours, would roll to the other end. Most pupils were able to judge that the likeliest result of throwing a handful of counters, each with a black and a white side, in the air would be a mixture of counters showing their black and their white sides, that the effect of shuffling a pack of cards would make it impossible to predict the position of any particular card, and that the most probable result of tilting a tray would be a mingling of marbles of two different colours. Most of the pupils also discriminated well between determined and random arrangements of the counters and of the marbles in the tray, and seemed to understand the different properties of these two kinds of arrangement. Since at the time of the pre-test these pupils had not been taught about randomness at all, we conclude that they had already built up for themselves a good idea of the effects of these familiar forms of randomisation, and certainly enough to provide a sufficient basis for learning from explicit instruction about, for example, the connection between fairness and randomness.

These results and this conclusion might seem at odds with the rather pessimistic picture drawn from previous research on pupils' understanding of randomisation. In fact, there may not be so great a conflict here. Most of the research that we were able to find on randomisation was about throwing dice, and it is possible that this particular form of randomising creates some specific problems for young pupils. For example, Watson and Chick (2001) report that some pupils produce the self-contradictory argument that it is equally possible to throw any of the numbers on a die but that 6 is harder to throw than the other numbers because in some games you have to throw a 6 in order to be allowed to enter the game and that sometimes takes a lot of throws. Nevertheless the pupils' high rate of success in our randomisation items indicates a promising start in their informal learning about randomness.

The pupils' scores in the pre-test items on the independence of successive events in a random sequence provided an even starker contrast with previous research and in particular with the results reported by Chiesi and Primi (2009). Our items were a direct replication of those devised by Chiesi and Primi and yet the pupils in our study made far more correct choices and far less positive or negative recency errors than any of the groups in the Chiesi and Primi study including their adult group of students. The only common pattern in our results and theirs was that the 9- to 10-year old pupils in our study and the 10-year old pupils in their study made more positive than negative recency errors in both studies, but the much small number of errors of either type in our study than in Chiesi and Primi's diminishes the significance of this similarity.

It is hard to explain the radical discrepancy between the two studies. One possible reason could be that we stressed very clearly, and perhaps more clearly than Chiesi and Primi, that the character who made the first four draws from the bag always put the marble he had just drawn back in the bag before the next draw, and we put an animation into our PowerPoint slides that represented this replacement. However, a failure to grasp that the marbles, all of the same colour, picked in the first four trials were immediately replaced, could only lead to negative recency errors, since the person not noticing the replacement would think that the number of marbles of this colour in the bag, had diminished by four and therefore would be less likely to be drawn again. So this would not explain why the 10-year olds in the Chiesi and Primi (2009) study made far more positive recency errors than the pupils in our study and far more positive than negative recency errors.

At any rate we are confident in our results for two reasons. One is that the pupils in the two groups who were not instructed about randomness maintained a high level of correct answers about the independence of successive events in the immediate and delayed post-tests as well as in the pre-test, which shows some reliability in the way that they dealt with questions about successive events. The other reason is that the majority of the pupils in our study was able to produce a relevant reason for their answers when asked to justify them. We conclude that, despite the negative claims made by Chiesi and Primi (2009), and to some extent by Fischbein and Gazit (1984) as well, most 9- to 10-year old pupils nevertheless do have a clear understanding that previous events in a truly random sequence have no bearing on what will happen next.

10.4.3 What the Pupils in the Two Intervention Groups Learned About Randomness

Our main aim in this study was to see whether it is possible to improve pupils' understanding of randomness through formal and explicit teaching. Our results suggest that the teaching given to the Randomness group did improve their understanding and reasoning about randomness to some extent, though not as strongly as we had predicted. The post-test scores of this group for the randomising items (shuffling cards, throwing counters, tilting a tray) were significantly higher than those of the Unseen Control group but were not different in statistical terms from those of the Directed Numbers group. Both interventions worked.

The unexpected beneficial effect of the directed numbers intervention suggests that this intervention might have helped children to think systematically about the effects of randomisation. The connection between the intervention and pupils' understanding of randomising needs to be explored in further research. One possibility (but we stress that is as yet an untested speculation) is that the connection is due to the directed numbers intervention laying great emphasis on the inverse relation between addition and subtraction, since the knowledge that the number of objects in a set remains the same if equal numbers of objects are added to it and subtracted from it is an essential part of being able to monitor the sample space. This idea can be tested in further intervention experiments which relate the understanding of inversion and the ability to monitor sample space in replacement and non-replacement situations.

Our failure to find significant effects of the randomness intervention on the number of correct choices in items on the independence of successive events in a randomised sequence was disappointing, but the justifications given by the pupils for their choices in these items was more encouraging. The greater improvement in the quality of these justifications by the pupils in the Randomness Intervention group than by the other pupils shows that teaching pupils about randomness helps them to reason about uncertain events. This difference, taken together with the generally high level of the choices made by the pupils in all three groups in the pre-test, suggests that pupils have acquired an effective but implicit understanding of the effects of randomisation by the age of 9–10 years, and that one of the main effects of instruction about randomness is to make this knowledge more explicit and therefore more open to instruction and to discussion in the classroom.

The Randomness intervention programme took the form of pairs of pupils solving problems that usually involved concrete material such as dice and cards or material presented on PowerPoint and then presenting their solutions to the rest of the group for discussions at the end of each session. This seemed to us to be a congenial as well as an effective way of improving pupils' understanding of randomness and of making this understanding more explicit. We cannot say whether other teaching approaches would also lead to improvements in pupils' understanding of randomness.

This study has established that 9–10 year old pupils are ready for instruction about randomness. The pupils' justifications for their responses have also shown that the instruction that we devised to teach them about randomisation did improve their explicit understanding of randomisation and of the independence of successive events in a random sequence.

Acknowledgements The project described in this chapter was funded by the Nuffield Foundation.

References

Batanero, C. (2016). Understanding randomness: Challenges for research and teaching. In K. Krainer & N. Vondrová (Eds.). *Proceedings of the Ninth Congress of the European Society for Research in Mathematics Education* (pp. 34–49). Prague: European Society for Research in Mathematics Education.

Batanero, C., Henry, M., & Parzysz, B. (2005). The nature of chance and probability. In G. Jones (Ed.), *Exploring probability in school: Challenges for teaching and learning* (pp. 15–37). New York: Springer.

Batanero, C., & Serrano, L. (1999). The meaning of randomness for secondary school students. *Journal for Research in Mathematics Education, 30*(5), 558–567.

Bennett, D. J. (1998). *Randomness*. Cambridge: Harvard University Press.

Bryant, P., & Nunes, T. (2012). *Children's understanding of probability: A literature review*. London: Nuffield Foundation.

Chiesi, F., & Primi, C. (2009). Recency effects in primary-age pupils and college students. *International Electronic Journal of Mathematics Education, 4*(3), 259–274.

Falk, R. (2016). *Many faces of the gambler's fallacy: Subjective randomness and its diverse manifestations*. Electronic book available on Amazon.

Falk, R., & Konold, C. (1997). Making sense of randomness: Implicit encoding as a basis for judgment. *Psychological Review, 104*(2), 301–318.

Fischbein, E., & Gazit, A. (1984). Does the teaching of probability improve probabilistic intuitions? *Educational Studies in Mathematics, 15*, 1–24.

Fischbein, E., & Schnarch, D. (1997). The evolution with age of probabilistic, intuitively based misconceptions. *Journal For Research In Mathematics Education, 28*(1), 96–105.

Green, D. R. (1989). Schools students' understanding of randomness. In R. Morris (Ed.), *Studies in mathematics education, the teaching of statistics* (Vol. 7, pp. 27–39). Paris: UNESCO.

Kuzmak, S. D., & Gelman, R. (1986). Young pupils' understanding of random phenomena. *Child Development, 57*(3), 559–566.

Piaget, J., & Inhelder, B. (1975). *The origin of the idea of chance in pupils*. London: Routledge and Kegan Paul.

Truran, K. (1994). Children's understanding of random generators. In J. P. Ponte & J. F. Matos (Eds.), *Proceeding of the 18th International Conference for the Psychology of Mathematics Education* (pp. 328–336). Lisbon: International Group for the Psychology of Mathematics Education.

Shtulman, A. (2009). The development of possibility judgement within and across domains. *Cognitive Development, 24*, 293–309.

Shtulman, A., & Carey, S. (2007). Impossible or improbable? How pupils reason about the possibility of extraordinary claims. *Child Development, 78*, 1015–1032.

Watson, J. M., & Chick, H. L. (2001). Factors influencing the outcomes of collaborative mathematical problems solving: An introduction. *Mathematical Thinking and Learning, 3*(2 & 3), 125–173.

Chapter 11
Understanding Children's Meanings of Randomness in Relation to Random Generators

Haneet Gandhi

Abstract This chapter describes an open-exploratory study with middle-grade Indian school students (aged 13–14 years) to understand their perceptions of randomness. Detailed discussions are provided of two groups of students who worked in small groups to choose, from a set of eight perfectly or partially symmetrical polyhedrons, a shape that according to them is optimal for playing the games of chance. The discussions bring to the fore different meanings that students ascribe to qualifying an object as a generator of randomness. The results indicate that students rely on physical characteristics of the object, as well as the size of sample spaces for making judgements. They also clinched to the notion of exhaustiveness of the sample space that explains, to some extent, the basis for their irrational beliefs against the cubical dice. When analysed from an alternative perspective, the responses exemplify a propensity or a disposition to the physical characteristics of the resources used in the experimental setting. The chapter concludes with a proposition of imbibing isomorphic resources as means for identifying, understanding and nurturing children's notions and beliefs.

Keywords Randomness · Propensity theory · Intuitions · Isomorphic resources

11.1 Introduction

It is almost impossible to talk about any work in probability without mentioning the notion of randomness, a construct that should perhaps be studied most carefully in the domain of probability. Since randomness is an abstract concept and difficult to define, a way to experience it is through an assortment of physical devices, resorting to dice, coins, cards, lotteries, and the like (Bennett 1998). While dealing with these resources, we test the ability of the device to generate outputs that show traits of

H. Gandhi (✉)
Faculty of Education, University of Delhi, 33 Chhatra Marg,
Delhi 110007, India
e-mail: haneetgandhi@gmail.com; hgandhi@cie.du.ac.in

C. Batanero and E. J. Chernoff (eds.), *Teaching and Learning Stochastics*,
ICME-13 Monographs, https://doi.org/10.1007/978-3-319-72871-1_11

being random, meaning thereby that we judge the randomness of an object by analyzing the sequence of outcomes that it produces (Batanero 2016). The sequence should not have any evident pattern and it should almost be impossible to predict its next outcome (Batanero and Serrano 1999). Thus, the testimony of outputs serves to check the randomness of an object.

In schools, children are asked to predict randomness by experiencing random mixtures, random distributions, random draws, and random sequences. In spite that, the word 'random' is never dealt with exclusively, seldom do we pause and reflect on how our children may have conceived the idea of randomness. Issues related to randomness are either ignored or taken for granted and consequentially, we move quickly to more sophisticated and challenging tasks such as calculating probabilities. Questions related to the object that create such sequences, and in particular, the physical and structural features of the object which were instrumental in generating random sequences, are hardly considered (Batanero 2016).

Eichler and Vogel (2012) analysed the functional and structural aspects of thinking processes and stated that children rely on data, objects and mental simulations in situations of uncertainty. While dealing with tasks related to probability, children form mental models of the situation structure through the characteristics of the physical objects present in the situation, and the data generated from an uncertainty process. They construct a mental simulation of the relevant aspects which help them in anticipating the possible results. The authors further go on to say that in situations where the generation of data is not available explicitly, children form mental simulations based on the physical characteristics of the objects that are made available. This view opens a new window to explore, which consists in analysing children's sense-making in a probabilistic situation when the data (frequentist approach) is not considered.

The available literature (Batanero and Serrano 1999; Green 1983; Piaget and Inhelder 1975; Pratt 1998; Pratt and Noss 2002) describes children's understanding of randomness from a frequentist lens relying on the nature of the outcomes, but there exists little research on knowing how children perceive the object that produces such outcomes. In this context, there are several questions which have been left unanswered, such as do children see randomness as entrenched in the object itself, or in the outcomes that it produces? How do they ascribe to randomness on their first encounter with the objects of randomness? What conceptions do they form while choosing a good random generator? How do children rationalise their decisions when selecting a random generator? Children's perceptions and understanding related to random generator itself often go unnoticed, unquestioned, and unexplored.

In this chapter, we present the results of a study that was part of a larger project on exploring the concept of probability in the Indian context. The episodic phase described in this chapter is based on the premise that by giving children opportunities to create randomness, we are able to understand their beliefs related to this construct. An exploratory approach was adopted wherein students were given open opportunities to explore and experiment with various symmetric polyhedrons, and to choose one for playing games of chance. It was assumed that in the process of choosing their random generator, the students would inadvertently bring their meanings of randomness. Consequently, a student-oriented approach to understanding what children would do in a probabilistic situation governed by an object of randomness was adopted with the aim of interpreting the different meanings that children ascribe to in order that they qualify an object as a generator of randomness.

11.2 Background

The study was based on a three-faceted foundation: (a) questioning the emphasis given to the analysis of the behaviour of sequences of outcomes in deciding the randomness of an object; (b) children's naïve, intuitive, and biased associations with certain objects that create randomness, especially the cubical die; and (c) historical evidence that supports the use of various artefacts as randomizers. In the following section, we give a brief account of how these considerations formed the background of the current inquiry.

11.2.1 Experiencing Randomness

Randomness as a construct is difficult to be defined. We attribute randomness to mixtures, distributions, draws and sequences, provided that their outcomes exhibit no pattern, and do not follow any fixed algorithm (Batanero 2016; Batanero and Serrano 1999). In other words, our claims of randomness are based on the expectations (or should we say non-expectations) of a pattern in their outcomes and to our inability to guess, a priori, the outcome that is about to appear.

In most studies that investigate pupils' notion of randomness (Batanero and Serrano 1999; Green 1983; Piaget and Inhelder 1975; Pratt 1998; Pratt and Noss 2002), a series of real or simulated outcomes were put to test in either judgmental or production based tasks (Bar-Hillel and Wagenaar 1991). In judgmental tasks, an already generated sequence of outcomes is examined for randomness (asking the participants, for example, "Which of the following series of outcomes actually correspond to the outcomes of a die?"), whereas in the production based tasks, the assumptions that one would make while creating a series of random outcomes are analysed (for example, "Produce a series of outcomes that you are likely to get from throwing a die"). Though judgemental tasks are a purer way of studying the perception of randomness, the basic biases in subjective perceptions of randomness were discovered by the production tasks (Bar-Hillel and Wagenaar 1991, p. 431). In either case, the randomness of an object is associated with the series of outcome that it produces or is likely to produce. If the sequence of outcomes 'behaves randomly,' the object is considered random.

One extensive study that classified randomness in observable terms, using a computer simulation, was conducted by Pratt (1998). In his Chance-Maker micro-world, children approached randomness from a frequentist approach.

Sequences of outcomes generated from the simulated world were analysed for short-term variations which owing to the lack of pattern exhibited four sub-constructs, thereby manifesting properties of randomness in unpredictability, lack of control, irregularity, and fairness. Although these sub-constructs are helpful in determining the randomness of an object, they were associated with the sequence of outcomes and not directly with the making of the random generator. There is scarce research associating randomness with the physical embodiments of the generator that creates it rather than with the study of sequences (Batanero 2016).

Anyway, hardly do we pause and reflect on the properties of the device that could have been instrumental in generating random sequences. In a way, we can see randomness being shadowed by the charm of the sequences that it generates and we still seem to be unsure of what people examine when they first encounter devices of probability. This, in turn, triggers a set of questions—How is randomness exhibited in the random generator? Which aspects of the generator would children invoke while testing it for randomness? In particular, we wished to explore how children perceive randomness in the configuration of the device itself in their initial encounter with the object.

11.2.2 Biases Related to the Cubical Dice

Studies that have contributed to understanding the notion of randomness in students of different ages (Batanero and Serrano 1999; Green 1983; Jones et al. 1999; Langrall and Mooney 2010; Metz 1998; Piaget and Inhelder 1975; Pratt 1998; Watson and Moritz 2003) have focused on different properties such as expected frequencies, lack of control, length of runs, lack of pattern, uncertainty or unpredictability as indicators for qualifying a sequence to be random. As part of these studies, students were provided with generators of randomness such as cubical dice, coins, and spinners; or other times were asked to classify a sequence of outcomes as random or not random. Their responses were analysed to understand their beliefs, conceptions, and misconceptions.

Concomitantly, studies have also brought to fore the irrational beliefs that children hold regarding the outcomes of a standard cubical die. For example, Truran's (1995) work established that children consider God, fate, mental or supernatural powers as being responsible for biased outcomes. Children's past experiences of playing with the cubical dice also lead them to believe that luck (Amir and Williams 1999), techniques of throwing the die (Green 1983; Pratt 2000), partiality towards particular numbers (Vidakovic et al. 1998; Watson and Moritz 2003), or difficulty in getting a six (Green 1983), are some of the reasons responsible for the (perceived) biased outcomes of the cubical dice.

Fischbein (1975) defines primary intuitions as cognitive attributes that are spontaneous, global, and self-evident to the believer. Since these beliefs are mostly derived from everyday experiences, they are difficult to be challenged and corrected when left on their own. He further adds that through systematic didactic situations such intuitions can be restructured, thereby emphasizing the role of appropriate pedagogic experiences in understanding, challenging, and modifying primary intuitions as a step for facilitating formal learning. Biases and misconceptions related to the cubical dice in particular and Fischbein's ideas of restructuring pedagogical resources led to the framework of the present study (discussed in detail in Sect. 11.3).

11.2.3 Randomisers in the Primitive Times

The third foundation for the study draws from the learning gathered from the historical anecdotes, which certified that people used a variety of objects as randomisers in primitive times. From these anecdotes, we know that the cube was not the only tool used in the past for making decisions in chance situations (Bennett 1998; du Sautoy 2008). Humans turned to an assortment of chance devices such as heel bones of hoofed animals (astragali), specially carved stones and leafed rosettes; spheres with twelve to fourteen knobs; shells, tetrahedral dice, rectangular prisms, triangular sticks, coins, and hexagrams, not only for playing games, but also for making vital decisions and soliciting divine direction (Bennett 1998; du Sautoy 2008). It is evident that they either believed in some divine power or relied on craftsmanship while selecting the objects. Most of the humanly crafted objects exhibited perfect symmetries, so it appears that the structural homogeneity of the shapes was given preference. Clearly, people had perceived a relationship between symmetry and randomness (du Sautoy 2008).

Thus, the background of the study was informed by the review of related literature which focuses on judging the randomness of an object based only on the nature of the runs of outcomes, and children's incomplete understanding of randomness as a construct as well as their biased opinions against a fair cubical die. In addition to all this, historical evidence attesting the use of symmetric objects for games of chance also proved as a reliable resource.

11.3 Framework of the Study

In an attempt to interpret and implement Fischbein's ideas into more practical forms, two perspectives of research have emerged. One focuses on *identifying* primary intuitions, heuristics, biases, and misconceptions that people hold in relation to probability (Kahneman and Tversky 1972; Konold 1989), while the other takes a *constructive-interpretative* path for understanding the reasons behind people's actions when they encounter stochastic situations (Nilsson 2004, 2007). Following the first paradigm, we can understand the notions that are already entrenched in the mind and may be difficult to unlearn or refine. The latter paradigm, on the other hand, enables us to understand how a concept is assimilated. The hallmark of this approach lies in grounding learning in situations in which it emerges.

Learning is an act of attaching meaning to external representations. Knowledge is embedded in contexts, and children's actions and reactions in their existing circumstances and situations symbolise their cognition and give an insight into the internal mental representations (Cobb et al. 1992). In other words, mathematical learning becomes a salient feature of the external representations and is characterised by students' actions in relation to those external resources. Since learning is an act of making connections between the external representations and the mental entities, by interpreting actions in the external world we can interpret internal mental representations. Therefore, in order to interpret learning it becomes important to study actions as they reflect cognition. Pratt (1998) further distinguishes between external and internal resources and elucidates how their interplay leads to the development of knowledge. Resources such as tools, peers' interactions, and experimental settings that are outside the individual form the external resources, and internal resources comprise all forms of intuitive and formal thinking that the child incorporates. The development of internal resources, however, depends upon the interaction with the external resources.

Creating a context for interpreting construction of knowledge was deemed useful for the present study. So we chose the constructive-interpretative approach wherein the tools and the small-group situation were counted upon as external resources. The associations that students made from these external resources enabled us to interpret their internal resources and develop an understanding in relation to randomness. We relied on children's actions in specifically designed probabilistic environments as gateways for discerning and interpreting their thoughts, intuitions, and possible reasons for their (mis)conceptions.

In addition to the above, we are also aware that children's notions about randomness are often shaped by their every-day experiences (Fischbein 1975) and "as long as we persist in assuming the inherent frailty of primary probabilistic intuitions, …we continue to design and research activities that only further confirm this assumption" (Abrahamson 2012, p. 870). Pitting children's intuitive ideas alone through complex stochastic resources does not guide their erroneous thinking. Rather, Abrahamson (2012) proposes designing epistemic resources that investigate the roots of children's intuitive thoughts and further nurtures these roots towards the targeted didactical content. "More generally, when a bias is observed in spontaneously devised models, correct models may in most cases be available, but experimental tricks are required to activate them" (Lecoutre 1992, p. 567). Children hold many unreasonable beliefs against the cubical die (as stated in Sect. 11.2), so to dig into the roots of their canonical erroneous beliefs, it is imperative to bring in fresh, unfamiliar tools as epistemic resources. It was hypothesised that in the process of interpreting children's work as they construct new internal resources while working on unconditioned resources, would open avenues for understanding the rationales that structure their intuitive ideas regarding randomness and random generators.

11.4 Methodology

The aim of this study was to identify students' association with unconventional resources of randomness in an exploratory setting in order to understand their meaning of randomness. In particular, it was hypothesized that an exploratory approach encourages freedom of expression and thus enhances the scope of understanding children's reasoning, judgement, and sources of (mis)conceptions related to randomness and random generators. In this study, we were interested in understanding students' dominant assumptions in considering symmetrical objects as generators of randomness.

11.4.1 Participants

The present study was conducted in a private school in Delhi (India). The sample of participants consisted of eight students from Grade 8 (age group 13–14 years) who were divided into two groups, with four students in each group. The results follow a case study approach aimed at drawing out, from the actions and discussions of the groups, the commonalities in the children's internal thoughts related to perceiving objects as generators of randomness.

In the Indian Mathematics Curriculum (recommended by NCERT 2005), since the concept of chance is introduced in Grade 7, we presumed in the present study that all the participants were familiar with elementary concepts related to probability. Most of the content of probability in Indian textbooks follow the classical interpretation (Gandhi 2015), which familiarises the middle-grade learners with the theoretical approaches of finding probabilities of simple events. As a part of their curriculum, the students were also exposed to the handling of typical materials and resources such as cubical dice and coins as random generators.

11.4.2 Data Sources and Experimental Setting

The tools used in the study comprised eight polyhedrons that exhibit either perfect or partial symmetry. Though Grade 7 and 8 students of India have been exposed to polyhedrons as a part of their geometry curriculum (NCERT 2005), they are unable to relate these to the topic of probability. The eight polyhedrons included four platonic solids that exhibit perfect symmetry: a regular tetrahedron, a cube, a regular octahedron, and a regular icosahedron. The other four partially symmetric polyhedrons consisted of two pyramids (one with a square base and the other with a regular hexagonal base), and two prisms (one having the top and bottom bases of regular triangles; and in the other, regular hexagons made the bases).

The edge lengths of all of the platonic solids were kept the same i.e. 3 cm. With the non-platonic polyhedrons, care was taken to ensure maximum regularity between lateral and base faces. The edges of at least one face of the non-platonic solids was kept at 3 cm. For example, the edges of the regular hexagon, which made the base of the hexagonal pyramid, were 3 cm. No markings were made on the polyhedrons. Most of these solids have, at some point, been used by humans in crude or in polished form for playing games or otherwise (Bennett 1998; du Sautoy 2008). This supported the selection of these polyhedrons.

In the experimental setting, both groups were made to sit apart from each other and work independently. The set of eight polyhedrons was provided to each group. The participants were encouraged to work within their small groups and decide which amongst the given polyhedrons, they would use for playing board games such as Ludo (Ludo is a very common board game played by Indian children in which a standard cube is used as a die for deciding the moves). After the groups had analysed and experimented with their chosen object, they were asked to state the reasons for their choice. The aim was to observe how students rationalised their decisions regarding their choice of the die and the inhibitions (if any) they carried against the standard die (the cube).

Participant observation emerges from the central concerns of symbolic interactionalism. To make meaning from social encounters, Rock (1999) identifies the potentialities of participant observation as it "centres on an interpretation of knowledge as praxis [and], not as an analytic a priori" (p. 6). In a social setting, "facts are principally produced in the symbolic work of social encounters. They are not 'given' but created" (Rock 1999, p. 15). The researcher of this study undertook the role of a participant observer, present in situ and intervening minimally only to seek clarifications and to make note of students' actions and discussions as they interacted in the social setting of their peers and experimental tools.

Since both groups were working simultaneously, it was not possible for the researcher to take notes for each of them so their working was video-recorded and later converted into transcripts. Transcripts of each group were read several times to get a sense of the whole, a gestalt. This provided a context for identifying the emergence of meanings of each group. Any act that seemed significant and expressed a meaning was taken as a unit (Hycner 1999). These units were then filtered in two categories: units of general meaning and units relevant to the research question (Hycner 1999). Thereafter, only the units categorised as relevant were considered. In the next stage, the relevant units were colour coded for any idea that reflected students' understanding of randomness. The units for which the colour codes of both groups matched were taken as representations of the consistencies in the rationales of the groups. Consistencies in their arguments led to the formulation of the conclusions of the present study.

As the setting was open and exploratory, the students were being given the flexibility to work at their own pace. Time was not a constraining factor, but it was noted by the researcher that both groups took nearly 70 min to arrive at their decision.

11.5 Findings and Results

This section deals with the presentation of research findings through the interpretation of students' actions, discussions and engagement with the various polyhedrons, by means of which they made sense of randomness as a construct. As stated earlier, the students were made to work in their respective groups and were encouraged to select a shape which they found most suitable for their dice.

The groups worked independently to choose their preferred piece. Both groups selected different pieces and their selection process elucidated certain similarities and some differences in their interpretations. The differences were found to be inconsequential and hence did not contribute much to the analysis.

In the sections below, the work of both groups is presented to highlight the threads that were found to be common in the actions and considerations of both groups. We identified how the children associated the physical characteristics of the shapes with the notions of fairness and unpredictability, and their perception of sample space while selecting an object for a die. A notion that helps explain possible reasons for children's biases against the outcomes of the cubical dice also emerged from the analysis of the actions.

11.5.1 Preliminary Explorations

Before making any decision, each group spent sufficient time investigating the physical features of all the shapes (approximately 40% of the total experimentation time was spent in these explorations). During their initial inspection, the children examined each shape by throwing, spinning, and rolling them. It was evident from the children's explorations that their focus was on judging the orientation and free motion of the shapes. In fact, the participants iterated this examination several times during the entire intervention phase. Given below are the accounts of each group as they selected their respective die.

After playing with all the pieces, Group One realised that physical motions such as throwing, rolling, and spinning, are not applicable to all types of solids, and different physical movements would be needed for different pieces. For instance, spinning works well with pyramids, while rolling is effective with platonic solids. The dilemma was now to decide on a mechanism that would ensure minimal external interference and thus the first round of elimination was done on the basis of the mechanisms that obstruct free movement. The students of this group articulated that rolling is the most appropriate action to ensure maximum free movement and inhibit manipulation. The act of rolling was thus preferred over spinning and throwing and this led to the exclusion of all pyramids except tetrahedron for any further deliberation. Certainly, minimizing the effect of friction and other physical or environmental interferences were considered as an essential criterion.

> Group One: [They toyed with the pieces by throwing, rolling, and spinning them] Rolling is good. [A little later, while spinning the pyramids] You can spin it slow or fast. We know where it will stop… If it stops at the corner (edges) then it will be difficult to decide the number.

In the second round of elimination, the children of Group One ruled out both the prisms (triangular prism and square prism or cuboid) as the lateral faces of these objects were more frequent to appear as compared to the top and bottom faces. They could conclude that the other two faces (top and bottom) "were a waste" and would never turn up. Finally, they had to make a choice between tetrahedron, octahedron, cube, and icosahedron. After several rounds of inspection, the group decided on the regular icosahedron. This choice was based purely on its ease of rolling and non-manoeuvrability. The group explained that as compared to other solids, the icosahedron rolls freely and thus can be manoeuvred least. They marked the faces of the icosahedron from 1 to 20.

> Group One: [While comparing tetrahedron and octahedron]: Which one should we take? This (tetrahedron) can't be rolled easily as this (octahedron). [One of the students picks the Icosahedron piece and rolls it on the table]. If I throw this (icosahedron) then it will not show the outcome that I have in mind… [Other students of the group also roll the tetrahedron, octahedron, cube and icosahedron several times. The examination of the solids lasted for about ten minutes]. This (icosahedron) rolls like a ball. It rolls nicely, doesn't stop.

Group Two, on the other hand, began by rejecting all partially symmetrical shapes (cuboid, hexagonal pyramid, triangular prism, and square pyramid). Of the remaining four shapes (tetrahedron, cube, octahedron, and icosahedron), they eliminated the tetrahedron and the icosahedron as being "extreme cases." Since a tetrahedron has only four faces, were deemed to be too few to play a game, and the icosahedron would also lead to "frustration" as one may have to wait for several turns to get the desired number. Both these options were thus ruled out during the preliminary inspection. The group selected the octahedron (having eight faces) as an optimal choice since it would add to the thrill of the game, being a median piece, and numbered the faces 1–8.

> Group Two: [While discarding tetrahedron and icosahedron]: In this case (tetrahedron) the game will be very slow and this (icosahedron) will make it fast.
>
> Researcher: What do you mean by 'slow'? How will it (icosahedron) make the game 'fast'?
>
> Group Two: It limits the opportunities between 1, 2, 3 and 4. The game would take a long time to finish, only four moves. This [pointing to the icosahedron] has twenty faces. It would make the game fast. The one who gets twenty will reach near the end and will win. There will be no thrill in the game. We may even have to wait long for some other number to come up. The game will be boring and frustrating.

For ease of readability, in the rest of the chapter, we will refer to Group One as the Icosahedron Group, and Group Two as the Octahedron Group.

11.5.2 Forming the Ideas Related to Randomness

In spite of choosing different objects, there were commonalities beneath the sense-making exercise of both groups. This section delineate the acts and discussions of the groups as they made connections with sample spaces, unpredictability, and fairness to fortify randomness of their die.

11.5.2.1 Sample Space

The construct of sample space is fundamental to all aspects related to probability. Horvath and Lehrer (1998) describe it as an act of recognising and listing all possible ways of obtaining an outcome. From the classicists' viewpoint, the sample space is built by enumerating all possible outcomes.

Previous studies have emphasised strengthening the idea of sample space as they form the backbone of probabilistic judgements (Benson and Jones 1999; Horvath and Lehrer 1998; Jones et al. 1999). Although these studies have contributed substantially to understanding the genesis of the idea of sample space in children of various age groups, little has been achieved in understanding how children correlate the sizes of sample spaces with probabilistic judgements. We found that for children of this age group, a paramount criterion for selection was the size of the sample spaces of their chosen die. Both groups ensured that their respective dice had larger sample spaces (as compared to the cube). Larger sample space meant wider possibilities of getting different numbers (8 and 20 possibilities, respectively), and having more faces (as compared to the cube) meant smaller chances of predicting the outcome ($\frac{1}{8}$ and $\frac{1}{20}$, respectively). Increased sample size also gave them the confidence that in their respective dice the probability of each outcome being smaller would result in greater uncertainty of knowing the number yet to occur. An example of their discussions reflecting this understanding follows:

> Icosahedron Group: This (Icosahedron) has the maximum number of faces, twenty. We have more numbers to put on each face so nobody will be able to know which number they will get on the next throw...The numbers are more. There will be more moves in one throw... we will, you know, have to make changes in the game too. It will be an interesting move...There are more options.

> Octahedron Group: There are eight faces and the probability of each face is $\frac{1}{8}$, which is very low. Range of outcomes (as compared to cube) will be more. It has more options... more variations.... More range of numbers.

The above discussions reaffirm the claims of Fischbein (1975), Konold (1989), and Langrall and Mooney (2010), that children have a natural intuition regarding proportionality and they tend to rely on deterministic approach even in events related to fortuitousness. The participants embarked upon proportional strategies for finding probability by enumerating the sample space and considering the ratio between the desired event and the size of the sample space. Both groups claimed that the probability of getting a face was equiprobable, a notion in line with the classical explanation of probability.

11.5.2.2 Unpredictability

Unpredictability is associated with non-determination of the outcome. The children in the present study associated unpredictability with the uniqueness of the shape of their die. Earlier studies (e.g. Batanero and Serrano 1999; Pratt 1998, 2000) regarded unpredictability and fairness as attributes to be considered in determining the randomness of the outcomes generated from the tools. The children in this study perceived these constructs as being inherent in the physical make of shapes and rather than thinking of the runs of the outcomes. They took a keen interest in studying the uniqueness in the shape and its influence on producing the outcomes.

The Icosahedron Group claimed that the presence of many faces would limit the chances of anticipation of the number to appear. Possessing a large number of faces would make it difficult to predict the number that would come next. The second conjecture was made with respect to its peculiar shape and icosahedron's semblance to a ball as it rolled freely, unobstructed by any external interference. Owing to this double semblance of shape and rolling property, the group ruled out the predictability factor, as well as to any a priori assumption about the resultant face on which the icosahedron would finally land.

Discussion between two students of the Icosahedron Group exemplifies this conclusion:

> Icosahedron Group (Student 1): There are so many faces that it will land on a different face each time. All faces are equilateral triangles, it can land on any of these.
>
> Icosahedron Group (Student 4): We can roll it many times… twenty different faces. How do you know which is coming?
>
> Icosahedron Group (Collectively): It rolls better… See… [Rolls the die]. It continues to roll, doesn't stop, even after it lands. It's a different number every time.

The group that chose the octahedron associated unpredictability with the instability that was apparent from the constitution of the solid. The group showed confidence in ensuring unpredictability of their die by quoting their observations related to the instability that the die showed upon rolling. They were observed mentioning that no matter how fast or slow they rolled the die, the face to show up would change just before the piece became stationary. The group experimented this by throwing and rolling the die several times and pointed that each time they threw the die on the working table, the octahedron would get unstable before landing and would turn slightly before finally coming to rest. This instability in landing resulted in different numbers coming up during and after landing, which gave them the confidence that the number to come up cannot be known in advance. No matter how skilled a person may get with the techniques of throwing or rolling, the physical structure of the octahedron would ridicule any act of manoeuvrability.

The following dialogue between the Octahedron Group illustrates their certainty in ensuring unpredictability of their die:

Octahedron Group (Student 3): In this diamond shape (octahedron) it is very difficult to do so. There are so many faces. It is so difficult to say which one will come up. Any of them can come up anytime.

Octahedron Group (Student 2): The shape can land anywhere. See this… [She throws the die in the air, allowing it to fall freely on the working table and then points to its wobbling state as it attained its balance before finally coming to rest]

Researcher: How does that help?

Octahedron Group (Collectively): It changes its direction every time.

[The group repeat the acts of rolling and throwing the die several times to convince the researcher of its instability]

Thus, we summarise that precautions to ensure the least influence of any external power were kept in mind by the children while selecting the dice and uncertainty of knowing the next outcome in advance was determined by mechanical analysis of the shapes. The peculiar make of the chosen shapes and their instability while landing were enough to convince the children that the selected dice could not be manoeuvred easily by human hands.

11.5.2.3 Fairness

Fairness in a die is associated with the same likelihood of each side coming up (Watson and Moritz 2003). While playing with dice, the fairness of a die is associated with each player's chances of winning; in other words, each should have the same, equal chance of winning. In the case of dice, it is presumed that if all the faces have an equal likelihood of showing up, it is fair. Often, while dealing with the cubical dice in the classroom, this aspect of randomness is taken for granted and is not given much emphasis.

With our sample groups, the children's choice of die revealed their assumptions about fairness. Their engagement with only perfectly symmetric platonic solids delineated their consideration of symmetry as a signifier of fairness. Although no exclusive discussion took place and neither was there any need for clarification, the students' choice of picking only perfectly symmetrical shapes in their initial stages of examination indicated, covertly, their notions of fairness. The groups affirmed that since their objects had congruent faces, the dice would land freely on any face without being effected by an external force. For them, since the faces of the respective dice were congruent, all faces had to show up once at least.

Icosahedron Group: [During initial explorative stages, while examining pyramids] … choose an object which is fair. The faces should be same… all faces should have the same area so that all of them have equal chances of coming up. The faces have to be similar (congruent). In these [referring to pyramids] the bases and other faces do not have equal areas. It will never land slantingly and these faces [referring to the square and hexagon at the bases] will never come.

[A little later, while zeroing in on the icosahedron polyhedron] Every face is same so each time the number that showed on the top-most face is different. You never know where it will stop… we cannot control the roll.

A similar reasoning was given by the Octahedron Group.

Octahedron Group: We should pick more of a nicer figure so that the probability of any face coming up is same.

We took these inspections as evidence that both groups were crediting fairness to the symmetry of the objects. The physical symmetry of the objects was also a powerful tool in ensuring free fall of the die after the act of throwing or rolling. It also hinted on the equiprobability notion.

11.5.3 Exhaustiveness of the Sample Space

A common thread that appeared through the discussions with both groups was their belief in the 'exhaustiveness of the sample space', that is, one would get each number on a die at least once during a game in a finite number of throws, irrespective of their order of appearance. The children intuited that irrespective of the order of occurrence of numbers, it was certain that each face would occur at least once before an already obtained number reoccurs. This exhibited their intuition of the occurrence or non-occurrence of events being cyclic and agreed that irrespective of the order of appearance, the occurrence of every number is an essential criterion.

An example of students' belief in the exhaustiveness of the sample space could be traced through the discussions that took place between two students of the Icosahedron Group, who claimed that all faces had to come up at least once:

Icosahedron Group (Student 2): In all, there are twenty numbers, and each of them will occur once in twenty throws. So, if four people are playing, the game will end early. They will get a 20…, or a 19… or a 15 once in twenty throws, and those players will lead to bigger jumps.

Icosahedron Group (Student 4): There are twenty numbers so one of the players will get a 20 within twenty throws. The next will get some other number, then other, then other. Like this, I will have to wait for a long time.

Icosahedron Group (Student 4): What if we change the game. Can twenty people play a game?

Icosahedron Group (Student 2): [Clarifying] Umm…Yes, that's possible. We will then have to change the rules of the game as well.

Icosahedron Group (Student 4): If we need a number, which has just appeared, we will have to wait for a long time, maybe for twenty throws.

Similarly, the Octahedron Group rejected the tetrahedron with an argument that the latter would make the game boring because of the limited number range. In addition, their comments on the likeliness of sequential occurrence of numbers in an icosahedron, making the game time-consuming and frustrating (refer Sect. 11.5.1

for discussions of Group Two) corroborated their belief on the principle of cyclic exhaustiveness of the sample space.

11.5.4 Prior Conditioning

As a concluding exercise, the students were asked to compare their observations with the cubical polyhedron and comment on the properties that make it a popular die for playing games. It must be noted that, though a cube was also given to the students in the set of polyhedrons, none of the groups selected it, perhaps owing to its fewer faces. It is likely that since a cube has fewer faces as compared to the other available polyhedrons, the children were more attracted to these other shapes. Even after they had had enough empirical experience with other symmetrical objects, the participants tenaciously clung to their beliefs that a cube would not give equal and fair results and that skilled hands can easily manoeuvre it to get desired outcomes. It is also possible that their prior experiences of playing with only the cubical die may have created such prejudices and that children are conditioned to using a cube (Watson and Moritz 2003).

11.6 Summary

In this chapter, students' reasoning while selecting objects as generators of randomness has been discussed. Analyses of experimentation and interactions of two sample groups helped in understanding how children make associations with physical characteristics of the objects to form notions of randomness. While substantiating their choices, the students not only experimented with randomness but also verified certain characteristics that qualified a device for being a generator of randomness.

As a step towards considering how children perceive randomness, we examined their (a) idea of randomness embedded in the object and (b) connections between the generator of randomness and the idea of randomness. From the working of both sample groups, we could draw three major findings: First, as an overview, we found children relying on the physical configuration of the objects while perceiving randomness. Their logic was largely based on the physical judgement of the objects that governed their understanding of randomness. It was noted that the children related randomness to the size of the sample space, unpredictability, fairness, and possibilities of non-manoeuvrability. Properties related to the perfectness of the shapes, their physical symmetry and orientation, the number of faces, and the mechanisms of using the dice (throwing, spinning, and rolling) were key components in selecting a die. Homogeneity in a shape was deemed to ensure equal chances of all faces showing up, indicating the idea of fairness; and unpredictability was the easiest to claim as it was only associated with non-manoeuvrability of the object.

Second, the work identified children's perceptions related to the size of the sample spaces of the random generators. The children considered the size of the sample space of each polyhedron, and the ones with larger sample spaces were selected. In fact, the children also connected other notions, such as fairness and unpredictability with the sample space. A larger sample space was deemed to ensure greater randomness. Notably here, in the absence of instructions, the children relied on the classicist approach and enumerated all possible outcomes (Horvath and Lehrer 1998) followed by the deterministic proportionality frame of measuring probability (Konold 1989; Fischbein 1975).

Although we had started with an intention to capture the reasons behind students' erroneous beliefs against the fair cubical dice, what we got as a product was much more profound. Our third result draws on a belief that explains, to some extent, the basis behind children's inhibitions against the standard cubical dice. The notion of exhaustiveness of the sample space was found dominant in the discussions of the groups as children intuited that each outcome of their die has to occur, but it will not reappear unless all outcomes have shown up at least once. While playing games with the cubical dice, the children hold a belief that in a cubical dice a number would appear only once in a cycle of six throws and that their opponent recently rolling a six eliminates their chances of rolling it again. They assume that they will have to wait for another round to start. It is possible that since in gaming situations the target is on winning the game in the least amount of time, children experience temporal gaps that lead to such misleading conclusions. Since the idea is driven by the context of competition, they are bound to perceive the dice as being biased towards their opponent. Perhaps, children's prior experiences of playing games with only cubical dice create such prejudices and conditioning (Watson and Moritz 2003).

We would also like to comment on the nature of embodiments that we considered while discerning children's logic of their erroneous beliefs. Since the aim here was to build a constructive, exploratory domain in which it was possible to work on intuitions and beliefs rather than approach the formal ideas as a separate domain grafted onto an activity (Cobb et al. 1992), we chose embodiments that were structurally isomorphic to the models that already existed in the children's cognitive domain. The symmetric polyhedrons with which the children had worked earlier in another area of mathematics but not in probabilistic situations served as isomorphic resources for capturing children's logic as they approached randomness. The intention was to put learners in situations where they could express their thoughts freely without having been conditioned. We suggest using isomorphic resources, set in open-constructive spaces as didactical opportunities for identifying and understanding the roots to children's intuitive, unreasonable beliefs and misconceptions, especially in probability.

11.7 Conclusion

Any didactic activity of probability should treat the "pre-articulated informal, embodied notions as epistemically bona fide cognitive content bearing directly on the inferential reasoning" (Abrahamson 2012, p. 871). Without disapproving of children's informal prior understanding related randomness, we probed the rationales that they give while perceiving it. As we delved in an exercise of making sense of children's notions, we realised that many of their existing intuitive beliefs related to the runs of outcomes were indeed a result of their mental associations with the physical construct of the objects. Their beliefs were influenced by the characteristic of the shapes; in so doing they grounded their logic on the physical attributes of the shapes impacting the outcomes. Clearly, children's reasoning could not be termed embryonic as they offered correct mathematical insights gained from their experiences of working with the objects thereby making sense of randomness from the perceptual judgement of the objects.

On the onset, the study initiated a quest for understanding children's reasoning as they make sense of randomness. During the course we could make profound connections with the propensity interpretation of probability expounded in the early writings of Popper (1959). The propensity interpretation of probability questioned the statistical approach of measuring probability and interpreted probability as a property entrenched in the experimental conditions. By doing so, one can define probability of a single event from its set of generating conditions. Physically existing dispositional properties of an experimental setting interact and combine to form generating conditions and it is these conditions which when taken holistically define the probability of a situation (Popper 1959). While analysing the objects for randomness, our students expressed propensities or dispositions to the structural characteristics of the objects (homogeneity, symmetry, number of faces) and to the external forces (gravity, friction, landing on a stabilised position). Insofar, propensity towards the physical aspects proposes a realistic account to understand the reasons behind the metaphysical and anthropomorphic meanings that children attach to probability. Often their so called intuitive inferences are made solely on the dispositions generated by the experimental arrangements.

Studies in probability education have been capturing responses as testimonies of biases, misconceptions or incomplete probabilistic understanding. Abrahamson (2009) and Chernoff (2009), on the other hand, clarified that often such incorrect responses may not actually be devoid of a correct mathematical proposition. It is only when such responses are viewed from a normative, sacrosanct probabilistic lens they seem meaningless and wrong. Rather than viewing things from a normative lens, there is a need to open our vistas and understand inferences from a different, rather prismatic lens. Our students' reasoning were not incongruent of a mathematically correct judgement and we could find several illustrations where they raised correct arguments. When we attended their responses from another perspective, as dispositions to the experimental arrangements, we could better explain the genesis of their inferences. The pre-empted incorrect responses when interpreted from an alternative perspective offer a prolific scope to unveil the meanings behind such conclusions.

11.8 Looking Ahead

Unlike other researches wherein the study of children's conception of randomness is based on the examination of the outcomes that the subject produces or in comparing the likelihood of different random sequences (Batanero 2016), this study correlates randomness to dispositions to physical properties of the object. The results of our research are based on a small set of students of specific age group experimenting with limited resources. Although the results add a dimension of acknowledging and interpreting children's tacit notions and perceptions through resources that were not too distant from their cognitive domain, replication of the findings with a broader set of isomorphic resources and with a larger group of children of various mathematical maturity levels can be the focus for further studies. As a follow-up, further research can be undertaken to study varied strategies that children adopt to make connections between random objects and other probabilistic judgements. A comparison of the relative frequencies of different symmetrical objects can also be undertaken to help get a more coherent view on how children draw parallels with various generators of randomness.

We also noticed in our research study that children only valued the faces of the solid as outcomes. Experimenting with other attributes of the dice, instead of faces, would be interesting. Similarly, many more connections can be made with historical resources to see children's conceptions of randomness. For instance, how ancient people used non-symmetrical objects such as an astragalus for taking probabilistic decisions. Arcavi et al. (1982) see potential in going back to history as it helps to re-examine and revisit the known and taken-for-granted mathematical ideas, which might serve as 'sources of insight' and enable us to unpack and unclog the 'automated processes.' Since many of the probabilistic constructs are subjected to counter-intuitions and irrational thoughts, engagement with such tools might enable us to dispel these beliefs.

Intuitions are often guided by varied interconnected experiences. There still remains a wide scope to understand and recognise how children reconcile their tacit knowledge and make sense of disciplinary knowledge. It is imperative to devise frameworks that can analyse intuitive inferences beyond the normative, sacrosanct lens and take into account the spectrum of experiences that influence the inferences. In probability, we need to design didactical resources that recognise children's prevailing ideas and intuitions, provide scope for interpreting their inferences, and make ways for formal analysis. By engaging with isomorphic resources, children internalise and construct mental representations that guide their intuitions and have the potential to challenge those notions which we assume to be given or known. The task of exposing the learners to the unconventional resources as a part of their everyday pedagogy remains a challenge.

References

Abrahamson, D. (2009). Orchestrating semiotic leaps from tacit to cultural quantitative reasoning—The case of anticipating experimental outcomes of a quasi-binomial random generator. *Cognition and Instruction, 27*(3), 175–224. https://doi.org/10.1080/07370000903014261.

Abrahamson, D. (2012). Seeing chance: Perceptual reasoning as an epistemic resource for grounding compound event spaces. *ZDM Mathematics Education, 44*(7), 869–881. https://doi.org/10.1007/s11858-012-0454-6.

Amir, G. S., & Williams, J. S. (1999). Cultural influences on children's probabilistic thinking. *Journal of Mathematical Behaviour, 18*(1), 85–107.

Arcavi, A., Bruckheimer, M., & Ben-Zvi, R. (1982). Maybe a mathematics teacher can profit from the study of history of mathematics. *For the Learning of Mathematics, 3*(1), 30–37.

Bar-Hillel, M., & Wagenaar, W. A. (1991). The perception of randomness. *Advances in Applied Mathematics, 12*(4), 428–454. https://doi.org/10.1016/0196-8858(91)90029-I.

Batanero, C. (2016). Understanding randomness: Challenges for research and teaching. In K. Krainer & N. Vondrová (Eds.). *Proceedings of the Ninth Congress of the European Society for Research in Mathematics Education* (pp. 34–49). Prague: European Society for Research in Mathematics Education.

Batanero, C., & Serrano, L. (1999). The meaning of randomness for secondary students. *Journal for Research in Mathematics Education, 30*(5), 558–567.

Bennett, D. J. (1998). *Randomness*. Cambridge: Harvard University Press.

Benson, C. T., & Jones, G. A. (1999). Assessing students' thinking in modelling probability contexts. *The Mathematics Educator, 4*(2), 1–21.

Cobb, P., Yackel, E., & Wood, T. (1992). A constructivist alternative to the representational view of mind in mathematics education. *Journal for Research in Mathematics Education, 23*(1), 2–33.

Chernoff, E. J. (2009). Sample space partitions: An investigative lens. *The Journal of Mathematical Behavior, 28*(1), 19–29. https://doi.org/10.1016/j.jmathb.2009.03.002.

du Sautoy, M. (2008). *Symmetry: A journey into the patterns of nature*. New York: Harper Collins.

Eichler, A., & Vogel, M. (2012). Basic modelling of uncertainty: Young students' mental models. *ZDM Mathematics Education, 44*(7), 841–854. https://doi.org/10.1007/s11858-012-0451-9.

Fischbein, E. (1975). *The intuitive sources of probabilistic thinking in children*. Dordrecht, The Netherlands: Reidel.

Gandhi, H. (2015). Traversing the epistemology of probability in Indian mathematics textbooks. In S. Chandrasekharan, S. Murthy, G. Banerjee, & A. Muralidhar (Eds.), *Proceedings of Sixth International Conference to review researches in Science, Technology and Mathematics Education* (pp. 195–202). Mumbai, India: Homi Bhabha Centre for Science Education.

Green, D. R. (1983). A survey of probability concepts in 3000 pupils aged 11–16 years. In D. R. Grey, P. Holmes, V. Barnett, & G. M. Constable (Eds.), *Proceedings of the First International Conference on Teaching Statistics* (Vol. 2, pp. 766–783). Sheffield, UK: Teaching Statistics Trust.

Horvath, J. K., & Lehrer, R. (1998). A model-based perspective on the development of children's understanding of chance and uncertainty. In S. P. Lajore (Ed.), *Reflections in statistics: Learning, teaching, and assessment in Grades K–12* (pp. 121–148). Marwah, NJ: Erlbaum.

Hycner, R. H. (1999). Some guidelines for the phenomenological analysis of interview data. In A. Bryman & R. G. Burgess (Eds.), *Qualitative research* (Vol. III, pp. 143–164). New Delhi, India: Sage.

Jones, G. A., Langrall, C. W., Thornton, C. A., & Mogill, A. T. (1999). Students' probabilistic thinking in instruction. *Journal for Research in Mathematics Education, 30*(5), 487–519.

Kahneman, D., & Tversky, A. (1972). Subjective probability: A judgement of representativeness. *Cognitive Psychology, 3*(3), 430–454.

Konold, C. (1989). Informal conceptions of probability. *Cognition and Instruction, 6*(1), 59–98. https://doi.org/10.1207/s1532690xci0601_3.

Langrall, C. W., & Mooney, E. S. (2010). Characteristics of elementary school students' probabilistic reasoning. In G. A. Jones (Ed.), *Exploring probability in school: Challenges for teaching and learning* (pp. 95–120). New York, NY: Springer.

Lecoutre, M. (1992). Cognitive models and problem spaces in "purely random" situations. *Educational Studies in Mathematics, 23*(6), 557–568. https://doi.org/10.1007/BF00540060.

Metz, K. E. (1998). Emergent understanding and attribution of randomness: Comparative analysis of reasoning of primary grade children and undergraduates. *Cognition and Instruction, 16*(3), 285–365. https://doi.org/10.1207/s1532690xci1603_3.

NCERT, National Council of Educational Research and Training. (2005). *Mathematics syllabus for classes at the middle Level, VI–VIII*. New Delhi, India: NCERT.

Nilsson, P. (2004). Students' ways of interpreting aspects of chance embedded in dice game. In M. Johnsen & A. Berit (Eds.), *Proceedings of the 28th conference of the International Group for the Psychology of Mathematics Education* (Vol. 3, pp. 425–432). Bergen, Norway.

Nilsson, P. (2007). Different ways in which students handle chance encounters in the explorative setting of a dice game. *Educational Studies in Mathematics, 66*(3), 293–315.

Piaget, J., & Inhelder, B. (1975). *The origin of chance in children*. New York: Norton (Original work published in 1951).

Popper, K. E. (1959). The propensity interpretation of probability. *The British Society for Philosophy of Science, 10*(37), 25–42.

Pratt, D. (1998). The coordination of meanings for randomness. *For the Learning of Mathematics, 18*(3), 2–11.

Pratt, D. (2000). Making sense of the total of two dice. *Journal for Research in Mathematics Education, 31*(5), 602–625.

Pratt, D., & Noss, R. (2002). The micro-evolution of mathematical knowledge: The case of randomness. *Journal of the Learning Sciences, 11*(4), 453–488.

Rock, P. (1999). Participant observation. In A. Bryman & R. G. Burgess (Eds.), *Qualitative research* (Vol. II, pp. 143–164). New Delhi, India: Sage.

Truran, K. (1995). Animism: A view of probability behaviour. In B. Atweh & S. Flavel (Eds.), *Proceedings of the 18th Conference of the Mathematics Education Research Group of Australasia* (pp. 623–630). Lismore, Australia: MERGA.

Vidakovic, D., Berenson, S., & Brandsma, J. (1998). Children's intuition of probabilistic concepts emerging from fair play. In L. Pereira-Mendoza, L. S. Kea, T. W. Kee, & W. Wong (Eds.), *Proceedings of the Fifth International Conference on Teaching Statistics* (Vol. 1, pp. 67–73). Voorburg: International Statistical Institute.

Watson, J. D., & Moritz, J. B. (2003). Fairness of dice: A longitudinal study of students' beliefs and strategies for making judgments. *Journal for Research in Mathematics Education, 34*(4), 270–304.

Chapter 12
Reasoning in Decision Making Under Uncertainty and Decisions of Risk in a Game of Chance

Ana Serradó Bayés

Abstract We present a design-based research study that aimed to enhance the stochastic reasoning of 48 Spanish students (Grade 7, age 12) when making decisions in situations involving uncertainty and risk during a game of chance task called Integer Addition Bingo. Grounded in a review of the literature, the task was designed with the aim of improving students' learning and reasoning about risk management through a process of understanding the random nature of the game, and operationalizing risk management. A qualitative retrospective analysis was performed to identify the structural static aspects of students' reasoning. Then, a retrospective analysis of the functional dynamic aspects of students' reasoning led to the identification of six different learning trajectories.

Keywords Probability · Risk · Mental models · Design-based research methodology · Lower secondary students

12.1 Introduction

Although it has traditionally been investigated in psychological studies, people's perception and quantification of risk is becoming a major topic of research in mathematics education. Studies in the field of psychology describe types of statistical innumeracy and the errors individuals tend to make when reasoning in situations involving risk, and emphasise the importance of stochastic numeracy for informed and accurate decision making in such situations (e.g. Gigerenzer 2002). On the contrary, the duality of innumeracy versus numeracy has been the traditional focus of research in probability education. For example, the dual-process theories of Kahneman and Frederick (2002) distinguish between the faster, intuitive dimension of decision making and reasoning, and the slower, more reflective, analytic type of

A. Serradó Bayés (✉)
Colegio La Salle-Buen Consejo, San Ignacio, 20, Puerta 4,
11510 Puerto Real, Cádiz, Spain
e-mail: ana.serrado@gm.uca.es

C. Batanero and E. J. Chernoff (eds.), *Teaching and Learning Stochastics*,
ICME-13 Monographs, https://doi.org/10.1007/978-3-319-72871-1_12

reasoning. This dual process only considers the person as an individual, excluding societal influences from his or her reasoning about risk. Borovnich and Kapadia (2011) consider both the individual and societal dimensions of the decision-making process and distinguish between decisions that are based on personal preference and decisions that are rationally bounded.

Influences of these dualities can be observed in some national curricular guidelines, such as those in Canada or in the United States. These guidelines recommend offering students opportunities to explore various situations involving risk and risk management, which may lead them to become more confident risk-takers. Meanwhile, other education systems, such as those of Germany, the United Kingdom, and Spain, are looking for ways to embed risk assessment, risk management, and decision making in situations involving risk into the K–12 school system (Russel 2015).

In Spain, where the present study took place, there is no reference to the concept of risk in the national mathematics curriculum; however, decision making is considered to be an important skill that should be developed during secondary education. This skill is represented in three grade bands: in middle school, it is recommended that Grade 7 and 8 students (ages 12–13) reflect on the decisions they have made and learn from them for similar future situations; in Grades 9 and 10 (ages 14–15), students are expected to learn to make rational decisions in situations involving uncertainty by considering the probabilities of various events; and in high school (Grades 11 and 12, ages 16–17), students are expected to compute probabilities and to use the results to make decisions in contexts related to the social sciences (Ministerio de Educación, Cultura y Deporte [MECD] 2015).

With regard to the relevant concepts that should be taught, Till (2014) suggests the concepts of proportions, conditional probabilities, and expected values as a basis for risk literacy in primary education (Grades 1–6, ages 6–12). Meanwhile, in high school (Grades 11 and 12, ages 17–18), Eichler and Vogel (2015) suggest that students should learn about conditional probability, Bayes' theorem, independence, absolute and relative risk, and distributions of probabilities (frequencies). However, there are no research recommendations **for** introducing risk in the middle school curriculum (Grades 7–10, ages 12–16). In order to consider the possibilities for integrating the topic of risk in middle school, we have designed, implemented, and retrospectively analysed a game of chance task called *Integer Addition Bingo* [IAB task]. This task wants to enhance students' reasoning about risk management through a systematic decision-making process called *operational risk management* [ORM]. In this chapter, we report on this experience with the aim of analysing:

a. How the game of chance task helped students to progress in their mental reasoning models about stochastic concepts.
b. The learning trajectories that emerged through implementation of the task in question.

12.2 Conceptualizing Operational Risk Management

When Blaise Pascal and Pierre de Fermat wrote about the gambling problem in 1654, they caused a paradigm shift in how humans perceived their world (Gigerenzer 1989). In particular, Pascal and Fermat expanded people's interest in quantifying the probability of future events so as to discern the difference between risk and uncertainty, a distinction that has been extensively analysed by Knight (1921). Knight (1921) argued that the difference lies in the predictive character of probability. According to Knight, if the underlying distribution of probabilities in a given situation is not known, any decisions regarding this situation can be said to be made under *uncertainty*; if it is known, the situation is said to involve *risk*. Gigerenzer (2002) also argues, from the perspective of psychology, that in situations in which a lack of empirical evidence makes it impossible or undesirable to assign numbers to the possible alternative outcomes, the term "uncertainty" should be used instead of "risk." The possibility of translating uncertainties into risks depends on having enough knowledge of the object or situation in question so to be able to assign or estimate probabilities to the possible outcomes.

Let us consider an object traditionally used in games of chance: a die. Etymologically, the word die comes from the Arabic word *al-zahr*, related to the word *khatar*, which means "to bet, or gamble." Around 1100 B.C.E., the word appeared as a root in the Spanish word *azar,* which then denoted a game of chance. The Spanish word azar would go on to evolve into the French word *hasard* and into the English word *hazard*, which has the same meaning as the original Spanish word.

The primary application of probability in games involving dice was determining the probability of loss. As a result, the definition of hazard was identified with the ability to explain why someone would take a given risk and with the harms that may be associated with this risk. Although the process of losing in a game of chance has multiple facets, including emotional, physical, psychologic, intellectual, and spiritual dimensions (Cline 2003), in this paper, we use the words *hazard* and *hazardous* in reference to the emotions involved in losing a game of chance.

In agreement with the aforementioned perspectives on the notion of hazard, Borovcnik (2015) states that "risk management is then any measurement to eliminate or control hazards at an acceptable level" (p. 5). According to this author, the concepts of risk and hazard are linked via probabilities or chance to the adverse events. If the consequences are immediate, this link between risk and hazard (as described in Borovcnik 2015) is obvious. In contrast, when the effect is delayed, the link is difficult to establish and it is difficult to provide evidence for this link. Although some objections have been raised to the subjective view for epistemologically distinguishing between risk and hazard (Borovcnik 2015), this distinction can be considered to be a starting point in establishing the connection between situations involving hazard and risk.

To sum up, historically, two difficulties emerged in the conceptualisation of risk. On the one hand is the documented difficulty of establishing connections between hazards and risks. On the other hand, as shown by Knight (1921), is the complexity of discerning between situations of uncertainty and situations of risk.

From a learning point of view, this difficulty can be viewed as a discontinuity in the transition from process to object (Gueudet et al. 2016). We suggest that discerning between situations of uncertainty and situations of risks involves a change in the students' mental structure. Following Gueudet et al. (2016), we conceive this conceptual change as a discontinuity in the transition from only being able to quantify uncertainty or risk to conceptually distinguishing situations of uncertainty from situations of risk. Consequently, we are interested in the discontinuity that may emerge in the transitions between primary education and high school in the process of quantifying the probability of events that arise when conceptualizing the risk of loss.

In order to facilitate this transition, teaching should attend simultaneously to developing the concepts of probability and developing strategies for decision-making and risk management. From a curricular point of view, we consider the possibility of decomposing the process of risk management into the components of probabilistic knowledge used by the students. This decomposition could help to analyse the possible changes when discerning between situations of uncertainty and situations of risk when making informed decisions in probabilistic contexts.

We have called this process systematic *operational risk management* [ORM]. We define operational risk management as a decision-making process of systematically evaluating possible courses of action, identifying risk and its benefits, and determining the best course of action for any given situation.

12.3 Conceptualizing Mental Models in an Operational Risk Management Process

The aim of the research described herein is to use the theory of mental models to study students' reasoning based on the personally-preferred and rationally-bounded decisions made when operationalizing risk management in a game of chance task. In agreement with Eichler and Vogel (2012), it is assumed that students cannot be ascribed to general cognitive levels, but that they may show situational cognitive levels—that is, in the context of a specific task.

12.3.1 Theoretical Framework of Mental Models

The *theory of mental models* takes into consideration: (a) the structural static aspect of thinking, which involves recognising the physical objects in a given situation and their characteristics (Johnson-Laird 1983); (b) the functional dynamic aspect of thinking, which refers to the construction of a mental model when coping with demands of a specific situation in order to simulate relevant aspects of the situation, such that these aspects come to be cognitively mastered (Seel 2001); (c) that the mental model may be changed, enriched, or modified while a student is working on representations of a situation (Schnotz and Bannert 1999); and (d) that mental models may show interpersonal differences in their structural and functional aspects (Schnotz 1994).

Traditionally, researchers have used the SOLO (Structure of Observed Learning Outcome) model of Biggs and Collins (1982) when coding students' responses as they reason about stochastic concepts. The SOLO model includes five levels of reasoning quality represented in students' responses: (a) *pre-structural response*, where there is no logical relation between the question and the response, or there is an attempt to link the question to the response by an irrelevant feature; (b) *uni-structural response*, where one relevant datum or concept related to the question is mentioned; (c) *multi-structural response*, where two or more relevant concepts or data related to the situation are mentioned, but the relevant features are not connected; (d) *relational response*, where two or more relevant concepts or data related to the situation are related through an overall concept or principle that accounts for the various isolated data in the multi-structural response, but the response is bounded by the data and concepts already taught; and (e) *extended abstract response*, where all of the relevant data and their interrelations are taken up and subsumed under a hypothetical abstract structure, which was not included in the original problem statement.

12.3.2 Literature on Stochastic Mental Models

In the literature, there have been different attempts to characterize students' reasoning about stochastic processes. Eichler and Vogel (2012) studied young students' mental modelling within simple situations of decision making under uncertainty, before the topic of stochastics had been formally addressed in school. They concluded that their work can serve as a basis to investigate the evolution of students' mental modelling in formal schooling. This is the aim of this chapter.

Sanchez and Orta (2015) explored middle school students' reasoning about variation when comparing two data sets and interpreting the dispersion (variation) in the data in terms of risk. Furthermore, Sánchez et al. (2016) investigated students' responses to two binomial tasks, one of prediction and another of distribution, to know how they express variability. The authors used SOLO model (Biggs and Collins 1982) to categorize students' answers in theoretical dogmatism and empirical commitment.

In the next section, we describe the methodology of the study in question.

12.4 Methodology

A design-based research methodology was adopted for the present study where two teachers, one of whom was also a researcher and the author of this chapter, undertook the process of introducing decision making and risk into the middle school curriculum with the aim enhancing both teachers' and students' understanding of decision-making processes in situations of uncertainty and situations of risk. We consider the three main features of this methodology: instructional design and planning, ongoing analysis of classroom events, and retrospective analysis (Cobb 2000). Cycles of interaction in the study involved the teacher and researcher in collaborative planning, implementing, evaluating, and revising a sequence of seven sessions of the game of chance task called Integer Addition Bingo.

12.4.1 Design of Integers Addition Bingo Task

Integer Addition Bingo [IAB] is a game of chance based on the game of bingo. In bingo, each player has a card arrangement filled with different numbers. Numbers are called out randomly, and players mark the numbers on their cards if they are called. Players compete against one another to be the first to have a winning arrangement.

The IAB cards have ten numbers from −10 to +10. Unlike in traditional bingo, the numbers are not called out. Rather, students watch as an applet (constructed using the Geogebra software) randomly generates two numbers from −5 to +5. Students mentally add the two numbers, and if the number result of the addition is on their card, they mark the result on their card. The first student to have marked all of the numbers on their card wins the game.

A task that stretched out over seven sessions was designed to help students develop an appreciation of the risk involved when playing IAB and to investigate how the cards should be constructed so as to minimize the risk of losing. With the previous theoretical analysis as a basis, the task aimed:

a. To involve students in systematic operational risk management [ORM] and to introduce them to decision-making processes. We expected the students to overcome the fuzzy distinction between risk management and hazard, as well as the distinction between decisions under uncertainty and decisions of risk.
b. To trigger an evolution in students' reasoning from decisions based on personal preference to decisions that are rationally bound; and
c. To further students' understanding of the concepts of proportion, distributions of probability and of frequencies, and decisions under uncertainty and of risk.

Students were involved in a sequential process of playing IAB, doing mathematics, and reasoning in three differentiated phases. The first phase aimed to familiarize students with the game, and to trigger subjective intuitions about randomness and possibility of appearance. Students first played with the same card of randomly-generated and ordered numbers (card A1: −7, −5, −2, +0, +1, +1, +3, +5, +9, +10). Next, students were asked to select between two cards (card B1: −7, −5, −2, +0, +1, +1, +3, +5, +9, +10 and card B2: −9, −6, −3, +0, +0, +0, +2, +2, +7, +10) and to explain their choice. Finally, each member of a team was given a different card to play with (card C1: +0, +0, +0, +0, +0, +0, +0, +0, +0, +0; card C2: −10, −9, −8, −7, −6, −5, −4, −3, −2, −1; card C3: −10, −8, −6, −4, −2, +2, +2, +4, +6, +8, +10; and card C4: 9, −8, −5, −5, +1, +3, +4, +10, +10, +10). These cards were designed with the aim of confronting students with the difference between events and sequences of events, and with the idea that some events are more likely to occur.

During the second phase, the task was designed to engage students in systematic operational risk management through: (a) constructing a first card, playing the game, analysing the distribution of frequencies, and reasoning about the criteria used to create the card and why they may have won or lost; (b) designing a second card, playing the game, determining the theoretical probabilities corresponding to the sums of two integers between −5 and 5, and graphically analysing the properties of the theoretical distribution; (c) constructing a third card, playing the game, comparing the distribution of frequencies and probabilities, and reasoning about the criteria used to create the card and why they may have won or lost; and (d) constructing the fourth and final card and playing the game.

In the last phase, students were asked to reflect on the decision-making process they had developed. Two questions were posed to students: "How would you construct the card to minimize your risk of losing?" and "What have you learned?" The latter question was designed with the aim of analysing the utility of the task from the perspective of the students and their perceptions about their own learning trajectories (Clements and Sarama 2004).

Figure 12.1 summarizes the task, describing the three phases and the consecutive actions of playing (P), comparing and constructing cards, analysing and comparing the distribution of frequencies and probabilities, and reasoning about the decision-making process carried out in order to determine how the risk of losing can be minimized.

12.4.2 Implementation and Retrospective Analysis of the Integers Addition Bingo Task

The task was implemented with 48 Grade 7 students (ages 12–14) in a Spanish middle school located in a low socio-economic coastal city. The students were divided into two groups, and each classroom group was also subdivided to create heterogeneous groups so as to promote deliberative dialogue in a collaborative learning environment (Serradó et al. 2015). Individual and collective research data was collected from audio and video recordings of classroom discussions, individual

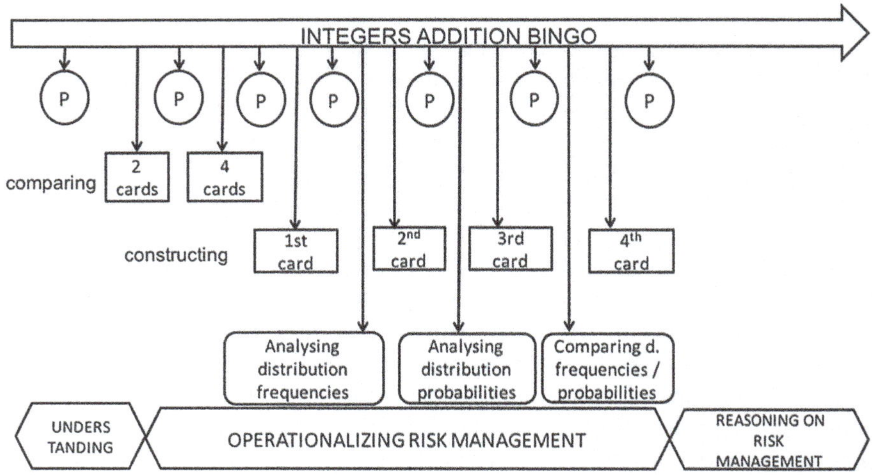

Fig. 12.1 Design of the "Integer Addition Bingo" task

and small-group work samples, and individual interviews (n = 5). The cooperative teams were coded based on the classroom (A or B), the particular team (numbered 1–6), and the team members (numbered 1–4). Although not all of the students attended the seven sessions, this chapter reports the results of the retrospective analysis of the work of the students who attended the successive phases. In particular, we report on the results from the 46 students who were present during the first phase of the task (understanding the game), and describe the results of the twelve teams who participated the second phase (operationalizing risk management). Finally, in the analysis of the third phase (reasoning about risk management), we report on the results of the 41 students who attended the seven sessions.

The iterative nature of design-based research has allowed for repeated opportunities to discuss and reflect on students' responses in each session of the task and to refine the approaches taken in each subsequent activity to develop students' reasoning and decision-making skills in stochastic situations. The research literature presented in the previous sections was used to design the game of chance task and to plan questions that prompted deliberate dialogue and sponsored the development of individual reasoning.

First, we present the results of a *qualitative static retrospective analysis* of the IAB task regarding the students' mental levels of reasoning when they are engaged in understanding the game, reasoning about the ORM process, and reasoning about risk management. This analysis focuses on the structural static aspect of thinking, which involves recognising the physical objects of the situation (Johnson-Laird 1983).

Subsequently, we analysed the students' opportunities and constrains for changing, enriching or modifying their reasoning when coping with a specific situation in which they are asked to make a decision. These changes provided information about the functional dynamic aspect of students' mental models. We refer to this analysis as *qualitative dynamic retrospective analysis*. These changes also allowed us to identify individual differences in relation to the hypothesized learning trajectory (Clements and Sarama 2004). Possible learning trajectories that emerged while students were engaged in the IAB task are discussed. Finally, the results concerning the distinction between decisions under uncertainty and decisions of risk are presented.

12.5 Decisions When Understanding the Game

In this section, the qualitative static retrospective analysis and the qualitative dynamic retrospective analysis of the students' reasoning while they were engaged in learning the game are presented. The analysis distinguishes between personal preference and rationally-bounded decisions.

12.5.1 Playing with One Card

During the first phase of understanding the game, students were engaged in playing IAB with Card A and were asked to individually reason whether the same numbers would come out in the next round. The aim of playing with one card was to unearth and develop students' subjective intuitions about the uncertainty inherent in the game and the random process underlying the game. A structural static aspect of the mental model that was expected to develop (Johnson-Laird 1983) was the characterization of the game as random. Nevertheless, nineteen students (out of 48) were unable to construct a random model of the situation because they considered the game to be deterministic. They reasoned that the next play would be determined by the numbers on the card (e.g., student B61: "Yes, because they are the same cards") or by the events that had already occurred (e.g., student B31: "Because if you know the numbers obtained, then they are going to be the same"). The rest of the students (29/48) referred to the uncertainty of the appearance of the numbers in the next IAB play. Differences in reasoning could be observed between the students. Some students reasoned about the appearance of a particular event (e.g., student A41: "No, because it cannot come out as the same number"), or the sequence resulting from a game (e.g., student B53: "No, because the numbers can be different").

Thus, a functional dynamic aspect (Seel 2001) of this random model was perceived in the reasoning of some students who need to cope with the demand of understanding the random events andrandom sequences of events that were generated by the IAB. In response to the question of whether the same numbers would occur in the next round, some students argued about the length of the game (student A34: "No, because it has been a long game play") or appearance of the events (student B13: "No, because the numbers are not ordered"). Other students, instead of analysing the particular events, considered the sequence of events.

A micro-analysis of the answers, as suggested by Gueudet et al. (2016), helped to distinguish between students who only recognised differences between sequences (student A62: "No, it has no relation, it can be others") and students who predicted the difference between sequences using the word "possible" (student B43: "Yes, it is possible, but it is a quite remote possibility") or the word "probability" (student A22: "it can be yes or no, it is a probability").

In both cases, students' reasoning can be categorized using dual-process theories (Kahneman and Frederick 2002). We categorize as fast the reasoning of those students who were unable to abstract the random nature of the IAB. In contrast, we consider that the reasoning of some students was slower because they contextually argued about the luck of the situation (student B54: "No, because that [the numbers] came out by chance"), the chance inherent in the game (e.g., student B32: "No, because that it is chance. [They are] not always the same"), and the randomness of the numbers generated (e.g., student B21: "No, it is random").

12.5.2 Decisions Made When Selecting Between Two Cards

This phase of understanding the game provided information about students' reasoning when choosing between two cards. Two possible models were analysed: decisions based on personal preference and rationally-bounded decisions (Borovnich and Kapadia 2011). As their justifications for their choice of card suggested, eight of the twelve teams chose a card based on personal preferences about the numbers on the cards (team B4: "The [numbers] that you prefer. The numbers"), whether the numbers are easy to count, their desire to win, or the perceived length of the game (e.g., team A3: "The first [card B1] because is quicker [to end]" and team B4: "The second [card B2] because you finish quickly"). These teams did not recognise the uncertainty of the game when making decisions and did not link their choice to the events or sequence of the events that occurred during the game. In contrast, two teams recognised the uncertainty of the game and, as a consequence, rationally bounded their decisions (team B6: "It would be the same for us because it is a game of chance"; team B1: "It would be the same for us, because we think that it is random").

Although the decisions made by two of the teams were rationally bounded, these were constricted by reasoning based exclusively on one card and analysing the number of zeros or repeated numbers (in particular, team A5 referred to the repetition of the number +1, and team B2 referred to the repetition of the numbers 0 and +2). We consider that these teams were able to analyse the individual events (data), but were unable to capture the meaning of the data distribution. Bakker and Gravemeijer (2004) describe this difficulty, arguing that novices in statistics tend to see individual values before understanding the distribution as a conceptual entity.

12.5.3 Decisions Made When Selecting Between Four Cards

Finally, in this first phase of understanding the IAB game, students were asked to play and select one of four cards (cards: C1, C2, C3, C4). The qualitative static retrospective analysis led us to conclude that some students did not transition from personally-preferred to rationally-bounded decisions (Borovnich and Kapadia 2011).

Teams A4, A5, A6, and B4 still justified their choices by referring to personal preference (e.g., ease of counting). Team A2 rationally bounded their reasoning by analysing the appearance of the number zero. Thus, the emotional factors involved in hazard (Cline 2003) restricted the vision of some students in the game of chance, who were consequently unable rationally bound their decisions.

Nevertheless, the second qualitative retrospective analysis provided a basis for understanding why the reasoning of some students evolved. Teams A1, A3, B3 and B5, who initially reasoned by expressing their personal preferences about, for example, the length of the game, later debated the pertinence of repeated numbers (team B5: "We would choose the card that has all 0 [card C1], because it is the card in which there is more repetition of numbers"). Meanwhile, the students of team B1 initially rationally bounded their choice of card based on the repetition of the numbers, and then informally predicted the possibilities of winning ("The card with more repeated numbers, because it has more possibilities of winning" team B1). Playing with four cards, then, provided students with information that allowed them to modify their reasoning (Seel 2001) in the construction of the random model, enriching the students' mental model (Schnotz and Bannert 1999) with the need of understanding the distribution of frequencies.

Summarizing this first phase, we consider that while understanding the game and playing with the first card, around 30% (14 of 46) of the students reasoned from a deterministic point of view. The rest of the students (70%, 32 of 46) reasoned by taking into account the uncertainty inherent in the events or sequences. Two of the students were able to anticipate and reason predictively about the possibility and probability of the same sequence occurring. The use of the phrase "it has more possibilities" can be interpreted as an attempt to qualitatively measure the uncertainty of the game. In accordance with the psychological studies of Gigerenzer (2002), this qualitative measure could be considered to be a step forward in the progression from statistical innumeracy to statistical numeracy. At this stage of the task, among these 32 students, only one of every six students were able to contextually reason about the role of luck, chance, or randomness in the generation of numbers during IAB.

When reasoning about selecting between two cards, many students' mental models were classified as pre-structural, according the SOLO model (Biggs and Collins 1982). In accordance with the findings of Borovnich and Kapadia (2011), we were again able to distinguish between students who made decisions based on personal preference and those whose decisions were rationally bounded. The teams whose responses were based on personal preferences did not link their choice to the events or sequence of events that occurred during the game. Meanwhile, the teams that made rationally-bounded choices based on an analysis of the number of zeros or repeated numbers were able to evolve their reasoning to the uni-structural level. Considering the theoretical framework, which establishes possible difficulties in conceptualizing risk management due its overlap with the notion of hazard (Borovcnik 2015), we conclude that those teams who contextualised their rationally-bounded reasoning via their feelings about the hazard of losing the game had difficulties in recognising that the numbers were randomly generated.

A micro-analysis of the transfer of knowledge when playing and reasoning with one and two cards could help to analyse the continuity or discontinuity (Gueudet et al. 2016) in the transition between the concept of event and that of sequence of events. This micro-analysis could be performed through the improvement of the task design, which may involve: (a) engaging students in a dialogue about the uncertainty of the randomly generated numbers; (b) reasoning about the difference between the numbers randomly generated by the IAB random generator and the random events obtained by the addition of the randomly-generated numbers; and (c) facilitating students' ability to discriminate between events and sequences of events. In accordance to Kahneman and Frederick's (2002) dual-process theories, these questions may help students abandon faster, intuitive decision making based on personal preference in favour of a slower, more reflective, analytic type of reasoning.

12.6 Decisions When Operationalizing Risk Management

The second phase of the task was designed to engage students in systematic operational risk management through the construction of four cards, reasoning about the criteria used to create the cards, playing the game, analysing the results (distribution of frequencies, distribution of theoretical probabilities, and comparing these distributions), and reasoning about how the winning card was constructed.

12.6.1 Decisions Made When Constructing the First Card

The qualitative static retrospective analysis informs us of subtle differences in student reasoning about how to construct cards that would be likely to win the game. Five of the teams (A1, A4, A5, A6, and B3) constructed their card by recalling which events were obtained with higher frequency (e.g., team A6: "Write the numbers that have appeared more"). Then, qualitative dynamic retrospective analysis revealed that analysing the graphical distribution of frequencies had provided students with a normative language (e.g., team B3: "Because we have looked at the numbers with higher frequency"). As suggested by Bakker and Gravemeijer (2004), it appears that some teams analysed the frequency and modal clumps of the sample data in order to make inferences about the density of the distribution. It is quite interesting that team B5 went beyond this analysis and looked at the relationship between positive and negative events (team B5: "Because there is the same amount of positive and negative [numbers]").

The qualitative static retrospective analysis of the reasoning of teams A2, A3, and B2 showed that these students predicted the relative likelihood of the event 0 (e.g. "the number zero is more likely to come out" team A2). Meanwhile, the qualitative dynamic retrospective analysis revealed that these teams were interested in understanding the variability of the individual events (e.g., "[we got] numbers with high variability, although these are not numbers with high [absolute] value" team B2).

Teams B1 and B4, when initially constructing their card, reasoned about the sequence of events and analysed the distribution of frequencies, which helped them to understand the random model involved (e.g., team B4: "Because the numbers have to come out randomly"). Finally, no functional dynamic evolution (Seel 2001) was observed in the reasoning of team B6, because the students' answers were still restricted to their understanding of the randomness involved in the game.

To sum up, using the SOLO model (Biggs and Collins 1982), we were able to identify interpersonal differences concerning students' structural and functional aspects of thinking (Schnotz 1994). Team B6 was still categorized as reasoning at the pre-structural level, because there was no evolution in their conception about the randomness of the game and because their reasoning was based on personal preferences. The reasoning of ten of the teams was again categorized as uni-structural because they looked exclusively at either the density, skewness, or spread of the distribution, or the variability of the data. The reasoning of two teams was clustered as multi-structural. In particular, the reasoning of team B2 was rationally bounded in reasoning about the density of the distribution and the variability of the individual events, and the reasoning of team B5 was bounded by reasoning about the density and the symmetry of the distribution.

12.6.2 Decisions Made When Constructing the Second Card

Students were then asked to construct a second card, explain their reasoning, analyse the distribution of frequencies, and justify why the winning card may have won. The qualitative static retrospective analysis classified the reasoning of ten of the teams as uni-structural. Analysing the structural static aspect of the students' reasoning (Johnson-Laird 1983), we observed that they were able to recognise the skewness of the distribution before and after the game. Analysis of the functional dynamic aspect of students reasoning (Seel 2001) provided us with information about the differences between teams in relation to the distribution of frequencies that they deemed as relevant in deciding how to construct a card more likely to win, such as: the previous winning card (team A5: "observe the previous [winning] card and change some numbers"), the table recording the absolute frequency of each simple event (team A3: “using the previous table, we have chosen those [numbers] that are repeated more often”), or the previous games (team A5: “We chose those [numbers] that appear more often [in the games]”). Although the members of team B2 enriched their reasoning through the analysis of the density of distribution and its variability, their reasoning was still categorized as multi-structural, as their decisions could have been enriched by taking variability into account (Sanchez and Orta 2015).

12.6.3 Decisions Made When Constructing the Third Card

Considering the importance of looking for ways to embed risk management and decision making into the K–12 curriculum (Russel 2015), students were asked to compute theoretical probabilities and analyse the properties of the distribution of probabilities, then to construct the third card and explain the criteria used to design the card so as to minimize the risk of losing. Then, students played using the constructed card, and were asked to graphically analyse the differences between the experimental (frequency) distribution and theoretical (probability) distribution. Finally, students explained why they thought their card would be the winning card. A possible learning trajectory was hypothesized (Clements and Sarama 2004) that may help students distinguish between experimental probability and theoretical probability.

A qualitative static retrospective analysis was performed with the aim of identifying the characteristics of the distribution of frequencies and probabilities that were used by the students in making their decisions (Johnson-Laird 1983). Seven of the teams explained that they had constructed the third card by exclusively considering the table or the distribution of frequencies. Meanwhile, five teams constructed their cards by considering the distribution of probabilities, analysing the density of the distribution of probabilities, and selecting the numbers with higher probability (e.g. team A4: “To choose the numbers with higher probability”). According to the SOLO model (Biggs and Collins 1982), the reasoning of these nine teams was uni-structural, because they were only able to analyse either the density of the distribution of frequencies or the probabilities, but not both. Meanwhile, the reasoning of team B2 was categorized as multi-structural because they were able to analyse the skewness and the symmetry of the distribution of probabilities (team B2: “[We chose] Small numbers, and as many negatives as positives”). Finally, the reasoning of team B6 was categorized as relational, because students were able to relate the frequency distribution of the events randomly generated by the IAB with the symmetry and the density of the theoretical distribution of probabilities (team B6: “because it’s luck and because we have combined the numbers +1, −1, +2, −2, +3, −3”).

The qualitative dynamic retrospective analysis then provided information about the evolution of the teams’ reasoning when coping with the demand of comparing both distributions, that is, the distribution of frequency and the distribution of probability. In this case, learners constructed a new mental model to reason about relevant aspects of the situation (Seel 2001). There was an evolution in reasoning on the part of team B6, because this team, initially categorized as uni-structural, progressed to multi-structural reasoning by basing their decisions on the density and symmetry of the distribution of frequencies and probabilities. An advancement in reasoning was also identified in team A5, who initially reasoned using only the distribution of frequencies; however, after comparing both distributions, they also reasoned using the distribution of probabilities.

12.6.4 Decisions Made When Constructing the Fourth Card

Neither static nor dynamic structural evolutions of mental levels of reasoning were observed during the process of constructing the last card.

To sum up, the first investigative cycle, which involved constructing the first card, playing the game to obtain data, analysing the distribution of frequencies, and using this information to construct the second card, provided students with the opportunity to reason about the density of the data and its skewness (Serradó 2016). The analysis of density provided students with the opportunity to evolve from a pre-structural to a uni-structural level, coinciding with the results presented by Sanchez and Orta (2015). The evolution from data to distribution, as proposed by Bakker and Gravemeijer (2004), was restricted by the lack of (informal or formal) knowledge about how to analyse the centre and spread of the distribution. Conversely, the analysis of the probability distribution guides students toward understanding properties of the data (Bakker and Gravemeijer 2004). More specifically, the analysis of skewness, the position of the mode, and the symmetry of the theoretical distribution allowed some students to reason about how they might increase their chances of winning and evolved their mental reasoning from the multi-structural to the relational level.

12.7 Decisions Under Uncertainty and of Risk

Individual differences were also identified when students were asked to generalize the situation and answer the question: "How would you construct the card to minimize the risk of losing?". These differences allowed us to distinguish between students whose decision making was based on uncertainty or on risk. Twenty-nine of the 41 students (70.7%) achieved a uni-structural level of reasoning. They based their decisions on recognizing the modal clumps of the distribution of frequencies or probabilities (e.g., student A33: "[select] the numbers with higher relative frequency"; student B22: "choose the numbers with higher probability"). Nine of 41 students (21.9%) achieved a multi-structural level of reasoning. They recognized and based their decisions on the numbers associated with the majority of the data in the distribution of frequencies, or on the central position of zero and the symmetry of the distribution of probabilities (e.g., student A54: "selecting numbers between −2 and +2"). Three of 41 students (7.3%) achieved a relational level of reasoning. Their decisions were based on the recognition of the modal clumps, the numbers associated with the majority of the data, and the symmetric distribution of probabilities (e.g., team B6: "writing + and − numbers with higher probability").

The 24 of 41 students (47%) who based their decisions exclusively on the analysis of the distribution of frequencies were categorized as reasoning at the uni-structural level, and 4 of 41 (9.7%) at the multi-structural level. We consider the decisions of these students as being made under uncertainty. Four of 41 students (9.7%) were categorized as reasoning at the uni-structural level because they only used a property of the distribution of probabilities in their reasoning. Six of 41 students (14.6%) were able to express their reasoning using two properties of the probability distribution. Only 3 of the 41 students (7.3%) were able to establish relationships between more than one property of the probability distribution (relational level).

The case of student B64 is particularly interesting. Initially, her evolution from pre-structural to multi-structural reasoning was restricted by the fact that she took the randomness of the numbers generated by the IAB app into consideration, but did not identify randomness in the sequence of events. Then, when she was confronted with the task of contrasting the distribution of frequencies and probabilities (presented graphically), her reasoning developed to a relational level. She reasoned about the modal clumps, the symmetry, and the density of the probability distribution. Finally, when she generalized the situation, her reasoning was categorized as relational, because she was able to relate different aspects of the distribution and understand the difference between probability and frequency (team B64: "one is what could happen and the other is what happened"). We consider that she was able to understand the predictive character of probability and to use it to minimize the risk of losing. She concluded that the utility of the task was to "learn to use probability to minimize the risk of losing."

To sum up, we have analysed the evolution of students' decisions under uncertainty to decisions of risk when operationalizing risk management. This operationalization was proposed as a means of surpassing two difficulties identified in the literature review. On the one hand, the overlap in the meaning of the notions of risk and hazard (Borovcnik 2015) restricted the advancement in the reasoning of team B6 until they analysed and compared the distribution of frequencies and probabilities. On the other hand, due the predictive character that the team B6 assigned to probabilities, they were able to distinguish between decisions under uncertainty and risk, despite the difficulties described by Knight (1921).

In particular, when frequency is used to reason about how to increase chances of winning in this task, the decisions made are characterized as decisions under uncertainty; meanwhile, when probability is applied to increase chances of winning, we consider the decisions made to be decisions of risk. Although Gigerenzer (2002) only describes the need for stochastic numeracy for informed and accurate risk decision making, the proposed distinction between decisions under uncertainty and of risk suggests that students need to appreciate the numerical properties of data, as well as those of the frequency distribution and the underlying probability distribution.

The observed difficulties were conceived as discontinuities in the learning process, which could hinder continuity in the learning process (Gueudet et al. 2016). The information obtained from the micro-analysis allowed for the identification of different learning trajectories and the development of different stochastic mental levels of reasoning among the students in the study.

12.8 Stochastical Mental Levels of Reasoning

As noted earlier, students cannot be ascribed to general cognitive levels, but situationally-provoked cognitive levels in the context of a specific task may be observed (Eichler and Vogel 2012). The SOLO model (Biggs and Collins 1982), which presents a generic description of mental levels of reasoning, can be particularized when students reason about stochastic concepts in the decision-making process of operationalizing risk management. These are:

a. *Pre-structural response*: Decisions in this category are based on personal preferences; there is no logical connection to the understanding of the uncertainty, randomness, or hazard involved in the IAB task.
b. *Uni-structural response*: Decisions in this category are said to be made "under uncertainty," and are rationally bounded by the analysis of the modal clumps of the distribution of frequencies.
c. *Multi-structural response:* Decisions in this category are either made "under uncertainty" (if they are rationally bounded on reasoning about the events associated with the majority of the data) or are decisions "of risk" (if they are rationally bounded by reasoning about the density and symmetry of the distribution).
d. *Relational response*: Decisions in this category can be classified as both "under uncertainty" (rationally-bounded reasoning about the events associated with the majority of the data in the distribution of frequencies) and "of risk" (rationally-bounded by reasoning about the symmetry of the distribution).

12.9 Learning Trajectories: Opportunities and Constrains

Sections 12.5 and 12.6 described the structural and functional elements of students' thinking and the evolution of mental models of reasoning about stochastic processes when making decisions in situations of emotional risk. The functional dynamic aspect analyses how learners construct a mental model in order to simulate relevant aspects of the situation to be cognitively mastered (Seel 2001). As the above analysis reveals, the mental model may be changed, enriched, or modified (Schnotz and Bannert 1999), for instance by involving students in learning trajectories such as systematic ORM. Although we have established ORM as a hypothetical learning trajectory, we have observed that students may follow different but equally valid learning trajectories (Clements and Sarama 2004). With regard to these models, students show individual differences in the details concerning the structural and functional aspects of their thinking (Schnotz 1994).

In concordance with the mental levels of reasoning about stochastic processes described in Sect. 12.8, six different learning trajectories (LT) can be identified.

Three learning trajectories suggest an evolution from pre-structural to uni-structural mental levels of reasoning. The decisions of these students mature from being based on personal preferences to being based on rationally bounded preferences. However, these students were unable to evolve to higher mental levels of reasoning due to: (a) a deterministic view of the game (LT1); (b) difficulties in discerning between the randomness of the generator and that of the events and the sequence of events (LT2); or (c) a lack of previous knowledge about the measures of centre in the frequency distribution (LT4). This result is consistent with Bakker and Gravemeijer's (2004) study on the progression from reasoning about individual data, in our case random events, to an understanding of the frequency distribution.

Two learning trajectories involved a progression from pre-structural to multi-structural reasoning. In particular, the LT3 of team A6 involved a progression from pre-structural to uni-structural reasoning when trying to understand the different lengths of the games and the sequences of events, and from uni-structural to multi-structural reasoning when comparing the distribution of frequencies and probabilities. The trajectory of team B2 (LT5) also suggests an initial evolution from pre-structural to uni-structural reasoning. A higher level of evolution was temporarily restricted by the students' inability to understand that the events and the sequences of events were random, just as the numbers generated by the random generator were random. Subsequently, the need for understanding the density and the skewness of the individual data (events) allowed them to mature from a uni-structural to a multi-structural mental level of reasoning.

Finally, the LT of team B6 (LT6) involved a progression from pre-structural to relational reasoning. When making decisions about the construction of the first and second card, the students' reasoning about the risk to lose was restricted by the emotional hazardous nature that the students attributed to the IAB generator; however, after constructing the fourth card and comparing the distribution of frequencies and the distribution of probabilities, they overcome the fuzzy distinction between emotional hazard and risk, in discerning between situations of uncertainty and of risk (Borovcnik 2015).

12.10 Conclusion

In summary, this chapter presented a design-based research study involving a game of chance task, IAB, which aimed to enhance students' reasoning when making decisions under uncertainty and decisions of risk. From a methodological point of view, the study involved three levels of complexity. First, at a macro-level, a review of the literature provided information about (a) the faster, intuitive dimension of decision making, and the slower, analytic type; (b) the distinction between individual and societal dimensions of decision-making, (c) the distinction between personal preference and decisions that are rationally bounded; and (d) the identification of two difficulties when trying to distinguish between rationally bound decisions *under uncertainty* and *of risk.*

We designed a task to firstly promoting the students' continue transition between personal preference and rationally-bound decisions; and, then inducing a continue transition between the quantification of the game uncertainty and of the risk to lose to the identification of uncertainty or of risk situations. The task engaged students in recognizing the uncertainty in the game, operationalizing risk management, and reasoning about risk management.

A qualitative static retrospective analysis of the IAB task was performed to identify the static aspects of students' reasoning about stochastic processes. We identified four mental levels of reasoning: (a) pre-structural, where decisions are based on personal preference; (b) uni-structural, where decisions are rationally bound in situations of uncertainty; (c) multi-structural, where decisions are rationally bound either in situations of uncertainty or in situations of risk; and (d) relational, when decisions are rationally bound both in situations of uncertainty and of risk.

Finally, six learning trajectories were identified. Three learning trajectories involved progression from pre-structural to uni-structural reasoning, which were restricted, respectively, by a deterministic view of the game, the difficulties in discerning between the randomness of the generator and that of the events and sequences of events, and a lack of previous knowledge about measures of centre for frequency distributions. Two learning trajectories involved an initial evolution from pre-structural to uni-structural reasoning and subsequently from the uni-structural to the multi-structural level. Finally, one learning trajectory involved a direct progression from a pre-structural to relational level of reasoning.

References

Bakker, A., & Gravemeijer, K. (2004). Learning to reason about distribution. In D. Ben-Zvi, & J. Garfield (Eds.), *The challenge of developing statistical literacy, reasoning and thinking* (pp. 147–168). Dordrecht: Kluwer.

Biggs, J. B., & Collins, K. F. (1982). *Evaluating the quality of learning: SOLO taxonomy.* New York: Academic Press.

Borovcnik, M. (2015). Risk and decision making: The "logic" of probability. *The Mathematics Enthusiast, 12*(1–3), 113–139.
Borovnich, M., & Kapadia, R. (2011). Determinants of decision making in risky situations. In *Proceedings of the 58th World Statistical Congress*. Dublin: International Statistics Institute. http://2011.isiproceedings.org/papers/950138.pdf. Accessed: 25 July 2016.
Clements, D. H., & Sarama, J. (2004). Learning trajectories in mathematics education. *Mathematical Thinking and Learning, 6*(2), 81–89. https://doi.org/10.1207/s15327833mtl0602_1.
Cline, P. B. (2003). Re-examining the risk paradox. In *Proceedings of the 2003 Wilderness Risk Management Conference*, State College. https://www.outdoored.com/sites/default/files/documents/files/wrmc_proceedings_03_examing_cline.pdf. Accessed: 11 November 2016.
Cobb, P. (2000). Conducting teaching experiments in collaboration with teachers. In A. E. Kelly, & R. A. Lesh (Eds.), *Handbook of research desing in mathematics and science education* (pp. 307–333). Mahwah, NJ: Lawrence Erlbaum Associates.
Eichler, A., & Vogel, M. (2012). Basic modelling of uncertainty: Young students' mental models. *ZDM Mathematics Education, 44*(7), 841–854. https://doi.org/10.1007/s11858-012-0451-9.
Eichler, A., & Vogel, M. (2015). Teaching risk in school. *The Mathematics Enthusiast, 12*(1–3), 168–183.
Gigerenzer, G. (1989). *The empire of chance: How probability changed science and everyday life.* New York: Cambridge University Press.
Gigerenzer, G. (2002). *Calculated risks: How to know when numbers decieve you.* New York: Simon & Schuster.
Gueudet, G., Bosch, M., diSessa, A. A., Kwon, O. N., & Verschaffel, L. (2016). *Transitions in mathematics education, ICME-13. Topical survey series.* New York: Springer Open.
Johnson-Laird, P. N. (1983). *Mental models.* Cambridge, MA: University Press.
Kahneman, D., & Frederick, S. (2002). Representativeness revisited: Attribute substitution in intuitive judgement. In T. Gilovich, D. Griffin, & D. Kahneman (Eds.), *Heuristics and biases: The psychology of intuitive judgement* (pp. 49–81). New York: Cambridge University Press.
Knight, F. H. (1921). *Risk, uncertainty and profit.* Boston, MA: Houghton Mifflin.
Ministerio de Educación, Cultura y Deporte. (2015). *Real Decreto 1105/2014, de 26 de diciembre, por el que se establece el currículo básico de la educación secundaria obligatoria y del bachillerato* (Royal decree establishing the basic curriculum for secondary and high school). Madrid: Author.
Russel, G. L. (2015). Risk education: Analysis of what is present and could be. *The Mathematics Enthusiast, 12*(1–3), 62–84.
Sánchez, E., Mercado, M, & García, J. (2016). La variabilidad en el razonamiento probabilístico informal de estudiantes de Bachillerato (Variability in the informal probabilistic reasoning in high school students). In J. A. Macías, J. L. González, M. T. Sánchez, P. Hernández, C. Fernández, F. J. Ruiz, T. Fernández, & A. Berciano (Eds.), *Investigación en Educación Matemática XX* (pp. 479–488). Málaga: Sociedad Española de Investigación en Educación Matemática.
Sanchez, E., & Orta, A. (2015). Levels of reasoning of middle school students about data dispersion in risk contexts. *The Mathematics Enthusiast, 12*(1–3), 275–289.
Schnotz, W. (1994). *Aufbau von Wissensstrukturen. Untersuchungen zur Kohärenzbildung beim Wissenserwerb mit Texten* (Construction of knowledge structures. Cultural cohesion when knowledge is acquired with texts). Weinheim, Germany: PVU.
Schnotz, W., & Bannert, M. (1999). Einflüsse der Visualisierungs- form auf die Konstruktion mentaler Modelle beim Text-und Bildverstehen (Influences of the visualization form when constructing the mental models with the understanding of text and picture). *Zeitschrift für Experimentelle Psychologie, 46*(3), 217–236.

Seel, N. M. (2001). Epistemology, situated cognition, and mental models: "Like a bridge over troubled water." *Instructional Science, 29*(4–5), 403–427.

Serradó, A., Vanegas, Y., & Giménez Rodríguez, J. (2015). Facilitating deliberate dialogue in mathematics classroom. In U. Gellert, J. Giménez, C. Hahn, & S. Kafoussi (Eds.), *Educational paths to mathematics. A C.I.E.A.E.M sourcebook* (pp. 289–303). New York: Springer.

Till, C. (2014). Fostering risk literacy in elementary school. *Mathematics Education, 9*(2), 83–96.

Chapter 13
Determinism and Empirical Commitment in the Probabilistic Reasoning of High School Students

Ernesto Sánchez, Jaime I García-García and Miguel Mercado

Abstract In this paper, we explore students' responses to two binomial tasks, one related to prediction and another to distribution, in an effort to understand how students express variability in their predictions before and after simulation activities. To collect data, a four-stage study was conducted with two student groups (one with instruction in probability and the other without it). The first and fourth stages consisted of administering a test that included questions about a binomial experiment, B(x, 2, ½). During the second and third stages, the students conducted simulations with manipulatives and with software. The SOLO taxonomy was used to analyse their progress in reasoning based on their responses to two questions on the test. By analysing the students' responses in a double-entry table to both questions, we were able to ascertain the difficulties that students face in integrating variability into their reasoning, despite their experience with simulation. Two patterns of student responses are salient from the set of answers: Determinism and Empirical commitment.

Keywords Determinism · Empirical commitment · Probabilistic reasoning

---

E. Sánchez (✉)
Departamento de Matemática Educativa, CINVESTAV-IPN,
Av. Instituto Politécnico Nacional 2508, Col. San Pedro Zacatenco,
07360 Del. Gustavo A. Madero, Mexico City, Mexico
e-mail: esanchez0155@gmail.com

M. Mercado
Colegio de Ciencias y Humanidades, UNAM, Mexico City, Mexico
e-mail: miguelmm@unam.mx

J. I García-García
Facultad de Matemáticas, Universidad Autónoma de Guerrero,
Av. Lázaro Cárdenas S/N, Ciudad Universitaria, Col. Haciendita,
Guerrero C.P. 39087 Chilpancingo, Mexico
e-mail: jaime_garcia1980@hotmail.com

C. Batanero and E. J. Chernoff (eds.), *Teaching and Learning Stochastics*,
ICME-13 Monographs, https://doi.org/10.1007/978-3-319-72871-1_13

13.1 Introduction

A general problem that learners face in the learning of statistics is establishing the connection between models and data. Thus, in the process of learning statistics, students should be presented with many tasks in which models and data come into play. A particular case of this problem consists of understanding the relationship between the classical and frequentist approaches to probability, which is not a minor task, considering the many subtleties of this connection (Batanero et al. 2016; Konold et al. 2011).

According to the classical interpretation, the probability of an event can be determined by assuming equal probability for each element of the sample space. This assumption is generally based on the analysis of the traits of the random generator. In contrast, according the frequentist interpretation, the probability of an event is estimated by analysing data collected through repetitions of an experiment. For this, it is assumed that the relative frequencies of any event converge around a constant number. While developing a theoretical model for natural or social phenomena can be arduous, in gambling situations (e.g., involving coins, dice, urns, or spinners) it is often easy to propose a theoretical model (i.e., a probability distribution) using the classical notion of probability. Moreover, as game situations are repeatable under similar conditions, it is also possible to generate data sets and apply the frequentist approach to estimate the probabilities empirically.

These characteristics, coupled with the availability of software that can be used to generate simulations, make game situations an appropriate means to study the relationship between (random) models and 'reality' (Chance et al. 2007). A core idea underlying this relationship is variability: since variability is unavoidable, it is crucial that students develop a sense of variability in order to understand both the model and the data collected from game situations. In this paper, we analyse how students deal with variability in the context of a simple binomial situation before and after simulation activities.

13.2 Conceptual Framework

In this section, we discuss relevant prior research and the motivation for the present study. At the end of the section, some relevant works to the present research are referenced.

13.2.1 Probabilistic Reasoning Beyond Mathematical Reasoning

Probabilistic reasoning overlaps with mathematical reasoning, but the former is not wholly encompassed by the latter. On the one hand, referring to the sample space and field of events of a probability space, Kolmogorov (1956) suggested: "What the elements of this set represent is of no importance in the purely mathematical development of the theory of probability" (p. 1). On the other hand, from the point of view of teaching and statistics reasoning, it is important to take into account the context in which the problems occurs.

In this study, we hypothesize that the non-mathematical aspects of probabilistic reasoning can be characterized with the help of the big ideas of probability proposed by Gal (2005) in his analysis of probabilistic literacy. The notions of randomness, variation, and independence, as well as the complementary ideas of predictability and uncertainty, are the underpinnings of the system of probabilistic ideas, and are not represented in any other mathematical domains. As suggested by Gal (2005): "Some aspects of these big ideas can be represented by mathematical symbols or statistical terms, but their essence cannot be fully captured by technical notations."

The focus of this research is to analyse how students reason with, or ignore variability when faced with a situation involving prediction/uncertainty where the underlying distribution is binomial $B(x, 2, ½)$. In this context, the variability of a sequence of experimental outcomes is expressed as the difference between the theoretical probability and the observed relative frequency of a target event. Such variability can also be expressed as the difference between observed absolute frequencies and expected (theoretical) frequencies, which can alternatively be expressed as a range in which we expect that observed frequencies would fall for a given target event.

13.2.2 Calculation Versus "Prediction"

For a long time, learning probability was synonymous with learning to compute probabilities. Basic problems in teaching according this approach include: "What is the probability of obtaining a white ball from an urn with three white and two black balls?" and "What is the probability of getting two heads by throwing two coins?" These problems become more interesting as sample spaces become more complex: for example, "What is the probability of getting exactly five heads when flipping a coin 10 times?"

Recently, as suggested by some curricula, teaching objectives have expanded to include understanding relationships between probability and inference, for example: "Develop and evaluate inferences and predictions that are based on data" (National Council of Teachers of Mathematics 2000, p. 324). In particular, new "prediction" problems (not to be confused with prediction problems related to regression) have emerged in the research that have been used to understand how students perceive and progress in their understanding of variability in chance settings (Watson and Kelly 2004; Watson et al. 2003). The kinds of probabilistic prediction problems that we would like to draw attention to are triggered by questions such as "What do you think will happen…?" and "What is your prediction about…?" These questions typically refer to an experience involving uncertainty where the probability distribution is known or can be approximated by simulation.

An important prediction problem of this kind arises when we are interested in the number of times than an event occurs after repeating many binomial trials, or in the proportion of a specific result in a sample. For example, Torok and Watson (2000) studied children's consideration of variation through the following Lollies task, which asks students to predict a result:

> Suppose we have a container with 100 lollies in it. 50 are red, 20 are yellow, and 30 are green. The lollies are all mixed up in the container.
>
> (A) Suppose you pick out 10 lollies. How many reds do you expect to get? (p. 168)

Several students of different levels respond "5" to this task, others suggest, "I cannot know" or "any number between 0 and 10 may happen" and only a minority try to reflect the underlying variability in their response.

Rational responses to prediction problems such as the lollies problem should be accompanied by what Gal (2005) calls a *statement of certainty*—that is, an estimate of the confidence with which we believe in a prediction: for example, "I think that one in four times I will get 5 red lollies." A statement of certainty is not necessarily a numerical estimate and may be merely subjective: for example, "I may get 5 red lollies" or "I think it is very likely that I will get between 3 and 7 red lollies."

A level of confidence in a prediction can be obtained by observing the frequency with which the prediction occurs in a simulation process. By initiating and observing such processes, students may learn to make statements of certainty to accompany predictions.

13.2.3 Simulation

Technology has transformed the practice and teaching of statistics to such an extent that it is hard to propose any innovation in teaching without suggesting how technological tools may support the learning process (Chance et al. 2007). For instance, Hoffman et al. (2014, p. 284) argue that "[a] contemporary course on stochastics, which adequately takes data and chance into account, is not feasible without the use of technology." One of the most important contributions of technology in supporting the teaching and learning of probability is that it allows students to perform simulation activities, in addition to offering different ways of generating, organising, representing, and processing data. Fathom is one example of educational software that offers such tools.

Most simulation activities involve four steps (Hoffman et al. 2014): (1) Constructing the model, (2) Identifying the events involved, (3) Repeating the experiment and collecting data, and (4) Analysing the data. The last two steps in particular allow students to observe data trends so that they can respond to problems of prediction by making statements of certainty. For example, if we consider the Lollies task and define the variable to be “the number of red lollies in a sample of 10 lollies extracted from the container,” we can obtain the relative frequency of different possible events via simulation (see Table 13.1). Statements of certainty, similar to those presented on the last column in this table, may be derived from the estimated probabilities (the exact theoretical probabilities are presented in the third column).

The way in which variability may be captured in statements of certainty is by expressing our confidence that the outcome belongs to a given range of results. Here, there is a tension between the desire to make a precise prediction and the desire to have confidence in the prediction. The more precise the prediction, the less confidence we can have in it, and vice versa.

13.2.4 Relationships Between the Law of Large Numbers and Prediction

Although a simulation process may help construct a simpler table than Table 13.1 without calculating the theoretical probabilities in question, the idea that the relative frequencies approximate theoretical probabilities is crucial. To give students the opportunity to reflect on this relationship, we propose a problem involving a simple binomial distribution. This problem asks students, on the one hand, to calculate the probabilities involved (1/4, 1/2, 1/4), and on the other hand, to develop their understanding of the process of simulation using the Fathom software. Students' reasoning about variability is elicited by asking them about their expectations for the frequency of results in a run of 1000 trials governed by the binomial distribution with parameters: $n = 1$ and $p = \frac{1}{2}$.

Table 13.1 Statements of certainty for the Lollies problem

Event	Relative frequencies obtained from simulation (1000 repetitions)	Theoretical probability	Statements of certainty
From 2 to 8	0.980	0.978	It is almost certain that the number of red lollies will be between 2 and 8
From 3 to 7	0.875	0.890	It is very likely that the number of red lollies will be between 3 and 7
From 4 to 6	0.639	0.656	It is likely that the number of red lollies will be between 4 and 6
Exactly 5	0.233	0.246	It is possible that the number of red lollies will be 5

At this point, we should clarify how the relationship between the theoretical probability of an event and its absolute and relative frequency, expressed by the law of large numbers, allows for the construction of a prediction range for a desired confidence level. Of course, we are not suggesting that students should learn the formal procedure below; rather, our goal is to demonstrate the process of obtaining prediction ranges and to highlight their importance in predictions.

For each value of the variable x of B(x, 2, ½), we can define a Bernoulli distribution, which takes on the value 1 if the event x occurs and 0 if it does not. Let $f_n(x)$ be the number of x observed in n independent repeated trials with a probability p of occurring success ($x = 1$) at each trial. Then, for any number $\varepsilon > 0$, it follows that (Parzen 1960, p. 229):

$$\lim_{n\to\infty} \mathrm{P}\left(\left|\frac{f_n(x)}{n} - p\right| \leq \varepsilon\right) = 1$$

This means that given $\delta > 0$ and an integer n, it is possible to find ε such that:

$$\mathrm{P}\left(\left|\frac{f_n(x)}{n} - p\right| \leq \varepsilon\right) > 1 - \delta \tag{13.1}$$

For example, taking $\delta = 0.20, n = 1000$, we have:

$$P\left(-\varepsilon \leq \frac{f_{1000}(x)}{1000} - p \leq \varepsilon\right) > 0.80.$$

Multiplying the inequality within the brackets by 1000 and adding 1000 p, we have

$$P(1000p - 1000\varepsilon \leq f_{1000}(x) \leq 1000p + 1000\varepsilon) > 0.80 \tag{13.2}$$

It should be noted that inequality (13.1) is equivalent to:

$$\mathrm{P}\left(\left|\frac{f_n(x)}{n} - p\right| > \varepsilon\right) \leq \delta.$$

To estimate ε, we use the following inequality, which is demonstrated in Parzen (1960, p. 230):

$$\mathrm{P}\left(\left|\frac{f_n(x)}{n} - p\right| > \varepsilon\right) \leq \frac{1}{4n\varepsilon^2}$$

Then, for $\delta = 0.20$, we have $0.20 = \frac{1}{4n\varepsilon^2}$, from which it follows that $\varepsilon \cong 0.035$. To sum up, and using expression (13.2):

- For $x = 0$ (also for $x = 2$) and $\delta = 0.20$, we have $215 \leq f_{1000}(x) \leq 285$;
- For $x = 1$ and $\delta = 0.20$, we have $465 \leq f_{1000}(x) \leq 535$.

Using simulation with the Fathom software, we can find better estimates for these ranges.

13.2.5 Some Related References

In response to Jones (2005), who warned, "there is a void in the research associated with the frequentist approach to probability" (p. 368), we note that there is already some literature related to the problem of articulating the relationships between the frequentist and classical approaches of probability, most of which employs educational software.

We mention three sources where more references to research on this issue can be found. Ireland and Watson (2009) documented eighth grade students' understanding of the relationships between several components involved in the transition from experimental (specific frequencies) to theoretical probabilities (abstract), which include the relationship between variability and sample size. Lee et al. (2010) explored the ways in which sixth grade students make inferences from data to determine whether a distribution is uniform or not. Konold et al. (2011) pointed out some pitfalls of the common teaching practice of introducing students at the same time to both "theoretical probability" and "experimental estimates of probability" (via relative frequencies). They argue and demonstrate that it is naive to believe that merely teaching both concepts leads students to establish the necessary connections. In this paper, we explore reasoning of students about connections between frequencies and theoretical probabilities from binomial distribution.

13.3 Method

A four-stage study was undertaken with two groups of high school students. The first group consisted of 37 students who had not previously taken a statistics and probability course, and the second group consisted of 66 students who had already taken an introductory statistics and probability course. During the first stage of the study, a probabilistic situation and several questions about the situation (in this paper, only two of these questions are reported) was presented to students. In the second session, students were guided in simulating the situation with manipulative materials and observing the results, and were then asked to respond to the same questions posed in the first session, considering their observations. In the third stage, the students were asked to simulate the situation using the Fathom software and to observe the results. Finally, in the fourth stage, the students were asked to answer the same questions that were posed during the first stage. The situation presented to the students was the following:

The Smith family consists of mother Ana, son Billy, and father Charles. Every night, the three of them gather to watch TV, but they never agree on which TV program to watch. Ana likes to watch movies, Billy prefers cartoons, and Charles likes the news. As there is only one TV at home, it would be easier for them to take turns. However, Billy proposes a game of chance. He suggests tossing two coins and the following conditions: Ana wins if the result is two heads; Billy wins if one coin lands on head and the other on tail; and Charles wins if the result is two tails.

After receiving the above information, the students were asked to reply to the following two questions:

Question 1. What do you think will happen if the coins are tossed 1000 times?
Number of times Ana wins: _____
Number of times Billy wins: _____
Number of times Charles wins: _____
Question 2. Assign the probability to each value of the variable X, (that is, determine the following probabilities):
Probability of A (X = 0) is equal to ____
Probability of B (X = 1) is equal to ____
Probability of C (X = 2) is equal to ____

The second question is common in the teaching of probability, as it requires either the application of the classical definition of probability or the application of the frequentist approach when data are available. On the contrary, the first question is a prediction/uncertainty question requiring highly sophisticated reasoning to answer, as it requires considering the probability distribution and having an understanding of variability.

Assuming that students are familiar with the binomial distribution, they are likely to predict that 250, 500, 250 is the most likely result, but if they perform the experiment, they would be more surprised by this outcome than by results such as 257, 510, 233. The event "250, 500, 250" is obtained by multiplying the number of trials, 1000, by the probability of each event, while an event such as "257, 510, 233" may be observed through experience. This situation gives rise to a conflict when it is assumed that the theory predicts outcomes, but a different outcome is observed in practice. This type of conflict is often resolved by using an expression such as "the event *X* or something similar" instead of simply "the event *X*."

In fact, a mathematical expression that encompasses the notion of "something similar" would provide a range of numbers around each expected frequency, together with the probability that it occurs; for example, $E = \{(a, b, c)$ $|a \in \mathbb{N}, \in \mathbb{N}, \in \mathbb{N} \nRightarrow 225 \leq a \leq 275, 475 \leq b \leq 525, 225 \leq c \leq 275\}$, where the probability of *E* is a little higher than 80% (calculated using results from a Fathom simulation). This is the mathematical solution to communicating the variability of the situation. However, it is difficult for students to model variability in this situation. Even when the language needed is accessible to the students, they may have not yet been introduced to the concepts that would allow them to express variability in this way. However, responses that approximate expected frequencies can be interpreted as attempts to represent variability in the absence of concepts that allow students to solve this problem more formally.

13.3.1 Procedure of Analysis

In analysing the data, we first classified the responses for each question in the pre- and post-questionnaires, using the SOLO taxonomy to identify and isolate some relevant aspects of the task solution. These aspects were identified by considering both the relevant theory and the features of observed responses. Combining these aspects, the categories Pre-, Uni-, Multi-structural, and Relational were defined in a way that is similar to the characterization proposed by Biggs and Collis (1991). Subsequently, frequencies of responses that were categorized at the Unistructural level or higher for both questions were classified in a two-way table.

Responses to question 1

In general, students responded to question 1 by giving three integer numbers, each of them representing the frequency of X for the values 0, 1, and 2. Such responses are incorrect because the probability that the observed frequencies match the three given numbers is very small, and hence they should not be expected to occur. However, some features of these responses can be highlighted—in particular, those related to the notions of distribution and variability, since they can help to identify the levels of sophistication of the students' responses. These aspects are:

1. *Coherence*: The sum of the three frequencies is 1000.
2. *Distribution pattern*: The frequency of the value $x = 1$ is greater than the frequency of values $x = 0$ and $x = 2$.
3. *Variability*: The frequencies of 0, 1, and 2 fall within the ranges 250 ± 25, 500 ± 25, and 250 ± 25, respectively. We also consider the lack of variability, namely.
4. *The frequencies of 0 and 2 are equal.* The values 0, 1, and 2 for x are assigned frequencies of exactly 250, 500 and 250, respectively. We add this condition in order to differentiate between those responses where the variability is not reflected and those that reflect variability.

Considering the above aspects, responses to question 1 are categorized according to SOLO level are presented in Table 13.2, with an example provided in the last column.

Responses to question 2

In question 2, students are also asked to assign three numbers. In this question, the correct answer is (¼, ½, ¼). The SOLO categories corresponding to different responses are as follows:

Table 13.2 Definitions of SOLO levels for question 1

Level	Conditions	Example
Prestructural	Does not satisfy 1, or Does not satisfy 2 and 3	0, 5, 9 334, 333, 333
Unistructural	Satisfies 1 and 2 but not 3, or Satisfies 1 and 3, but not 2, or Satisfies 4	250, 500, 250 100, 200, 700 325, 350, 325
Multistructural	Satisfies 1 and 2, but neither 3 nor 4	102, 705, 193
Relational	Satisfies 1, 2, and 3 but not 4, or Give the expected frequencies "250, 500, 250" with a statement of certainty or expressing variability	235, 507, 258 The result is likely to be numbers around 250, 500, 250

- Responses are classified as *Prestructural*: (1) if they refer to elements not related to the situation or are incomprehensible, or (2) when the answer consists of three numbers whose sum is not 1 or 100% and which are not proportional to (1, 2, 1).
- Responses that meet one of the following three conditions are classified as *Unistructural*: (1) Students assign values whose sum is not 1 or 100 (generally integers), but the values are proportional to (1, 2, 1). (2) The response values sum to 100 or 1000, but they are not proportional to (1, 2, 1). (3) The answers are given as three decimal or fractional numbers whose sum is 1, but the probability of $x = 1$ is not greater than the probability of $x = 0$ or of $x = 2$.
- The responses that are classified in the *Multistructural* level meet any of the following conditions: (1) They are expressed as decimal numbers, fractions or percentages whose sum is 1 or 100%, with higher probability assigned to $x = 1$ than to $x = 0$ and $x = 2$, but with values that are far from the theoretical probabilities (¼, ½, ¼). (2) The given values (usually expressed as integers, but not necessarily) sum to 100 and greater probability is assigned to $x = 1$, but the "%" symbol is omitted. (3) Students express the theoretical probabilities and indicate that these probabilities are just an approximation: for example, "(about ¼, about ½, about ¼)."
- The responses classified in the *Relational* level are probability distributions, which probabilities are obtained via the Laplacian definition or via the frequentist approach.

Responses to both questions

The SOLO classification was used to evaluate the quality of each response to the two questions independently. However, a joint analysis of responses to questions 1 and 2 can be valuable for understanding how students address variability. We thus classified each pair of responses given by the students and then only took into account those students who provided responses to both questions that were classified at the Unistructural level or higher. The responses of these pairs of students were classified in one of four categories:

Determinism. The response to question 1 is (250, 500, 250) and the response to question 2 is obtained by the classical approach to probability: (¼, ½, ¼).

Connecting theoretical probabilities and data: Instead of providing the response (250, 500, 250) to question 1, the student provides three numbers reflecting that they know the distribution and showing also variability; the response to question 2 is obtained by the classical approach: (¼, ½, ¼).

Blurred relationship between theory and data: The response to question 1 is (250, 500, 250) and the response to question 2 is obtained by the frequentist approach: for example, (0.243, 0.496, 0.261).

Empirical commitment. The response to question 1 is not (250, 500, 250)—that is, it consists of three numbers reflecting variability and distribution, and the response to question 2 is obtained by the frequentist approach: for example, (0.244, 0.498, 0.258).

The frequencies of these categories can be presented in a contingency table after categorizing the responses to each question according to two complementary classes (see Table 13.3). With respect to the responses to question 1, one class contains responses in which the expected frequencies are (250, 500, 250), and the other contains the remaining responses. With respect to question 2, we distinguish responses in which the probabilities were obtained by the classical approach from those obtained by frequentist approach.

We interpret each category as follows:

The students with responses classified in the *determinism* category begin by assigning theoretical probabilities to the events in question, and then compute a prediction by simply multiplying the probability of each event by the number of trials (in this case, 1000). Students' whose responses after the simulation activities are classified in this cell are assumed to have learned little from the experiments and therefore avoid or ignore the variability inherent in the situation in their responses. Some of the students, being familiar with the classical approach to probability, may believe that a prediction problem is only a mathematical or theoretical issue, and do not consider the results of a real situation.

Table 13.3 Students responses to both questions

	Question 1	
Question 2	Expected frequencies (250, 500, 250)	Non-expected frequencies (others)
Classical approach	Determinism	Connecting theoretical probabilities and data
Frequentist approach	Blurred relationship between theory and data	Empirical commitment

Students whose responses are classified in the *empirical commitment* category ignore the theoretical distribution and propose a distribution based solely on empirical results. For them, there is no underlying probability distribution in the situation, since they only propose a description based on what they observe during the simulation or from their previous experience. Even though they provide three different numbers as expected frequencies in their answers to question 1, they do not consider variability, since they do not have the theoretical distribution as a reference for assessing the differences.

Students who connect theory and data give better quality responses: that is, when they accept that a theoretical distribution underlies the situation, but realize that in practice, the results usually differ from the theoretical expected values, and accept that such variability cannot be eliminated. Such responses are an indication that they have properly understood the notion of variability and are beginning to express this understanding. These students seem to understand the law of large numbers and that frequencies approximate theoretical probabilities. However, it is necessary that they translate the expected frequencies to relative frequencies and compare them with the corresponding probabilities.

Finally, responses in the *blurred relationship between theory and data* are odd, because we assume that a student who answers the first question with 250, 500, 250 does so because he/she knows the probability distribution (¼, ½, ¼). This is likely obtained through the classical approach; however, it is strange that at the same time, he/she answers the second question by proposing approximate probabilities, for example (0.244, 0.498, 0.258).

13.4 Results

In Tables 13.4 and 13.5, we present the responses to questions 1 and 2 for pre- and post-questionnaires. Table 13.4 presents the results from the group of students who had never taken a statistics or probability course, while Table 13.5 presents the results from the group of students who had previously taken a statistics and probability course.

Table 13.4 Percentage of responses according to SOLO categories of group 1 (n = 37)

Level	Question 1		Question 2	
	Pre-test	Post-test	Pre-test	Post-test
Prestructural	51	17	34	11
Unistructural	34	37	21	19
Multistructural	9	20	13	27
Relational	6	26	32	43
Total	100	100	100	100

It is worth noting that for both questions and for both groups (Tables 13.4 and 13.5), the number of responses in the post-test classified in the two lower SOLO levels is lower than the corresponding number in pre-test. This means that, in general, the quality of the post-test responses is better than the quality of the pre-test responses. Our interpretation of these results is that students achieved a better understanding of the problems because of the learning activities. For example, a greater number of students propose frequencies whose sum is 1000 and probabilities whose sum is 1 (or percentage values that sum to 100). Many learned to associate 0, 1, and 2 with the events "Ana wins," "Billy wins," and "Charles wins." They also learned that the event $x = 1$ is more likely than $x = 0$ and $x = 2$, and that the probability of the latter two events is approximately equal. However, it seems very difficult for them to articulate their understanding of the distribution and their intuitive sense of variability.

Table 13.6 presents frequencies from students who had not previously taken a probability and statistics course for the categories defined by combining the responses to both questions. It is important to remember that only those pairs of responses that were both classified in the Unistructural or level or higher in the SOLO classification are taken into account; for this reason, there are only 13 responses from the pre-test and 28 from the post-test, although there were 37 students in total.

Table 13.5 Percentages of responses according to SOLO categories of group 2 ($n = 66$)

Level	Question 1		Question 2	
	Pre-test	Post-test	Pre-test	Post-test
Prestructural	52	27	48	4
Unistructural	45	50	15	16
Multistructural	3	0	0	12
Relational	0	23	37	68
Total	100	100	100	100

Table 13.6 Frequencies of responses classified in each category of group 1

Categories	Number of responses	
	Pre-test	Post-test
Blurred relationship between theory and data	4	3
Determinism	3	6
Empirical commitment	2	4
Connecting theoretical probabilities and data	4	15
Total	13	28

The number of answers classified in the determinism category (3 on the pre-test, 6 on the post-test) as well as those classified in empirical commitment category (2 on the pre-test, 4 on the post-test) doubled from pre-test to post-test. However, we note that in the post-test, variability was considered in 15 of the responses, while in the pre-test it was only considered in 4; that is, 11 additional students were able to determine the probability distribution and to make a prediction that tried to express variability in the post-test.

The simulation activities thus promoted understanding of the nature of the problem and allowed the students to perceive trends in the results—that is, the shape of the distribution. It is likely that their observations motivated them to search for an explanation, which many of them found in determining the theoretical distribution. That is, the students identified the elements of the sample space and the favourable cases that corresponded to each value of the variable. This is why 18 new students provided the distribution in the post-test. This hypothesis is plausible due to the simplicity of the distribution, which has a sample space that consists of only four elements that become apparent in the simulation program.

In Table 13.7, we present data from students who had previously taken a course in probability and statistics. As in the previous analysis, it is important to remember that here, only those pairs of responses that were both classified in the Unistructural level or higher in the SOLO classification are taken into account. for this reason, only 25 pairs of responses in the pre-test and 41 in the post-test are considered, rather than the original 66 (the total number of students).

It should be noted that 8 pairs of responses are classified in the empirical commitment category. Since these students had already taken a course on probability and statistics, their responses suggest that they may think that the classical and frequentist approaches to probability are mutually exclusive, and that they must therefore make a choice between them. Perhaps these students prefer the second approach because they feel that it better reflects reality (Konold et al. 2011). We also note that most pairs of answers, both in the pre-test (18) and the post-test (21), are classified in the determinism category: such students determine the correct distribution (¼, ½, ¼) and predict frequencies of 250, 500, 250, without referring to variability. It is unlikely that the students did not realise that the simulation results do not produce exactly the frequencies 250, 500, 250, so one of the following beliefs may explain their response:

Table 13.7 Frequencies of responses classified in each category of group 2

Component	Number of responses	
	Pre-test	Post-test
Blurred relationship between theory and data	7	7
Determinism	18	21
Empirical commitment	0	8
Connecting theoretical probabilities and data	0	5
Total	25	41

- In mathematics, there is always a unique and exact answer; expressions such as "more or less," "approximately," etc., are not mathematical.
- What matters in the result is the theoretical response (in this case, the expected frequencies).

Such beliefs are reinforced by poor development of language for expressing variability. Our suggestion is that to develop such language, students should learn to combine ranges and probabilities to make *statements of certainty*.

13.5 Discussion and Conclusion

Because of the simulation activities, many students achieved a better understanding of the problems, reflected by the fact that a greater number their responses were classified in higher SOLO levels. However, most of the students had difficulty integrating variability into their responses, which may be attributed to the students not having developed an appropriate language to express variability or to certain beliefs about the form of correct responses in mathematics. Prior to the simulation activities, students who knew or determined the theoretical distribution were prone to predict the expected frequencies (determinism) while those who did not know or could not build the distribution used only the frequentist approach (empirical commitment). After the activities, many students persisted in responding as if it was possible to get the expected frequencies; but several of them added a commentary such as "It is a theoretical result, but the actual result can be different" or "approximately…" etc. Students who had not taken a probability course mostly adopted the frequentist approach but some of them constructed the distribution (¼, ½, ¼) after the activities.

The educational statistical software served as a tool to get results by provided a new form of representation for the probabilistic objects under consideration. For instance, it allowed students to provide statements of certainty such as "it is very probable that frequencies of 0, 1 and 2, are around 250, 500, 250 respectively". These results would be practically inaccessible without such a tool because in the teaching approach without simulation activities, students do not have the opportunity to perceive patterns of the results.

The foregoing analysis is primarily descriptive, because the concepts employed (for example, determinism and empirical commitment) were designed to fit the context described herein. Determinism means interpreting theoretical propositions about chance as if they were causal statements or as if they predicted results that can only occur in one way. This misconception is generally due to students' inability to address the uncertainty, randomness, or variability present in a situation. It can also be observed in responses to other problems or situations. For example, "If the probability of rolling five with a die is 1/6, then exactly one five will be obtained when rolling a die six times," or "A random sample is a miniature replica of the population." This type of thinking is not unusual and is promoted by the conception of mathematics as a set of definitions and procedures that are unrelated to reality.

On the other hand, empirical commitment suggests a belief in only that which can be observed. Faced with the difficulties of conceptualizing frequencies and the variability of results, students are limited to describing them. Here again, the same thinking can be observed in other situations. It is related, for instance, to the outcome approach (Konold 1989), whereby an answer to a probability question is assessed as correct or incorrect depending on whether it is coincident or not with the observed results.

In order to sponsor deeper understanding of the notion of variability, the teaching and learning of probabilistic reasoning must integrate models and data. It should involve conceptualization and not just description, since concepts cannot be seen in reality; however, it is naive to think that students will acquire understanding of the underlying distribution simply by performing simulation activities and observing their results. In teaching probability, the classical and frequentist approaches should therefore be treated under the more general framework of integrating theory and data, and avoiding neglecting variability for the sake of simplicity.

Acknowledgements Research Projects EDU2016-74848-P (AEI, FEDER), EDU2013-41141-P (MINECO), and P254301 (CONACYT).

References

Batanero, C., Chernoff, E., Engel, J., Lee, H., & Sanchez, E. (2016). *Research on teaching and learning probability*. ICME-13. Topical survey series. New York: Springer.

Biggs, J. B., & Collis, K. F. (1991). Multimodal learning and the quality of intelligent behavior. In H. A. H. Rowe, (Ed.), *Intelligence: Reconceptualization and measurement* (pp. 57–76). Hillsdale, NJ, USA: Lawrence Erlbaum Associates Publishers and Australian Council for Educational Research.

Chance, B., Ben-Zvi, D., Garfield, J., & Medina, E. (2007). The role of technology in improving student learning of statistics. *Technology Innovations in Statistics Education, 1*(1), 1–26.

Gal, I. (2005). Towards "probability literacy" for all citizens: Building blocks and instructional dilemma. In G. A. Jones (Ed.), *Exploring probability in school: Challenges for teaching and learning* (pp. 39–63). New York: Springer.

Hoffman, T., Maxara, C., Meyfarth, T., & Prömmel, A. (2014). Using the software FATHOM for learning and teaching statistics in Germany – A review on the research activities of Rolf Biehler's working group over the past ten years. In T. Wassong et al. (Eds.), *Mit Werkzeugen Mathematik und Stochastik Lernen.* Heidelberg: Springer.

Ireland, S., & Watson, J. (2009). Building a connection between experimental and theoretical aspects of probability. *International Electronic Journal of Mathematics Education, 4*(3), 339–370.

Jones, G. A. (2005). Reflections. In G. A. Jones (Ed.), *Exploring probability in school. Challenges for teaching and learning* (pp. 367–372). New York: Springer.

Kolmogorov, A. N. (1956). *Foundations of the theory of probability.* New York: Chelsea Publishing Company.

Konold, C. (1989). Informal conceptions of probability. *Cognition and Instruction, 6*, 59–98.

Konold, C., Madden, S., Pollatsek, A., Pfannkuch, M., Wild, C., Ziedins, I., et al. (2011). Conceptual challenges in coordinating theoretical and data-centered estimates of probability. *Mathematical Thinking and Learning, 13*(1 and 2), 68–86.

Lee, H. S., Angotti, R. L., & Tarr, J. E. (2010). Making comparisons between observed data and expected outcomes: Students' informal hypothesis testing with probability simulation tools. *Statistics Education Research Journal, 9*(1), 68–96.

National Council of Teachers of Mathematics. (2000). *Principles and standards for school mathematics.* Reston, VA: Author.

Parzen, E. (1960). *Modern probability theory and its applications.* New York: Wiley.

Torok. R. & Watson, J. (2000). Development of the concept of statistical variation: An exploratory study. *Mathematics Education Research Journal, 12*(2), 147–169.

Watson, J, M., & Kelly, B. A. (2004). Statistical variation in a chance setting: A two-year study. *Educational Studies in Mathematics, 57*(1), 121–144.

Watson, J, M., Kelly, B. A., Callingham, R. A., & Shaughnessy, M. (2003). The measurement of school students' understanding of statistical variation. *International Journal of Mathematical Education Science and Technology, 34*(1), 1–19.

Chapter 14
Students' Reasoning About Sample Space and Probabilities of Compound Events

Pedro Landín and Jesús Salinas

Abstract A probabilistic reasoning hierarchy for the concepts of sample space and the probability of a compound event is proposed as a tool to describe high-school students' performance when solving two problems involving binomial experiments, before and after a period of teaching. As expected, the responses to the post-tests indicate an improvement in the levels of reasoning in comparison with the pre-test results. In the pre-test, list and tables were the most frequently used procedures to count sample space elements and to compute the probabilities of compound events. In the post-test, the most common strategies were enumeration, tree diagrams, and use of the product rule. Tree diagrams were a useful tool in improving the students' probabilistic reasoning.

Keywords Probabilistic reasoning hierarchy · Sample space · Classical probability · Compound event · Binomial experiment

14.1 Introduction

Learning about sample space and probability of compound events[1] in high school is a multi-factorial problem. On the one hand, the concept of probability is of great importance in scientific thinking (Steinbring 1986). The topics of sample space and

[1]We consider compound events to be events such as "*A* or *B*," "*A* and *B*," and "the complement of *A*." In this chapter, we use compound events consisting of pairs or trios of outcomes (e.g., Girl-Girl, Even-Even-Even).

P. Landín (✉) · J. Salinas
Colegio de Ciencias y Humanidades, UNAM, Av. Universidad 3000,
2° Piso, Ciudad Universitaria, Mexico City, Mexico
e-mail: pedrorubenl@yahoo.com.mx

J. Salinas
e-mail: jes54@unam.mx

C. Batanero and E. J. Chernoff (eds.), *Teaching and Learning Stochastics*,
ICME-13 Monographs, https://doi.org/10.1007/978-3-319-72871-1_14

probability of compound events are included in the high school curriculum in several Western countries (Jones et al. 2007, p. 914), and are considered to be fundamental stochastic concepts (e.g., Heitele 1975; Jones et al. 1997). On the other hand, students' difficulties in computing frequencies and probabilities of compound events is a fact that is widely documented in the literature (e.g., Chernoff and Zazkis 2011; Fischbein and Schnarch 1997; Kahneman and Tversky 1972; Lecoutre 1992; Polaki 2005). According to Batanero et al. (2016), some distinctive features of probability are its different approaches (e.g., classical, frequentist, subjective) and that "counterintuitive results also appear in elementary probability" (p. 9). The complexity of the notion of probability and of related concepts (e.g., combinatorics and algebra) leads to difficulties experienced by students in computing probabilities of compound events, even at college and high school levels (Keren 1984; Rubel 2007; Shaughnessy and Ciancetta 2002).

However, according to Jones et al. (1997), "research evidence with regard to sample space is conflicting and highlights the need to study children's thinking on this construct more comprehensively, and within a probability context" (p. 105). Furthermore, in reference to computing probabilities of compound events, Rubel (2007) stated "there remains a need for further exploration of school students' strategies, and reasoning" (p. 533). In response to this need, our objective in this chapter is to document the reasoning and mathematical objects used by high school students when tasked with determining the sample space or calculating probabilities of compound events in a binomial experiment.[2]

14.2 Literature Review

In this section, we provide an overview of research on the assignment of probability to compound events in binomial random experiments. We first describe the research conducted during the period between 1972 and 1997, during which researchers found that many subjects of different ages use heuristics and show biases in their probabilistic reasoning. We then describe the teaching experiments conducted by other researchers in an effort to describe hierarchies in students' probabilistic reasoning.

[2]According to Yates et al. (2003, p. 439), a binomial experimental setting is a situation where the following conditions are satisfied: (1) Each observation (trial) falls into one of just two categories ("success" or "failure"); (2) there is a fixed number of independent observations; and (3) the probability of success (p) is the same for each observation.

14.2.1 Heuristics and Biases in Probabilistic Reasoning

In the 1970s, Kahneman and Tversky (1972) performed a study involving 1500 Israeli students in Grades 10, 11, and 12 (aged 15–18). According to Kahneman and Tversky, a person uses the *representativeness heuristic* when "the likelihood of an event is evaluated by the degree to which it is representative of the major characteristics of the process or population from which it originated" (p. 451). Some behaviours and beliefs, such as the negative recency strategy or the gambler's fallacy (the belief that, for example, if a coin has been flipped three times and three heads have been observed, the result of the fourth flip is more likely to be tails) can also be explained by this heuristic. Tversky and Kahneman (1983) also studied a heuristic linked to compound events in a study involving more than 3000 subjects of different backgrounds who were tasked with dozens of problems. The study found that the probability of an event $A \cap B$ appears, under certain conditions, to be higher than the probability of A, which Tversky and Kahneman refer to as the *conjunction fallacy*.

In a study by Lecoutre (1992), more than 1000 students with different backgrounds in probability were tasked with a problem about two dice being thrown simultaneously. The researchers found that most of the students believed that obtaining two sixes had the same likelihood as obtaining a five and a six because both outcomes were a matter of chance. They referred to this belief as the *equiprobability bias*, defined more generally as reasoning according to which "the two results to compare are considered to be equiprobable because it is a matter of chance" (p. 561).

Working with college students and students in Grades 5, 7, 9, 11, Fischbein and Schnarch (1997) studied the evolution of probabilistic misconceptions with age. They designed a questionnaire that included eight well-known problems from the literature, four of which were related to compound events. Among their results, they found that the frequency of misconceptions decreased with age for problems that assessed representativeness and the negative recency effect. In contrast, they found that the equiprobability bias was frequent and stable across ages. It must be noted that the authors requested direct responses, which expressed "immediate intuition" (p. 101) and did not offer justification.

The aforementioned studies followed a psychological approach that involves observing and describing participants' responses. However, some researchers (e.g., Bill et al. 2009; Chernoff and Zazkis 2011; Jones et al. 1997; Polaki 2005; Sánchez and Landín 2014; Rubel 2007) have conducted teaching experiments to describe students' reasoning strategies using a variety of theoretical frameworks.

14.2.2 Hierarchies for Students' Probabilistic Reasoning

Jones et al. (1997) formulated the Probabilistic Reasoning Framework, which distinguishes, in order of increasing sophistication, four levels of reasoning: subjective, transitional, informal quantitative, and numeric. This framework incorporates four key constructs: sample space, probability of an event, probability comparisons, and conditional probability. Their study tested third-grade children three times (at the beginning, middle, and end of the school year).

Polaki (2005) reported the strategies used by elementary and middle school students in listing the results of simple and compound random experiments. These strategies included the following (listed in order of increasing sophistication): (a) arbitrary or incomplete lists for simple events, (b) trial and error strategies, (c) partially generative strategies for compound events, and (d) generative strategies for compound events. In addition, Polaki described certain thinking tendencies of elementary school students in formulating probability predictions for simple and compound random experiments. Some of the tendencies observed were: (a) subjective responses, (b) use of informal quantitative phrases to describe probabilities, and (c) linking predictions to the composition of the sample space and order probabilities. Polaki concluded that "most of the students experienced more success in listing sets of outcomes than in making probability predictions for compound random experiments" (2005, p. 212).

Rubel (2007) reported on a study involving students from Grades 5 to 11 in a private school ($n = 173$). The students completed a probability test that included tasks adapted from the research literature; they were asked to justify their responses. The author reported the results from a simple events task that involved drawing from an urn and three compound events tasks that involved coin flipping. The correct and incorrect responses were described and their justifications were analysed to provide a description of the students' strategies, reasoning, and underlying cognitive models. Among the results, Rubel reported that most of the students in the 9th and 11th grades used the notion of sample space to correctly calculate the probability that one will get one "head" and one "tail" in flipping two coins.

Bill et al. (2009) reported on a teaching experiment with 19 Grade 10 students during which a binomial problem was examined through the use of three methods: the binomial formula, Pascal's Triangle, and a Fathom simulation. Using both physical and virtual coins, the students explored the distinction between a coin sequence and a coin combination (where order is not important) as a foundation for a 5-item true/false test task. The data were collected through a pre-test and a post-test applied under traditional exam conditions and were used to evaluate the development of the students' comprehension of binomial distribution using the SOLO (Structure of Observed Learning Outcome) taxonomy. In the lowest level of the taxonomy, the computation is incorrect, while in the highest level the student explicitly compares different methods and provide justification for the method chosen.

In another study, Chernoff and Zazkis (2011) asked prospective high school mathematics teachers to solve two tasks involving compound event probability. In the first task, the teachers were asked: "In a family of three children, what is the probability that there are two boys and a girl?" (p. 15). The second task was a version of the classic problem of an interrupted game: Two players are playing a game of chance involving coins. If the game is interrupted before the agreed-upon winning score is reached, how should the stakes be split between them? The authors defined a sample set as any set of all possible outcomes, which do not have to be equiprobable, and suggested "a desirable pedagogical approach in which the solution builds on the set of outcomes as identified by learners and serves as a bridge towards mathematical convention" (p. 15).

Sánchez and Landín (2014) described high school students' responses to binomial probability problems. A pre-test and a post-test were administered before and after a course to assess the influence of instruction. The authors constructed and used a reasoning hierarchy to classify the students' responses. They stated that understanding of the classical definition of probability, the multiplication rule, and the binomial probability formula were indicators of transition between levels.

Mooney et al. (2014) synthesised key characteristics of probability frameworks, based on SOLO. Such frameworks were built by two research teams: one led by Graham Jones and another led by Jane Watson. The Probabilistic Thinking Framework (Mooney et al. 2014) considers levels of growth (prestructural, unistructural, multistructural, and relational) in probabilistic thinking, which moves from being non-mathematical or personalized to quantitative and proportional. The framework seeks to summarise research results of the past 20 years conducted with elementary and middle school students. The authors suggest that "researchers have not characterized thinking at the extended abstract level, which would bridge into the formal mode of functioning" (p. 499).

Considering the significance of the topics sample space and probability of a compound event, their inclusion in the high-school curriculum, the difficulties experienced in learning such topics, and the gap found in the literature review, the aim of this study is to explore the following research question:

- How can we describe high-school students' reasoning in determining a sample space and in computing the probability of a compound event of a binomial random experiment, before and after a teaching and learning process?

Below we describe the conceptual framework used in this research.

14.3 Conceptual Framework

Many frameworks of probabilistic reasoning have been based on the neo-Piagetian SOLO model (Biggs and Collis 1982). In this taxonomy, Biggs and Collis (1991) incorporate five modes of representation, which illustrate levels of abstraction that subjects may reach (without necessarily working with every task in a given level) and which are assumed to constitute the basis of cognitive development. These five modes and the age at which they typically emerge are: sensorimotor (from birth), ikonic (beginning at 18 months), concrete-symbolic (beginning at around 6 years), formal (beginning at around 14 years), and post-formal (beginning at around 20 years).

According to Biggs and Collis (1991), every mode includes learning cycles constituted by unistructural, multistructural, and relational levels, which are defined according to the structural complexity of the responses to a task. They defined the SOLO levels as follows: unistructural, when the response includes only one relevant component (fact, concept, or process) of the task or idea; multistructural, when the response includes several unrelated relevant features; and relational, when the learner integrates several parts in a coherent way. There are two additional levels: at the prestructural level the learner is distracted or misled by an irrelevant aspect, and at the extended abstract level, "the learner generalizes the structure to take in new and more abstract features" (Biggs and Collis 1991, p. 65). Biggs and Collis (1982) suggest that these levels can be used to categorize responses and "are not meant as labels to tag students" (p. 23). The SOLO taxonomy has been used to develop different frameworks of probabilistic reasoning for students at the concrete-symbolic mode.

A second component of our framework is constituted by some concepts developed by the Onto-Semiotic Approach (OSA) to mathematics education, which will be useful in analysing the mathematical objects used by the students in solving the problems presented. This theoretical framework asumes an ontology of mathematical objects that considers mathematics as: (a) a socially shared problem-solving activity; (b) a symbolic language; and (c) a logically organised conceptual system. According to OSA, a *mathematical practice* is any action or manifestation that a subject carries out to solve mathematical problems, communicate the solution to other people, and validate and/or generalise the solution to other contexts and problems (Godino et al. 2007, p. 129).

When carrying out a mathematical practice, a subject works with a set of primary mathematical objects: *problem situations, language, concepts, properties, procedures,* and *arguments. Situations* (problems, exercises, etc.) are tasks that induce mathematical activity, while *language* constitutes terms, expressions, notations, and graphs. *Procedures* are operations, algorithms, or techniques carried out to solve a mathematical task; *concepts* are given through definitions or descriptions; and *properties* are in the form of statements or propositions. Finally, an *argument* (deductive, inductive, etc.) is used to explain or validate propositions and procedures (Godino 2002, pp. 245–246).

In sum, we have chosen the SOLO taxonomy to develop the probabilistic reasoning hierarchy used to analyse and to categorize the responses to the given questionnaires. The reason for this choice is that this model addresses the level of structural complexity of responses and this research seeks to characterize probabilistic reasoning in the formal mode of representation. Additionally, some elements of OSA will be used to describe the mathematical objects that students use when solving problems or tasks.

14.4 Methodology

Below, we describe the participants involved and instruments used, the hierarchy proposed, and the procedure used to classify students' responses according this hierarchy.

14.4.1 Participants and Instruments

The Mexican high school curriculum is very heterogeneous. Most of the institutions providing high school education include mandatory courses on algebra, Euclidian geometry, trigonometry, analytic geometry, and pre-calculus, while statistics and probability courses are typically optional. Therefore, some students in Mexico complete their high school studies without ever having taken a course on this subject.

The sample consisted of 28 high school students (ages 17–18, Grade 12) from a Mexican public school who were enrolled in a probability course. At this age, students begin thinking in the formal mode, which "both incorporates and trascends particular circumstances" (Biggs and Collis 1991, p. 63). The course focused on classical probability and problem solving, and content included basic combinatorial operations, use of tree diagrams to determine the sample space of compound random experiments, random variables, probability distributions (binomial and normal), and sampling distributions. Two weekly sessions lasting about 90 min each were taught from January to April in 2015.

Questionnaires of eight items each were used to assess the students' reasoning before and after instruction at the beginning, middle, and end of the course (one pre-test and two post-tests) under traditional assessment conditions. By the time the pre-test took place, the group had already studied descriptive statistics topics (for one and two variables) as well as sample space and probability of events in simple random experiments. The first post-test was administered in the middle of the course after the students had studied compound random experiments and probability distributions of discrete random variables, with the aim of analysing the influence of teaching on students' reasoning. The second post-test was administered at the end of the course to study the stability of students' knowledge.

Probability problems involving the concept of sample space and the measure of probability in the context of binomial experiments $\left(p = q = \frac{1}{2}\right)$ with two or three stages were included in the questionnaires. None of the problems used in the questionnaires were used in the teaching program. Below, we present four of the problems included in the questionnaires and the procedure used to analyse the students' responses. Problems 2a and 2b are isomorphic (i.e., they have the same mathematical structure) and involve a compound experiment with two stages $(n = 2)$, while Problems 3a and 3b are isomorphic and include a compound experiment with three stages $(n = 3)$.

Problems 2a (involving birth sequences) and 3a (involving coin sequences) were included in the pre-test and the first post-test questionnaires because their contexts are frequent in the research literature. Problem 2a was adapted from Moore (2007), while Problem 3a was taken from a Mexican textbook. Problems 2b (involving colour sequences) and 3b (involving die sequences) were only included in the second post-test.

Problem 2a (Moore 2007, p. 268). A couple plans to have two children. Suppose that a boy and a girl are equally likely to be born and that the gender of one child does not influence the gender of the other.

a. Write down all of the possible gender arrangements for two children.
b. Supposing that the results listed in (a) are equally likely, what is the probability that two girls are born?
c. What is the probability of having exactly one child of each gender?

Problem 3a (Pastor 1998, p. 76). You toss a $1 coin, a $2 coin, and a $5 coin in the air and observe the sides they land on.

a. Determine the number of possible outcomes of tossing the three coins, taking into account that each coin has heads on one side and tails on the other.
b. What is the probability that you will get no tails when you toss the three coins?
c. What is the probability that you will get one tail when you toss the three coins?
d. What is the probability that you will get two tails when you toss the three coins?
e. What is the probability that you will get three tails when you toss the three coins?

Problem 2b Two marbles, one white and one black, are put into a bag. If two drawings with replacement are performed:

a. Make a list of the possible colour outcomes in the experiment.
b. Knowing that the outcomes listed in (a) are equally probable, calculate the probability of drawing two white marbles.
c. Calculate the probability of drawing exactly one marble of each colour.

Problem 3b Consider the following task: You toss three six-sided dice, then observe whether the outcome is an even or an odd number for each die.

a. Determine the number of possible outcomes when tossing the three dice, considering whether an even or an odd number is observed on each die.
b. Determine the probability of obtaining no even numbers when tossing the three dice.
c. Determine the probability of obtaining exactly one even number when tossing the three dice.
d. Determine the probability of obtaining exactly two even numbers when tossing the three dice.
e. Determine the probability of obtaining exactly three even numbers when tossing the three dice.

14.4.2 Hierarchy

According to Hart (1981), a hierarchy "implies a string of abilities/levels/stages/concepts which are ordered from simple to complex" (p. 205). Hart (1981) explains that a hierarchy can refer to learning, teaching, or a logical sequence; however, these three aspects must be closely linked. In Table 14.1, we present the components of knowledge or mathematical objects considered necessary to understand or construct the sample space of a binomial experiment and to calculate the probability of an event in such an experiment.

In Table 14.2, we present the description of our proposed probabilistic reasoning hierarchy, which includes learning descriptors for the concepts of sample space of a compound experiment and the probability of a compound event. In developing this hierarchy, we considered the elements from the Table 14.1, the reasoning hierarchy developed by Sánchez and Landín (2014), the responses provided by the students in the sample of this research and the Probabilistic Thinking Framework (Mooney et al. 2014). In Sects. 4.3.1 and 4.3.2, we present examples of responses for each category.

14.4.3 Procedure

Based on this probabilistic reasoning hierarchy, we independently categorized the responses from all twenty-eight participants collected during the three assessments. Coding discrepancies between researchers were discussed until a consensus was reached (inter-jugde agreement; Biggs and Collis 1982). We analysed the responses in the effort to validate and adjust the hierarchy we propose here. In addition, the mathematical objects used to solve each problem were described and analysed according to three of the OSA categories: concepts, procedures, and language.

Below, the procedure used to categorize responses to two- and three-stage problems is described. The responses shown were selected because they were either frequent responses or because they were representative of a particular reasoning level.

Table 14.1 Components of knowledge needed to understand or construct the sample space and probability of events in a binomial experiment

1. Complementarity of the elements of the sample space
2. Listing the elements of the sample space systematically
3. Fundamental product rule for counting (logical multiplication of properties in Nunes et al. 2014)
4. Laplace's rule (probability is the proportion of favourable to all possible cases; Batanero et al. 2016)
5. Addition and product rules of probability
6. Binomial probability function

Table 14.2 Probabilistic reasoning hierarchy for the concepts of sample space and probability of a compound event in a binomial experiment

Probabilistic reasoning level	Characteristics of the responses
Prestructural	The sample space of the compound experiment is omitted. The responses are idiosyncratic, influenced by cognitive biases (such as the equiprobability bias or the conjunction fallacy) or heuristics (such as the representativeness heuristic). The responses are incoherent or contain many mistakes
Unistructural	The responses show incomplete sample spaces with or without the aid of lists or tree diagrams for two- or three-stage experiments, as well as a tendency to fall back on subjective probabilistic reasoning. The students correctly use Laplace's definition of probability, or they use the product rule but commit errors or omit calculations
Multistructural	The responses show systematic sample spaces for experiments with two or more stages, supported by tree diagrams or lists of results. The subjects count and use proportions or probabilities when judging probabilistic situations. The classic definition of probability is used in the correct responses. The addition rule or the product rule is used to obtain partial or correct responses
Relational	Sample spaces for experiments with two or more stages are systematically generated with or without a list of results. Sample spaces and probabilities are connected. Probabilities for complex situations, including situations involving non-equiprobable outcomes, are determined. The addition rule, the product rule of probabilities, or a binomial random variable is identified and used

14.4.3.1 Procedure to Categorize the Responses for Two-Stage Problems

Figures 14.1, 14.2, 14.3 and 14.4 reproduce examples of the responses for Problem 2a in the first post-test, categorized according to probabilistic reasoning level.

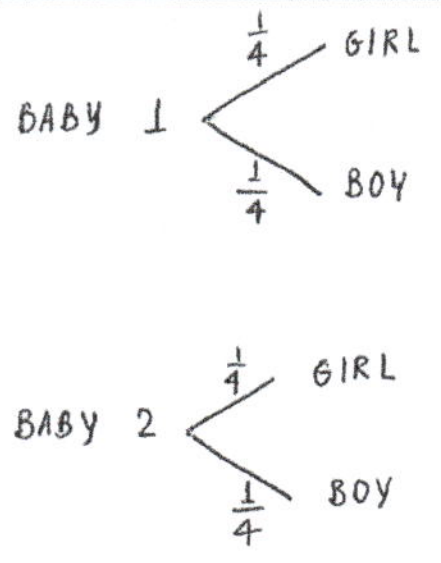

Probability that two girls are born

$\frac{2}{4} = \frac{1}{2} =$ 0.5 possibilities that two girls are born

Probability of having exactly one child of each gender

$\frac{1}{4} =$ 0.25 of having a girl and

$\frac{1}{4} =$ 0.25 of having a boy

Fig. 14.1 Response to Problem 2a categorized at the prestructural level

Girl-girl, Girl-boy, Boy-boy

$P(b)=\frac{1}{3}$ $P(c)=\frac{1}{3}$

Fig. 14.2 Response to Problem 2a categorized at the unistructural level

(boy-boy) (boy-girl), (girl-girl) (girl-boy)

P (two girls are born)= $1\ out\ of\ 4 = \frac{1}{4} = 25\%$

P (of having exactly one child of each gender)=(boy-girl)(girl-boy) = $\frac{2}{4} = 50\%$

Fig. 14.3 Response to Problem 2a categorized at the multistructural level

The response in Fig. 14.1 was categorized at the prestructural reasoning level because the student does not determine the sample space for the compound experiment, which should involve compound events (boy-girl, girl-girl, etc.). Rather, this student only considers simple events linked to each simple experiment. Moreover, the probabilities requested are incorrect because the student obtained those probabilities by dividing the number of girls (two) by the total number of possible outcomes. This response can be described as being at the prestructural (subjective) level in the Probabilistic Thinking Framework (Mooney et al. 2014).

The response in Fig. 14.2 was very common. We categorized this response as unistructural, since the sample set (Chernoff and Zazkis 2011) includes three compound events. Because the student does not consider the order of boys and girls in the outcomes, the probabilities calculated using Laplace's rule are incorrect. This student might have reasoned according to the equiprobability bias (Lecoutre 1992) if he or she considers the event Girl-Boy equivalent to the event "exactly one child of each gender," which can in fact occur in two ways: Boy-Girl and Girl-Boy.

The response in Fig. 14.3 was categorized at the multistructural reasoning level because the student determined a complete sample space, used Laplace's definition of probability, and correctly calculated the probabilities of the events "two girls" and "exactly one child of each gender." This response shows a systematic sample space and is described as multistructural in the Probabilistic Thinking Framework (Mooney et al. 2014, p. 502).

Finally, the response in Fig. 14.4 was categorized at the relational reasoning level. It includes a tree diagram, a complete sample space, and calculations for the requested probabilities based on the product rule and the addition rule. Using the product rule, either for counting or for computing probabilities, is suggestive of a higher degree of abstraction, and allows the student to calculate the cardinality of considerably large sample spaces and to determine probabilities of events without listing all possible outcomes, whether they are equiprobable or not. The student was able to compute conditional probabilities in replacement situation, a response classified in the upper level (relational) in the Probabilistic Thinking Framework (Mooney et al. 2014).

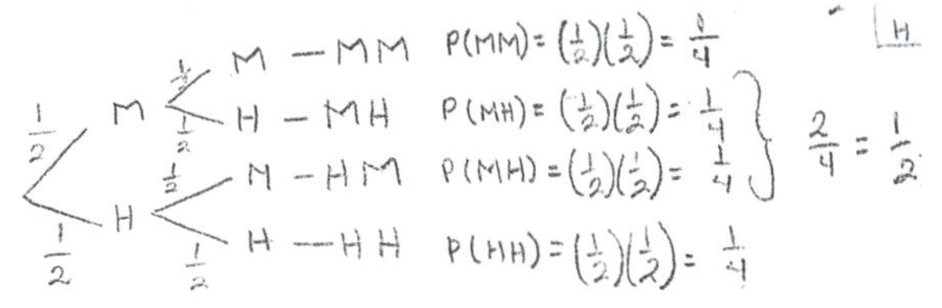

Fig. 14.4 Response to Problem 2a categorized at the relational level

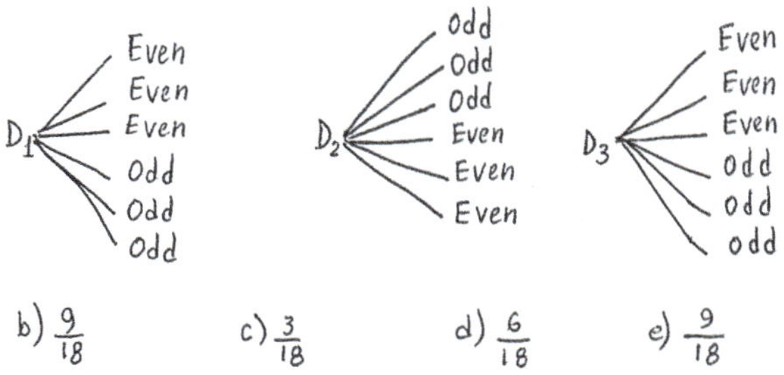

Fig. 14.5 Response to Problem 3b categorized at the prestructural level

14.4.3.2 Procedure to Categorize the Responses to Three-Stage Problems

Figures 14.5, 14.6, 14.7 and 14.8 reproduce examples of responses to Problem 3b in the second post-test, categorized by probabilistic reasoning level. The response in Fig. 14.5 was categorized at the prestructural reasoning level because the student did not list the sample space for the compound experiment. Instead, the student considers the simple sample space that corresponds to each die, and then only considers 18 simple events in his sample set. Moreover, the probabilities requested are incorrect: The student assigns the probability $\frac{3}{18}$ to the event "obtaining exactly one even number," which is the number of even faces on a die divided by the 18 simple events listed. Such responses were considered to be prestructural in the Probabilistic Thinking Framework (Mooney et al. 2014). This student follows the same line of reasoning in all the sections of the Problem 3b.

The response in Fig. 14.6 was categorized as unistructural, since the student's response includes six compound events; still, the student is unable to determine the sample space, which includes eight elements. The reason is that the enumeration lacks a system (Batanero et al. 1997), even though some elements of the sample space are enumerated. As a result, the probabilities calculated using Laplace's rule for (b), (c), (d), and (e) are incorrect: the student applies Laplace's rule but bases calculations on an incomplete sample space. Note, however, that the application of this rule is correct for the set considered by the student. Finally, the response in Fig. 14.6 can be considered as transitional from subjective thinking to quantitative thinking (Mooney et al. 2014, p. 501).

a) Ω={even, odd}

Even-even-even

Even-even-odd

Even-odd-even

Odd-odd-odd

Odd-even-odd

Odd-odd-even

Total= 6 possible outcomes

b) P(no even-numbers)= $\frac{1}{6}$

c) P(one even-number)= $\frac{2}{6}$

d) P(two even-numbers)= $\frac{2}{6}$

e) P(three even-numbers)= $\frac{1}{6}$

$\frac{6}{6} = 1$

Fig. 14.6 Response to Problem 3b categorized at the unistructural level

D_1={1,2,3,4,5,6} D_2={1,2,3,4,5,6} D_3={1,2,3,4,5,6}

a) The possible outcomes are 6x6x6= 216

b) The probability of obtaining no even numbers= 3x3x3= $\frac{27}{216}$

c) The probability of obtaining exactly one even number= $\frac{1}{216}$

d) The probability of obtaining exactly two even numbers= $\frac{2}{216}$

e) The probability of obtaining exactly three even numbers= 3x3x3= $\frac{27}{216}$

Fig. 14.7 Response to Problem 3b categorized at the multistructural level

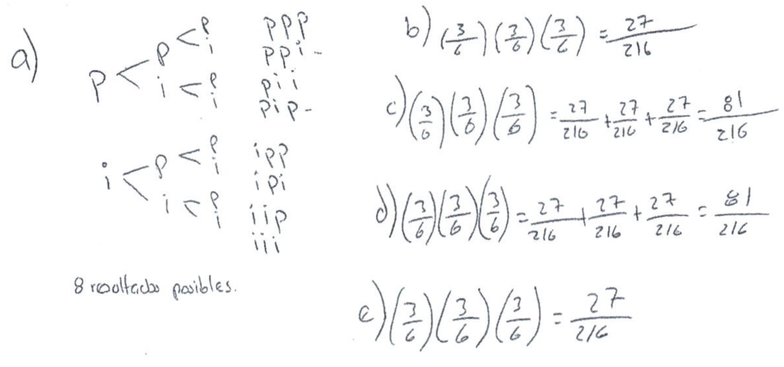

Fig. 14.8 Response to Problem 3b categorized at the relational level

The response in Fig. 14.7 was categorized at the multistructural level of reasoning because the student makes use of a fundamental counting principle: the product rule. This is a characteristic in multistructural level in the Probabilistic Thinking Framework (Mooney et al. 2014). Using this principle, the student obtains the correct total number of possible outcomes in the sample space. Using the same strategy, the student correctly calculates the number of favourable events and the probabilities requested in sections (b) and (e) of the problem.

However, the probabilities assigned to events (c) and (d) are incorrect. For instance, the student's response to (c) is $\frac{1}{216}$. The correct response can be obtained by considering that an even number can occur in the events EOO (Even, Odd, Odd), OEO, OOE, and that each of these events can occur in 27 different ways. As a result, the requested probability is $\frac{81}{216} = \frac{3}{8}$.

The response in Fig. 14.8 was categorized at the relational reasoning level because the student systematically constructed a sample space of eight compound events using a tree diagram. Additionally, probabilities are correctly assigned to the required events using the multiplication and addition rules of probability. From such results, the student might have deduced that the tree diagram synthesised 216 possible equiprobable outcomes from tossing three dice.

14.5 Results

Below, we present the results obtained in the three tests, classified according to the hierarchy previously described, as well as a summary of the mathematical objects needed to solve the problems.

14.5.1 Responses and Mathematical Objects Brought into Play When Solving Two-Stage Problems

In Table 14.3, we summarise the responses to Problems 2a and 2b in the different questionnaires. In the pre-test, only 29% of the responses were categorized at the multistructural or relational level of probabilistic reasoning. In the first post-test, 54% of the responses were categorized at the multistructural or relational level, and importantly, in the second post-test, over 64% of the students' responses were categorized at the multistructural or relational level. Thus, as Bill et al. (2009) observed in their study, we observe through the responses to the questionnaires continuous progress in probabilistic reasoning, which we attribute to the teaching the students received.

In Table 14.4, we present the mathematical objects identified in students' responses to questions 2a and 2b in the different questionnaires. Each student may use more than one mathematical object and, therefore, the sum of individual columns may be greater than 28 (the total number of students). In this table, we observe that in comparison to the pre-test, the students used tree diagrams more frequently in listing the sample space in the post-tests. In all of the tests, on average, nine of the students used Laplace's rule and obtained correct probabilities; another set of nine students used Laplace's rule but obtained an incorrect value of probability. In the post-tests, an average of eight students used the addition rule and/or the product rule of probability to answer sections (b) and (c). The binomial distribution was rarely used, and only appeared in the first post-test (used in two cases).

Table 14.3 Responses to two-stage Problems 2a and 2b by probabilistic reasoning level and time of application

Probabilistic reasoning level	Time of application		
	Pre-test	First post-test	Second post-test
Relational	0	7	2
Multistructural	8	8	16
Unistructural	12	11	8
Prestructural	8	1	2
No response	0	1	0
Total	28	28	28

Table 14.4 Mathematical objects brought into play when solving two-stage problems

Mathematical object	Problem 2a		Problem 2b
Strategies or procedures	Pre-test	First post-test	Second post-test
Enumerating or counting the sample space (four compound events)	12	17	22
a. With tree diagram	1	10	13
b. Without tree diagram	11	7	9
Laplace's rule	20	14	19
a. With tree diagram	1	3	9
b. Without tree diagram	19	11	10
Addition and product rule of probability	0	9	7
a. With tree diagram	0	8	4
b. Without tree diagram	0	1	3
Binomial probability function	0	2	0
Mistaken arithmetic operations or mistaken probabilities	15	11	8
Language or representations			
Verbal expressions	6	2	3
Lists or tables	27	14	13
Tree diagrams	1	12	13
Algebraic notation	0	2	1

With respect to the procedures employed, the enumeration strategy (Batanero et al. 1994) stands out as a frequently used strategy in the pre-test, expressed through lists or tables used to determine the sample space and to calculate probability. This method was also used by most students in the study of Rubel (2007). In the post-tests, we note the tendency to use tree diagrams to enumerate the elements of the sample space, the calculation of probabilities with Laplace's rule, and the use of the addition and product rules of probability. All of these elements of knowledge were highlighted in Table 14.1 as necessary to understand binomial probabilities, and have been considered by different authors (e.g. Nunes et al. 2014). The number of arithmetic operations and probabilities that are incorrect decrease throughout the three assessment points.

14.5.2 Responses and Mathematical Objects Brought into Play When Solving Three-Stage Problems

In Table 14.5, we summarise the reasoning levels of students' responses to Problems 3a and 3b. In the pre-test, only 18% of the responses to Problem 3a were categorized at the multistructural level and there were no responses at the relational level. In the first post-test, 46% of the responses to Problem 3b were categorized at the multistructural or relational level, and in the second post-test, 32% of the responses were categorized at either the multistructural or the relational level. Therefore, there is a clear pattern of increase in sophistication of reasoning from the pre-test to the first post-test, similar to results observed by Bill et al. (2009), but not from the first to the second post-test.

In Table 14.6, we present the mathematical objects identified in the students' responses (presented in Table 14.1 as components of knowledge needed to understand binomial probabilities). We observe that an average of 14 students (50%) listed the sample space in the post-tests, while a third of them did in the pre-test. Across all of the tests, an average of 10 students used Laplace's rule to determine the requested probabilities. In the post-tests, on average, seven students used the addition rule and the product rule of probability to correctly calculate the probabilities.

With respect to the procedures and representations used by the students, 43% of the students employed enumeration in the pre-test via lists or tables to construct the sample space or calculate the requested probabilities. In the post-tests, most of the students used lists, tables, or tree diagrams to enumerate the elements in the sample space or to calculate requested probabilities. Six students (21%) represented the random variable using algebraic notation in the first post-test and one in the second post-test, and three students used the product rule for counting the elements of sample space and calculating the probabilities.

Table 14.5 Responses to three-stage Problems 3a and 3b by probabilistic reasoning level and time of application

Probabilistic reasoning level	Time of application		
	Pre-test	First post-test	Second post-test
Relational	0	7	2
Multistructural	5	6	7
Unistructural	9	6	4
Prestructural	10	6	11
No response	4	3	4
Total	28	28	28

Table 14.6 Mathematical objects brought into play when solving three-stage problems

Mathematical object	Problem 3a		Problem 3b
Strategies or procedures	Pre-test	First post-test	Second post-test
Enumerating or counting the sample space (eight or 216 compound events)	8	16	9
a. With tree diagram	0	12	3
b. Without tree diagram	8	4	6
Laplace's rule	11	8	10
a. With tree diagram	0	3	2
b. Without tree diagram	11	5	8
Addition and product rule of probability	0	10	3
a. With tree diagram	0	9	2
b. Without tree diagram	0	1	1
Mistaken arithmetic operations or mistaken probabilities	18	9	15
Language or representations			
Verbal expressions	0	0	3
Lists or tables	12	5	12
Tree diagrams	0	12	3
Algebraic notation	0	6	1

In general, three-stage problems were more difficult for students than two-stage problems since the levels of reasoning in the responses to Problems 2 were higher than those in the responses to Problems 3. This is to be expected because, according to Nunes et al. (2014), in a logical multiplication of properties "when the number of attributes increases, the level of difficulty of the task increases" (p. 440). Moreover, Problem 3b used in the post-test was more abstract than Problem 3a used in the pre-test (Problem 3b has a sample space with 216 possible outcomes, which can be summarized by 8 compound outcomes). These results might have also been influenced by the fact that the probability distribution of discrete variables was taught in the first two months of the course, while the probability distribution of continuous variables was taught during months three and four. Students might have therefore forgotten certain knowledge from the beginning of the course, which may have been reflected on the post-tests.

14.6 Discussion and Conclusion

In this paper, we proposed a hierarchy of probabilistic reasoning for the concepts of sample space and probability of a compound event as a tool to describe high school students' responses to problems involving binomial distributions, which involve such concepts. We based our analysis of students' responses on the SOLO model, as well as on the reasoning hierarchy by Sánchez and Landín (2014), the responses provided by the students in the sample of this research and, the Probabilistic Thinking Framework (Mooney et al. 2014). Through this hierarchy, we seek to contribute to characterising probabilistic reasoning in the formal mode of representation when working with binomial probability problems and to complement the aforementioned research.

Another aim in our research was analysing the evolution of students' reasoning at three different times in an instructional setting. We observed improvements in responses over time: Regarding the structural complexity of the students' responses to the problems, the number of responses coded in the unistructural reasoning level decreased as time progressed, while the number of responses categorized in the multistructural reasoning level increased. This shows improvement with instruction, and suggests stability in knowledge after teaching.

In analysing the students' procedures, we observed that the tree diagram was an important instrument in improving probabilistic reasoning because it helps students to easily represent the stages of a compound experiment (Batanero et al. 1994), systematically list the elements of a sample space, assign probabilities to compound events, and "prepare the intuitive ground for theorems concerning the addition and multiplication of probabilities" (Fischbein 1975, p. 94).

In the post-tests, about a quarter of the students used mathematical objects such as the addition rule and the product rule of probability to correctly compute the probabilities; however, a few students were able to use these rules to count the events in the sample space without the help of tree diagrams. Moreover, only on the first post-test were some students able to identify and represent the random variable through algebraic notation and use of the binomial function. We also observed that some students produced errors due to an unsystematic listing of outcomes (e.g., using trial and error strategies and arbitrary or incomplete listings for the simple events), in agreement with the work of Polaki (2005).

In summary, the tree diagram was a productive tool both in counting the elements of the compound sample space and in representing the structure of the situation so as to determine the correct solution to the problem (Fischbein 1975). We therefore suggest that teachers emphasize the teaching of tree diagrams to solve tasks involving the enumeration of a sample space and the computation of the probability of compound events. Adquiring competence in these tasks will help students advance from subjective reasoning in binomial experiments, where heuristics such as representativeness (Kahneman and Tversky 1972) and biases such as equiprobability (Lecoutre 1992) are common, and toward quantitative or proportional reasoning (Mooney et al. 2014), where the students use the rules of probability to obtain correct responses.

Acknowledgements We thank referees and editors for their helpful suggestions to improve this chapter.

References

Batanero, C., Chernoff, E. J., Engel, J., Lee, H. S, & Sánchez, E. (2016). *Research on teaching and learning probability*, ICME-13. Topical Survey series. New York: Springer. https://doi.org/10.1007/978-3-319-31625-3_1.

Batanero, C., Godino, J. D., & Navarro-Pelayo, V. (1994). *Razonamiento combinatorio* (Combinatorial reasoning). Madrid: Síntesis.

Batanero, C., Navarro-Pelayo, V., & Godino, J. D. (1997). Effect of the implicit combinatorial model on combinatorial reasoning in secondary school pupils. *Educational Studies in Mathematics, 32*(2), 181–199.

Biggs, J., & Collis, K. (1982). *Evaluating the quality of learning: The SOLO taxonomy (structure of the observed learning outcome).* New York: Academic Press Inc.

Biggs, J. B., & Collis, K. F. (1991). Multimodal learning and the quality of intelligent behaviour. In H. A. Rowe (Ed.), *Intelligence: Reconceptualization and measurement* (pp. 57–76). Hillsdale, NJ: Lawrence Erlbaum.

Bill, A., Watson, J., & Gayton, P. (2009). *Guessing answers to pass a 5-item true false test: Solving a binomial problem three different ways.* In R. Hunter, B. Bicknell, & T. Burgess (Eds.), *Crossing divides: Proceedings of the 32nd annual conference of the Mathematics Education Research Group of Australasia,* Vol. 1, (pp. 57–64). Palmerston North, NZ: MERGA.

Chernoff, E. J., & Zazkis, R. (2011). From personal to conventional probabilities: From sample set to sample space. *Educational Studies in Mathematics, 77*(1), 15–33.

Fischbein, E. (1975). *The intuitive sources of probabilistic thinking in children.* Dordrecht, The Netherlands: Reidel.

Fischbein, E., & Schnarch, D. (1997). The evolution with age of probabilistic, intuitively based misconceptions. *Journal for Research in Mathematics Education, 28*(1), 96–105.

Godino, J. D. (2002). Un enfoque ontológico y semiótico de la cognición matemática (An ontological and semiotic approach to mathematical cognition). *Recherches en Didactique des Mathématiques, 22*(2–3), 237–284.

Godino, J. D., Batanero, C., & Font, F. (2007). The ontosemiotic approach to research in mathematics education. *ZDM. The International Journal on Mathematics Education, 39*(1–2), 127–135.

Hart, K. (1981). Hierarchies in mathematics education. *Educational Studies in Mathematics, 12*(2), 205-218.

Heitele, D. (1975). An epistemological view on fundamental stochastic ideas. *Educational Studies in Mathematics, 6*(2), 187–205.

Jones, G. A., Langrall, C. W., & Mooney, E. S. (2007). Research in probability. In F. K. Lester, Jr. (Ed.), *Second handbook of research on mathematics teaching and learning* (pp. 909–955). Charlotte, NC: Information Age & NCTM.

Jones, G. A., Langrall, C. W., Thornton, C. A., & Mogill, T. (1997). A framework for assessing and nurturing young children's thinking in probability. *Educational Studies in Mathematics, 32* (2), 101–125.

Kahneman, D., & Tversky, A. (1972). Subjective probability: A judgment of representativeness. *Cognitive Psychology, 3*(3), 430–454.

Keren, G. (1984). On the importance of identifying the correct "problem space." *Cognition, 16*(2), 121–128.

Lecoutre, M. (1992). Cognitive models and problem spaces in "purely random" situations. *Educational Studies in Mathematics, 23*(6), 557–568.

Mooney, E., Langrall, C., & Hertel, J. (2014). A practitional perspective on probabilistic thinking models and frameworks. In E. Chernoff, & B. Sriraman (Eds.), *Probabilistic thinking: Presenting plural perspectives* (pp. 495–507). New York: Springer.

Moore, D. (2007). *The basic practice of statistics* (4th ed.). New York: W. H. Freeman.

Nunes, T., Bryant, P., Evans, D., Gottardis, L., & Terlektsi, M. E. (2014). The cognitive demands of understanding the sample space. *ZDM. The International Journal on Mathematics Education, 46*, 437–448.

Pastor, G. (1998). *Estadística básica* (Basic statistics). México: Trillas.

Polaki, M. (2005). Dealing with compound events. In G. A. Jones (Ed.), *Exploring probability in school: Challenges for teaching and learning* (pp. 191–214). New York: Springer.

Rubel, L. H. (2007). Middle school and high school students' probabilistic reasoning on coin tasks. *Journal for Research in Mathematics Education, 38*(5), 531–556.

Sánchez, E., & Landín, P. (2014). Levels of probabilistic reasoning of high school students about binomial problems. In E. Chernoff, & B. Sriraman (Eds.), *Probabilistic thinking: Presenting plural perspectives* (pp. 581–597). New York: Springer.

Shaughnessy, J. M., & Ciancetta, M. (2002). Students' understanding of variability in a probability environment. In B. Phillips (Ed.), *Proceedings of the Sixth International Conference on the Teaching of Statistics,* Capetown, South Africa [CD-ROM]. Voorburg, The Netherlands: International Statistical Institute.

Steinbring, H. (1986). Mathematical concepts in didactical situations as complex systems: The case of probability. In C. Marjorie (Ed.), *Proceedings of the Fifth International Congress on Mathematical Education* (pp. 56–88). Berlin: Birkhäuser.

Tversky, A., & Kahneman, D. (1983). Extensional versus intuitive reasoning: the conjunction fallacy in probability judgment. *Psychological Review 90*(4), 293–315.

Yates, D., Moore, D., & Starnes, D. (2003). *The practice of statistics* (2nd ed.). New York: W. H. Freeman and Company.

Chapter 15
The Six Loses: Risky Decisions Between Probabilistic Reasoning and Gut Feeling

Joachim Engel and Arne Orthwein

Abstract The paper reflects on students' intuitive strategies in a game of chance and contrasts their understanding with a normative probabilistic point of view. The game involves selecting optimal strategies outweighing potential gains with small probabilities with more probable losses. The paper investigates the reasoning of 46 students, analyses the rationales of their preferred strategies and their perceptions when dealing with risk. It sheds light on how their reasoning with risk is related to their knowledge about probability and how their thinking includes an evaluation of potential consequences of their choices. Responses are analysed in the light of well documented fallacies in probabilistic reasoning. Hence, the paper provides insight into students' thinking between probabilistic reasoning, assessing risks and gut feelings.

Keywords Risk · Intuition · Probabilistic reasoning

15.1 Introduction

In the dice game "The six loses" a fair die is being tossed repeatedly and the up face is recorded. As long as no 6 occurs, the rolled numbers are added up to a score. The game can be stopped at any time; then the obtained score is the prize (e.g. as money in Euros) won. However, if a 6 shows up, everything is lost and the game is over. If, for example, the die shows the consecutive numbers 4, 2, 3, 3 and then you stop, the gain is 12 Euro. With the sequence 2, 1, 6 you go empty handed, because you did not stop after the second roll and the 6 in the third roll destroys all your previous gains in this sequence.

J. Engel (✉) · A. Orthwein
Department of Mathematics and Computer Science, Ludwigsburg University of Education, Reuteallee 46, 71634 Ludwigsburg, Germany
e-mail: engel@ph-ludwigsburg.de

A. Orthwein
e-mail: arne-orthwein@arcor.de

C. Batanero and E. J. Chernoff (eds.), *Teaching and Learning Stochastics*, ICME-13 Monographs, https://doi.org/10.1007/978-3-319-72871-1_15

Imagine you play this game often. Which strategy for stopping the game should be employed to get on average over many games a score as high as possible?

This game of chance has been popularized by the Italo-American magician John Scarne under the name Pigs (Neller et al. 2006). A variant can be played online http://cs.gettysburg.edu/projects/pig. The underpinnings of this game relate to important topics in stochastic thinking and the game has proven to be quite motivating for learners as it appears as a toy model for stock market speculation, recalling the old saying on Wall Street that the market is driven by just two emotions: *fear and greed.* In the framework of decision theory and risk analysis the question of a good strategy in the game "The six loses" can be seen as balancing a potentially higher gain through rolling the die more often with an increasing risk of not gaining anything because of a six showing up in a longer sequence of throws.

Mathematically, the expected gain is a concept that reflects an empirical average when the game is played infinitely often. Adhering to a particular strategy, the actual average gain over finitely many games is not a fixed number but a random variable, with a certain expected value and a standard deviation. Under finitely many repetitions of the game, it is reasonable also to take considerations of variability into account. Various strategies do not only differ with respect to their expected gain but also with respect to their standard deviation (see Sect. 15.6). Therefore, reasonable strategies for risky decisions may consider both, the expected gain and the standard deviation involved.

We presented this game to college students and inquired about their choice of strategy. While the students were familiar with basic concepts of probability such as the notion of the expected value and standard deviation, they lacked the formal probabilistic knowledge for a full mathematical analysis of this problem. Therefore, they were challenged to reason informally about risks involved in continuing to roll the die, based on their limited knowledge about probability, their intuition and gut feeling. We present and discuss students' answers followed by a normative presentation of an optimal stopping strategy.

One main goal of the study was to investigate if and what kind of probabilistic reasoning the students displayed in explaining their preferred strategy. A second goal was to investigate students' perception and dealing with risk: how reasoning with risk is related to their knowledge about probability and if their thinking also included an evaluation of potential consequences of their choices.

Risk Literacy is understood as the ability to deal with uncertainty in an informed way (Gigerenzer 2014). Beyond understanding and dealing with probabilities, risk literacy implies an evaluation of the consequences various uncertain outcomes may have. As claimed by Borovcnik (2015, 2016), the concepts of probability and risk are closely interrelated; hence literacy in probability is entrenched by literacy in risk and vice versa. While in today's school curricula risk is rarely an explicit topic of probability instruction, it is very desirable to explore in which ways the two topics of probability and risk connect in the thinking of students. Fischbein (1975) seminal research established the importance of formal instruction for learners to overcome primary intuitions about probability, but little is known about how much and what kind of *formal* instruction is needed to raise risk literacy and if it depends on some level of previously acquired probability literacy.

15.2 The Dice Game

The game "The six loses" was presented to 46 second-year students preparing to be secondary school mathematics teachers. The students had some basic knowledge of probability from previous instruction but were not formally introduced to the concept of risk.

Figure 15.1 informs about the tasks the students were asked to do. After devising a strategy and noting down a rationale for the proposed approach students were asked to actually play the game at least ten times in a row, sticking exactly to their recommended strategy. The suggested number of repetitions was chosen as a compromise to provide some empirical evidence regarding the efficacy of the proposed strategy and practical considerations of time restraints including the danger of students' losing interest in the game by too many repeats. The repetitions provided an opportunity to challenge students' original intuition with concrete data.

Game of dice: A fair die is thrown repeatedly. As long as no 6 occurs, the up face numbers are added up to a score. The game can be stopped at any time. Then the obtained score is your gain (in Euros). However, if a 6 occurs, everything is lost. Which strategy for stopping do you recommend in order to maximize your average gain, provided you are playing the game very often?

1. Note down which strategy you choose in order to maximize your average gain when playing this game very often.
2. Explain, why you think this is a good strategy.
3. Now take a die and play the game at least 10 times following exactly your proposed strategy.
4. After testing your recommended strategy, would you still recommend this strategy? Why or why not?
5. If you favour a change of strategy, what is your preferred choice?

Fig. 15.1 Tasks given to the students inquiring a written response

Appropriate probability literacy implies an appreciation of data, but also an awareness of random variability inherent in the low number of repetitions (Gal 2005), hence would generalize the findings of 10 outcomes only with great care. Therefore, an additional goal was to investigate (1) if the students' choice of proposed strategy has been influenced by concrete data (in case data and original strategy are dissonant) and (2) if they were able to critically evaluate the data by being aware of random variability due to the small sample size.

Finally, after having played the game, students were asked to reconsider their originally proposed strategy and reason if they still stick to the original plan or favour a possible change of strategy. All these tasks were performed individually and students noted down their answers in writing while the researchers were present just to administer the tasks. The following week various students' answers were discussed in class and suggested solutions were tested via simulations based on a very large number of repetitions, thus providing insight into the normative solution (see Sect. 15.6).

In previous class meetings students were introduced to the concept of probability and its different notions (classic, frequentist, subjective), learned about computing probabilities in a multi-stage situation via tree diagrams and were introduced to the notion of the expected value as a theoretical average of repeated outcomes that is based on the law of large numbers.

Thus, it could be expected that the participating students were capable of calculating the probability of, say, "no six" occurring in 1, 2, 3, … repeated rolls of a die. In Sect. 15.6 we present a mathematical analysis of the optimal solution based on the notion of conditional expectation. However, the expected gain as a valid criterion for a good decision is based on an infinite number of games. In a finite world, considerations of the expected gain have to be complemented by taking variability of potential outcomes into account. The mathematical background for these considerations and techniques was not available to the students, thus their judgment about preferred strategies was based on intuition. Assessing the problem required them to recognize that with an increasing number of rolls the probability of a 6 (and hence losing everything "earned" so far) increases, but also the potential gain in terms of raising the score may get larger. Hence, a decision in a typical risk-benefit trade-off situation had to be made.

Therefore, their reasoning sheds some light on their perception and rationalizing about risks. It also provides insight into students' probabilistic thinking, in particular their understanding of the independence of consecutive throws of a die (*Does the likelihood of a 6 stay the same or does it increase if a 6 hasn't shown up in the rolls before?*) and their frequentist interpretation of the results after they have played the game under the proposed strategy several times. On the one hand, the concrete experience of playing the game may provide a valuable background to confirm or reject the originally proposed strategy. On the other hand, as the number of games played was rather low ("at least 10"), students may run the risk of overgeneralization and be subject to the *small sample fallacy* by considering their small sample experience as representative.

15.3 Fear of Losing or Greed for More: Balancing Risk and Benefit

Each additional roll of the die offers the possibility of higher gains, but also increases the chance of losing everything that has been gained so far. Therefore, a reasonable strategy has to involve some trade-off between an increasing risk of not gaining anything and of obtaining a higher score. While the probability of "no 6" (and hence maintaining a winning score) in k consecutive throws of a die approaches zero at a rate of $(5/6)^k$, the potential gain increases with each new roll of die. With a moderate number of, say, 7 rolls, either you win a substantial amount with a rather small probability of $(5/6)^7 = 27.9\%$, or you win nothing with a high probability of 72.1%. On the other hand, if you roll the die only twice, the winning probability is about 69.44%, but the gain will be modest. Therefore, the situation can be seen within the framework of risk assessment.

Risk in decision theory is defined as a product of a probability p times a measure for the (dis-)utility of an event (see, e.g. Edwards and Tversky 1967). In the classical situation, it is about evaluating a loss of a particular resource (money, health, time, energy …) that occurs with small probability and compare it with consequences of other events that may happen with certain other probabilities. One perspective at the present case is that you may obtain a score (gain money) with small probabilities as a "positive risk" or benefit and the question is which strategy maximizes your benefit. An alternative perspective is to look at the situation from the viewpoint of consecutive decisions: Assuming, you already achieved a score of $s > 0$ points through previous rolls of the die and you have to decide to stop or continue playing. When you continue, you can increase your score with probability 5/6 by up to 5 points resulting in an (equally likely) score of $s + 1$, $s + 2$, …, or $s + 5$. But with probability 1/6 you'll lose all the s points you have gained so far.

Hence risk-averse people may decide for very few rolls only, while risk-prone players will try to challenge their luck and opt for many more rolls hoping for a high score. The perspective of sequential decision making highlights the fact that risk is not only about probabilities of losing a resource, but also it depends on the utility of the good that you are losing.

15.4 Heuristics, Misconceptions and Gut Feelings

Because students had some experience and knowledge about probability, they may activate this knowledge, i.e. by computing the probability of "No six" in k consecutive rolls and recognize its geometrically decreasing values. However, they are not expected to perform a complete (large sample) probabilistic normative analysis of the game. From a practical point this mathematical analysis, even if available, would be of limited value as the game is being played only a finite number of times and not infinitely often as assumed when referring to the expected value as single criterion.

Therefore, besides rational considerations students will have to rely on their intuition. They may be tempted to predict the result of their next toss by relying on previous experience. While on normative grounds of probability all possible outcomes remain equally probable, regardless the results of the past rolls, students may resort to alternate reasons, pseudo-causalities and search for patterns. As an example, some may believe that the sixth throw of the die must result in a six if not shown in any of the five rolls before.

The literature is full of evidence that humans have difficulties grasping the independence of events (e.g., Engel and Sedlmeier 2005). Furthermore, students oftentimes deem small samples as representative. If you have 10 times a six when rolling a die 60 times, then they believe to get exactly one six in 6 throws. For some people, rolling a die is charged with strong emotions due to losing or winning in popular games like Ludo which may lead to biased perceptions.

Students encounter stochastic situations in their day-to-day life. But the problem is that these experiences are often not reflected which may result in biases. Decisions in risky situations are often strongly influenced by emotions and based on gut feelings and personality traits. Taking a wider "real-world" perspective, the subjective probability assigned to threatening events is usually highly exaggerated. Gigerenzer (2007) evaluates the quality of decisions that are strongly influenced by gut feelings and concludes that a mixture between analysis and gut decisions might be superior to pure analytic decisions. Gigerenzer discusses several heuristics in responding to risky situation that may also apply to our game. A student may recall similar situations when rolling dice and argue (representative heuristics): "In the past I did well to stop at the score I achieved now, therefore I stop". Or "Experience tells me that after four roles the next roll results most likely in a six". Related to these considerations is the less-is-more effect (Gigerezer 2007). "When I don't know much about probabilities and dice, then I may quickly conclude that it may be advantageous to roll the die 5 times, because the 6th roll ends with a six."

Another prevalent intuition is the "Take-the-Best" heuristic (Gigerenzer 2007): Base your decision on the single one criterion that appears best or accessible to you. In our context, one may compute the probability of "No six" in 3, 4 or 5 rolls as accessible computation and recognize that with 3 rolls this probability remains still under the benchmark of 1/2, i.e. the odds for "No six" are bigger than 1 by 3 throws while they become smaller than 1 if the die is rolled a fourth time. This is related to the "attribute substitution" of heuristic judgements (Kahneman and Frederick 2005), which is guided by the general hypothesis: When confronted with a difficult question, people may answer an easier one instead and are often unaware of the substitution. Applied to the current situation: "An evaluation of expected values may be too difficult therefore I look only at probabilities or some other simple computation I can perform".

Another consideration is based on the 1/n (or "nothing matters") heuristics. Because the outcome is unpredictable, the outcome is not influenced by the strategy. Any strategy is equally good or bad. Therefore, before starting the actual game, I roll the die to determine the number of times (between 1 and 6) to roll the die in the game.

15.5 Students' Responses

The written answers of the participants were analysed by the authors and in a first step categorized according to strategies proposed. In a second step the responses were studied with respect to the degree of exposed probabilistic thinking including heuristics and fallacies and the level of informal reasoning with risk.

Strategy 1: Roll the die r-times.
Strategy 2: Roll the die until you have at least a score of s points.
Strategy 3: A mixture of strategy 1 and strategy 2 in the form: Roll the die at least r times. But if your score is below s, then continue rolling the die one or two more times.

Strategy 1: Roll r times (37 students)
The majority of the students chose strategy 1 ("Roll the die r times") with different recommendations for the number r of rolls. All but one stuck to that strategy after they played the game ten times or more, some with recommending a change of the number of throws. 21 students favoured only 3 rolls before, 17 students after playing the game. Table 15.1 presents details. 20 students stuck to the same number of rolls (diagonal in Table 15.1), while 17 students changed the number of recommended rolls, mostly by being more daring, i.e. increasing the number of rolls.

Some students based their reasoning on computing the probability of "no six" in several consecutive throws and recognized the (geometrically) decreasing probabilities of "no six". However, very few students balanced this observation with the increasing potential gain when rolling more often. Hence, they resorted to an easier or partial problem that they knew how to solve without being aware that they only address a part of the original problem (Kahneman and Frederick 2005). Some of these answers also could be attributed to the outcome approach (Konold 1989, p. 62), because with four or more throws the odds of getting nothing due to a rolled six are greater than 1 and increase rapidly with the number of throws, as the following quotes document:

Table 15.1 Choice of the number of rolls before and after playing the game at least 10 times by students who recommended the "Roll the die r times" strategy

Before rolling the die	After rolling the die						
	1	2	3	4	5	6	7
1							
2		2	1				
3	1		16	3	1		
4				1	5	1	
5					1	4	
6							
7						1	

- "The probability of no six still has to be above 50%. This happens, if I roll three times".
- "Probability of three times 'no six' is about 60% (5/6)3 = 57.87%".
- "The probability of getting a 6 increases rapidly the more often I roll".

Some students based their recommendation just on intuition without reasoning about any probabilities involved:

- "If I throw more often, the risk of getting a six gets bigger; 4 times appears as a reasonable compromise".

The probability of a six is always the same, regardless how often one throws the die. Ignoring this fact, the following arguments are indicative of gambler's fallacy:

- "Roll 5 times, because on average every sixth roll results in a six".
- "Roll 5 times, because the probability for each side up is 1/6 because of 6 faces. Stop one throw before 6. Therefore roll 5 times".

Several students were discouraged to roll the die 5 times because the actual experiment they performed resulted in many zero-gains or, vice versa, others stuck to the result because of some "lucky" games. Not a single student questioned the random variability inherent in such a small sample size. Thus, they overgeneralized the experience of playing the game 10 times:

- "With 5 rolls I lost in 6 out of 10 cases. This is more than one half, therefore you should throw the die less than 5 times".
- "I won only 3 out of 10 times, therefore I'd rather change and roll less frequently".
- "In 9 out of 10 games I won with the "roll 3 times" strategy. Therefore, I would stick to it".

A few students recognized that the decision is about balancing possible gains with the risk involved in more throws:

- "With three rolls, there is quite a lot to win, an additional throw would diminish the probability of keeping my gain".
- "A reasonable compromise between high gain and probability of winning".

Some rather idiosyncratic explanations were the following:

- "The die has 5 sides [arguing for 5 throws]".
- "The probability of "no six" is 5/6. Hence I roll 5 times".
- "Because 3 is half of 6 [reasoning for three throws]".

Strategy 2: Rolling until a score of s is achieved (4 students)
Another intuitively reasonable strategy is to roll until a certain score is obtained. But none of the four students who opted for that strategy had strong arguments for their choice. At first, the low number of students (4 out of 46) choosing strategy 2 may be surprising, but it was obviously more difficult to come up with a convincing reason for a particular choice for the attainable score s that defines this stopping

rule. The students' irrelevant computations may, to say the least, be classified as an example for attribute substitution (Kahneman and Frederick 2005) or regarded as idiosyncratic.

- "Roll until a score of at least 15, because 1 + 2 + 3 + 4 + 5 = 15" (two students).
- "Roll until 15: risk and probability are always the same. Therefore, it doesn't matter if I go by the number of rolls or the score obtained".
- "Roll until a score of 20, because either I win 20 or I win nothing".

It is remarkable that none of the student changed from strategy 1 to strategy 2 or vice versa after actually having rolled the dice. Everyone stuck to their original strategy, some "fine-tuned" the parameters of the selected strategy.

Strategy 3: Mixing 1 and 2 (5 students)
A very reasonable strategy consists in relating the number of rolls to the score already obtained. This mixture of strategy 1 and 2 may well be indicative of a certain level of risk literacy because it requires an assessment of how much is at stake to lose before deciding for another roll. Only four students have chosen this strategy initially, one other student changed to that strategy after the hands-on experience with rolling the die.

- "Stop after 3 rolls if there is one roll with at least 4 points [otherwise continue]".
- "If I get high numbers at the beginning, I would secure the gain. Is the score low, however, I would dare to continue".
- "At least 3 throws, at most 5; stop if above 10 points".
- "Roll 4 to 5 times, but stop at 20 points".

Just one out of 46 students changed to strategy 3 after her initial choice was a fixed number of 5 rolls. Noteworthy is her reasoning or 5 rolls which is based on the expected waiting times for the first 6:

- "Consider when – on average – the first 6 occurs".

After testing this strategy, this student departed from the fixed number of 5 rolls and altered her preference by stating:

- "Roll 4 times, but I would take into consideration how much is at stake to lose".

While balancing gain and the probability of loss has been most obvious in strategy 3, several students who chose strategy 1 or 2 also connected their possible win or loss with the likelihood of rolling a six.

- "Rolling 5 times is optimal. Even if you get a six, you still get [on average] a high score".
- "You always run the risk of rolling a six, but if 5 times [rolling] works, at least the score is high".
- "Rolling 3-times is a good mixture between high gain and the probability of winning".

Nine students who had chosen strategy 1 brought forward similar arguments. They were aware of the fact that an increased probability of a total loss may be compensated by a potentially higher score and, vice versa, avoiding a risky decision may result in a low score.

Some students perceived the risk in additional rolls of the die, but did not reason with probabilities and based their decision on personal beliefs and dispositions instead:

- "Because here is much luck involved, I wouldn't risk more than 4 rolls".
- "Due to my personality to prefer the safe way, 3 rolls with 57% appear too risky to lose the score earned so far".
- "My own experience tells me that a 1 is followed by a 6 most of the time".

A majority of twenty-eight (out of the total of 46) students never took explicitly into account that the obtainable score depends on how often the die is rolled, neither before nor after actually playing the game. They stayed focused on reasoning with the probability of losing only [i.e. $p = (5/6)^k$]. A group of fifteen students realized *after* the concrete experience of playing the game that not only the probabilities but also different outcomes in dependence on the number of rolls have to be taken into account. Only the remaining three students reflected from the very beginning (before actually rolling any dice) that a successful strategy involves balancing probabilities with potential outcomes.

15.6 The Normative Solution for Maximizing the Expected Gain

Two intuitively plausible strategies are: "roll *r* times" and "roll until the score is at least *s*". Probability calculations allow us to compute the expected gains and to determine optimal choices for the parameters *r* and *s* under these strategies.

Strategy 1: Roll r times

An easy access is obtained through the formula for the conditional expectation (see, e.g. Feller 1967, p. 223). Let X denote a discrete random variable assuming values $x_1, x_2, \ldots, x_k$, and A be an event. The conditional expectation of X under the condition A is defined as

$$E(X|A) = \sum_k x_k P(X = x_k|A).$$

Then, provided A_n form a partition of the sample space (i.e. $\bigcup_n A_n = \Omega, A_i \cap A_j = \phi$ for $i \neq j$), it holds that

$$E(X) = \sum_n E(X|A_n) \cdot P(A_n)$$

Now let A_r denote the event: “No 6 in r consecutive rolls”, and let the random variable S_r denote the score after r rolls. Then it holds

$$(1) \quad E(S_r) = E(S_r|A_r) \cdot P(A_r) + E\left(S_r|\overline{A}_r\right) \cdot P\left(\overline{A}_r\right) = E(S_r|A_r) \cdot P(A_r) = 3r\left(\frac{5}{6}\right)^r,$$

because the expected gain in each of r throws is 3 provided no 6 is thrown and $S_r = 0$ otherwise, i.e. $E(S_r|A_r) = 3r,\ E\left(S_r|\overline{A}_r\right) = 0$.

In order to find the value of r maximizing this expression, consider the quotient

$$\frac{E(S_{r+1})}{E(S_r)} = \frac{r+1}{r} \cdot \frac{5}{6}$$

This quotient is exactly greater (smaller) than 1, if $r < 5$ (resp. $r > 5$).

Furthermore, we have $E(S_5) = 15 \cdot \left(\frac{5}{6}\right)^5 = 18 \cdot \left(\frac{5}{6}\right)^6 = E(S_6) = 6.028$.

Alternatively, one may look at this situation as a random walk on the integers with a move of a random number of units to the right (from 1 to 5 with probability 1/6 each) and back to the origin (with probability 1/6). The question then becomes: After how many jumps is it better to stop the walk to maximise the walker’s average distance to the origin?

Since the problem involves averaging, a natural “didactical reduction is to replace the original random walk by the associated (deterministic) barycentric walk, i.e. the walk of the barycentre of the mass system representing the original walk at each step. Otherwise said, it is the walk performed by the expected position of the walker, step after step. This deterministic walk heads right from the origin, reaches a maximum distance from the origin and then goes back to the origin (where it lands after an infinite time). This follows immediately from the deterministic equation of motion $3r(5/6)^r$.

When rolling the die $r = 5$ times, we can expect an average score of 6.028, we’ll end up with a score of 0 with a probability of $1-(5/6)^5 = 59.81\%$. In about 6 out of 10 games we will gain nothing, while the score in the other games will be distributed between 5 and 25. Henze (2011, p. 5) also calculates the distribution of the random variable S_r. Table 15.2 summarizes the expected value and standard deviation of S_r for various values of r, Fig. 15.2 shows the distribution for $r = 5$. Notice, that while a value of $r = 5$ and $r = 6$ maximizes the expected value the standard deviation increases monotonically with r. This has implications for the risk assessment. While the expected value can be considered as a theoretical average over infinitely many repetitions of the game, in real life the game is played only a finite number of times. Then a higher variability increases the probability of gaining less (and gaining more) than expected.

Table 15.2 Expected gain and standard deviation under the strategy "roll r times"

r	$E(S_r)$	$s(S_r)$	r	$E(S_r)$	$s(S_r)$
2	4.167	3.228	6	6.028	8.729
3	5.208	4.819	7	5.861	9.630
4	5.787	6.310	8	5.582	10.321
5	6.028	7.623	9	5.233	10.867

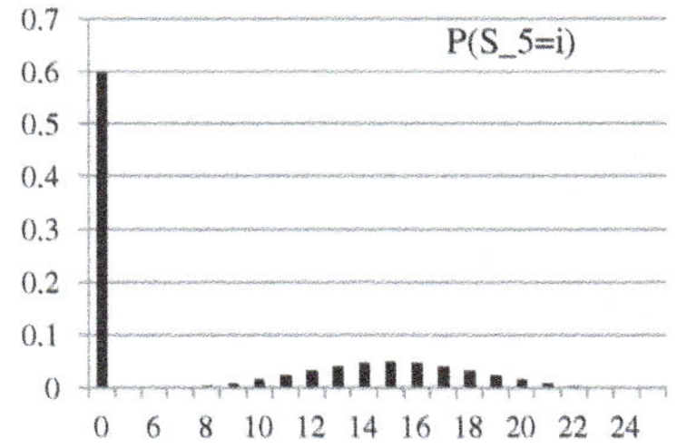

Fig. 15.2 Probability distribution of gains under the strategy *roll 5-times*, P(S_5 = i) denotes the probability that random variable S_5 assumes the value i

Strategy 2: Roll until a score of at least s is obtained
Assume, you already have a score of s and roll once more. Let Y_s denote your gain after the next roll. Y_s assumes the values $s + 1$, $s + 2$, …, $s + 5$ and 0, all with probability 1/6, therefore we have

$$(1) \quad E(Y_s) = \frac{1}{6}\sum_{j=1}^{5}(s+j) = \frac{5s+15}{6}.$$

To continue rolling the die pays off if and only if $E(Y_s) > s$ which is equivalent to $s < 15$. Hence the optimal rule is to stop rolling when a minimum score of 15 is achieved.

Because of $E(Y_s) = s \Leftrightarrow s = 15$ waiting till 16 will not deteriorate your score. Therefore, the best choice within Strategy 2 is "continue playing until your score is 15 (or more); otherwise stop and cash in your gain".

Knizia[1] (1999) describes the analysis for the "stop at 15" policy, where each roll is viewed as a bet: The true odds of such a bet are 1–5. If you ask yourself how much you should risk, you need to know much there is to gain. A successful throw produces one of the numbers 1, 2, 3, 4 and 5. On average, you will gain 3 points. If you put 15 points at stake this brings the odds to 3–15, that is 1–5, and makes a fair game. Whenever you accumulated points less than 15, you should continue throwing, because the odds are in your favour.

[1]Knizia's game is based on 1 instead of 6 being the losing number. Here his considerations are adapted to 6 being the loser.

Table 15.3 Expected gain and standard deviation under the strategy: "roll up to a score of at least s"

s	$E(Y_s)$	$s(Y_s)$	s	$E(Y_s)$	$s(Y_s)$
8	5.318	4.667	14	6.130	7.531
9	5.550	5.188	15	6.154	7.934
10	5.744	5.710	16	6.154	8.311
11	5.896	6.212	17	6.133	8.662
12	6.003	6.663	18	6.094	8.993
13	6.081	7.105	19	6.040	9.585

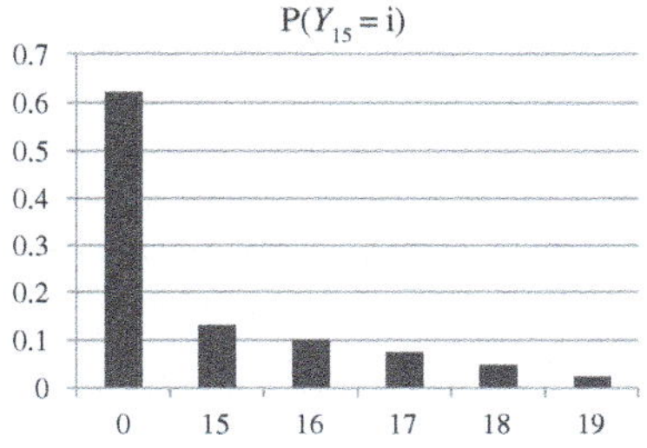

Fig. 15.3 Probability distribution of gains under the strategy *roll until at least a score of 15*, with $P(Y_{15} = i)$ denoting the probability that random variable Y_{15} assumes the value i

The formal derivation of the expected gain under this rule is beyond the scope of this paper. Henze (2011, p. 3) suggests a recursive procedure resulting in Table 15.3. Figure 15.3 gives the distribution for s = 15.

Furthermore, the probability of gaining nothing at all under this strategy is 62.22% (Henze 2011, p. 5), hence even slightly higher than under the optimal roll-k-times-strategy. Henze and Vehling (2017) derive optimal solutions for a variant of the game and outline how to introduce this result for the upper secondary school level.

Roters (1998) could prove that the Strategy 2 to stop when obtaining a score of at least 15 is not only superior to Strategy 1, but it is the overall optimal strategy for maximizing the expected value of the game.

15.7 Conclusions

The students' reasoning in the game "The six loses" revealed many typical and well studied fallacies, biases and preconceptions (Batanero et al. 2016), most notably the gambler's fallacy, the small sample fallacy, the equiprobability bias (everything being equal) and the attribute substitution (availability heuristics). Most students followed an outcome approach (Konold 1989, p. 62) and neglected the fact that a good strategy balances a potentially higher gain through rolling the die more often with an increasing risk. However, the study also showed a sizeable effect on student's awareness of risk through reflecting on their concrete hands-on experience. After playing the game several times, many participants reconsidered their original strategy by recognizing the fact that with more rolls the potential gain increases and that a good strategy is characterized by balancing the probability of loss with the chance for higher scores. Good instruction has to take student's understanding, intuition and heuristics as starting point, not the normative solution. The game "The six loses" has the potential of providing valuable experience to challenge students' thinking about risk.

References

Batanero, C., Chernoff, E., Engel, J., Lee, H., & Sanchez, E. (2016). *Research on teaching and learning probability. ICME-13. Topical Survey series*. New York: Springer.

Borovcnik, M. (2015). Risk and decision making: The "logic" of probability. *The Mathematics Enthusiast, 12*(1), 113–139.

Borovcnik, M. (2016). Probabilistic thinking and probability literacy in the context of risk. *Educação Matemática Pesquisa, 18*(3), 1491–1516.

Edwards, W., & Tversky, A. (Eds.). (1967). *Decision making: Selected readings*. London: Penguin Books.

Engel, J., & Sedlmeier, P. (2005). On middle-school students' comprehension of randomness and chance variability in data. *Zentralblatt für Didaktik der Mathematik, 3*, 168–179.

Feller, W. (1967). *An introduction to probability theory and its applications* (3rd ed.). New York: Wiley.

Fischbein, E. (1975). *The intuitive source of probability thinking in children*. Dordrecht, The Netherlands: Reidel.

Gal, I. (2005). Towards "probability literacy" for all citizens: Building blocks and instructional dilemmas. In G. A. Jones (Ed.), *Exploring probability in school. Challenges for teaching and learning*. New York: Springer.

Gigerenzer, G. (2007). *Gut feelings: The intelligence of the unconscious*. New York: Viking.

Gigerenzer, G. (2014). *Risk savvy: How to make good decisions*. New York: Penguin Books.

Henze, N. (2011). Zwischen Angst und Gier: Die Sechs verliert [Between fear and greed: The six loses]. *Stochastik in der Schule, 31*(2), 2–5.

Henze, N., & Vehling, R. (2017). Eine möglichst große Augensumme, aber bitte ohne Sechs! [A score as large as possible, but without a six, please!]. *Stochastik in der Schule, 32*(2), 1–10.

Kahneman, D., & Frederick, S. (2005). A model of heuristic judgement. In K. Holyoak & B. Morrison (Eds.), *The Cambridge handbook of thinking and reasoning* (pp. 267–293). Cambridge: Cambridge University Press.

Knizia, R. (1999). *Dice games properly explained*. Lower Kingswood, UK: Elliot Right-Way Books.

Konold, C. (1989). Informal conceptions of probability. *Cognition and Instruction, 6*, 59–98.

Neller, T., Presser, C., Russell, I., & Markov, Z. (2006). Pedagogical possibilities for the dice game Pig. *Journal of Computing Sciences in Colleges, 21*(6), 149–161.

Roters, M. (1998). Optimal stopping in a dice game. *Journal of Applied Probability, 35*, 229–235.

Part III
Education of Teachers

Chapter 16
Comparing the Relative Probabilities of Events

Egan J. Chernoff, Ilona Vashchyshyn and Heidi Neufeld

Abstract The purpose of this article is to contribute to the research investigating the use of logical fallacies, in particular the fallacy of composition, to account for normatively incorrect responses given by prospective teachers to relative probability comparisons. Our results respond to certain assumptions made regarding research on relative probability comparisons of coin flip sequences, which have suggested that participants were actually comparing events rather than sequences, and demonstrates that even when presented with events, the majority of respondents still give normatively incorrect responses. As with all research in this area, abductive reasoning is employed to substantiate our claim that the fallacy of composition is the most probable explanation of respondents reasoning.

Keywords Heuristics · Biases · Informal conceptions · Logical fallacies Probability

16.1 Introduction

Since 2007, there has been a rather systematic investigation into prospective elementary, middle, and high school mathematics teachers' relative probability comparisons, specifically as related to coin flip sequences (e.g., given a sequence of coin flips, participants are typically asked to determine which sequence is more or less likely). The varied results stemming from this investigation have been presented at a variety of mathematics education conferences. Without fail, the question and answer period for each and every presentation always includes one particular line of reasoning. In particular, the following is inevitably overhead during the question-and-answer period following the conference presentations in question:

E. J. Chernoff (✉) · I. Vashchyshyn · H. Neufeld
University of Saskatchewan, 28 Campus Drive, Saskatoon, SK S7N0X1, Canada
e-mail: egan.chernoff@usask.ca

C. Batanero and E. J. Chernoff (eds.), *Teaching and Learning Stochastics*,
ICME-13 Monographs, https://doi.org/10.1007/978-3-319-72871-1_16

> Yes, your results (and analysis) demonstrate that the participants deem [using H for head and T for tail] HHHHT as less likely than HHTHT, but they're *probably* comparing the event [i.e., a set of outcomes] four-heads-and-one-tail to the event three-heads-and-two-tails, in which case, then, 'HHHHT' *is* less likely than 'HHTHT'. This means that, although their answer is incorrect for the question you gave them, they're not necessarily incorrect if they're comparing events instead of sequences.

Responses to this line of reasoning typically make references to research literature that has: controlled the ratio of heads to tails in the sequences presented (e.g., Chernoff 2009); investigated what happens should sequences be interpreted as events (e.g., Abrahamson 2009); and investigated the idea that students are answering questions that may not have been asked (e.g., Chernoff and Mamolo 2015). More often than not, however, the presenter is *not* asked to respond or comment on this line of reasoning. Instead, attendees leave the presentation with the air of security that participants were, in fact, comparing events instead of sequences, and doing so correctly. Their conviction is strong enough for them to leave believing not only that participants are comparing events and not sequences, but also that any research conducted would simply confirm their convictions. However, this research has now been conducted (e.g., Chernoff 2012a) and suggests that these convictions are unfounded.

Investigations into relative probability comparisons, especially as related to coin flip sequences, have become an established thread in the last 50 years of probability education research (see, e.g., Abrahamson 2009; Batanero and Serrano 1999; Chernoff 2009, 2012a, b; Chernoff and Russell 2012; Falk 1981; Konold et al. 1993; Shaughnessy 1977). Speaking in general terms, the aforementioned research is dedicated to accounting for normatively incorrect relative probability comparisons. Should a research participant deem, incorrectly, that one sequence of coin flips is more or less likely than another, there are now a plethora of established theories and models to account for the "error," which include the *representativeness heuristic* (see Tversky and Kahneman 1974; Shaughnessy 1977), the *equiprobability bias* (Lecoutre 1992), the *outcome approach* (Konold et al. 1993), *sample space partitions* (Chernoff 2009), *attribute substitution* (Chernoff 2012b; Kahneman and Frederick 2002), *inadvertent metonymy* (Abrahamson 2009), and others. As we shall argue, despite the differences in the theories and models accounting for normatively incorrect relative probability comparisons, they all have something in common.

16.2 A Brief Appraisal of the Research Literature

Given the potential readers of this book (researchers in probability education and educators interested in probability), we feel justified in taking certain liberties with this appraisal of the research literature. While we offer below a brief overview of a number of theories, models, and frameworks that have been used to account for normatively incorrect responses to probabilistic tasks, for more detailed and exhaustive analyses we direct the reader of this chapter to comprehensive summaries of the theories given elsewhere (see, for example, Batanero et al. 2016; Borovcnik and Bentz 1991; Garfield and Ahlgren 1988; Jones et al. 2007; Jones and Thornton 2005; Kahneman et al. 1982; Shaughnessy 1992).

16.2.1 Heuristic and Informal Reasoning

In mathematics education and certain psychology research, relative probability investigations are categorized, in a broad sense, into two particular categories. Although, as detailed above, different individuals account for incorrect responses differently, two types of reasoning—heuristic (e.g., Tversky and Kahneman 1974) and informal (e.g., Konold 1989)—have come to dominate research literature.

16.2.1.1 Representativeness Heuristic

In the early 1970s, psychologists Tversky and Kahneman conducted a series of studies that led them to conclude that people generally rely on a limited number of heuristic principles that "reduce the complex tasks of assessing probabilities and predicting values to simpler judgmental operations" (1974, p. 1124). In particular, based on the results from multiple surveys that asked individuals to compare sequences of random events, they identified a type of reasoning that was coined the representativeness heuristic, whereby one "evaluates the probability of an uncertain event, or a sample, by the degree to which it is: (i) similar in essential properties to its parent population; and (ii) reflects the salient features of the process by which it is generated" (Kahneman and Tversky 1972, p. 431). For example, when considering a sequence of tosses of a coin, Tversky and Kahneman found that subjects deemed the sequence HTHTTH was more likely to occur than HHHHTH (where H stands for "heads" and T stands for "tails") because the ratio of the number of heads to the number of tails was not close enough to one. As Chernoff and Russell (2012) explain, the representative heuristic, at the time, provided a new model for interpreting normatively incorrect responses to comparisons of relative likelihood, and as a result, heuristic reasoning became a dominant lens for research investigating comparisons of relative likelihood.

Since the seminal work of Tversky and Kahneman in the 1970s, a number of researchers have addressed issues associated with the inferential nature of Tversky and Kahneman's research. For instance, as Shaughnessy (1992) notes, "there was no attempt made [by Tversky and Kahneman] to probe the thinking of any of these subjects" (p. 473). Consequently, Shaughnessy (1977) brought two new elements to the task: First, an equally likely response option was provided. Second, the participants were asked to provide a reason for each of their responses, enabling the researcher to gain some insight into the thinking process of the participants as they responded to the questions. With the new justification element of the task,

Shaughnessy was able to determine that subjects did, in fact, find that a sequence such as BBBGGG (where B stands for "boy" and G for "girl," with the sequence representing birth order) was not representative of randomness and that the sequence BBBBGB was not a representative ratio of boys to girls.

16.2.1.2 Outcome Approach

Building upon Shaughnessy's (1977) task modifications (i.e., including an equally likely response option and asking for a response justification), Konold et al. (1993) introduced a different version of the relative likelihood task than had been seen in the past, and coined another mode of informal reasoning the outcome approach: "a model of informal reasoning under conditions of uncertainty" (Konold 1989, p. 59) according which the participants interpret a question as asking them to predict what will actually happen instead of taking probability into consideration. The outcome approach became, as had the representativeness heuristic earlier, a new model to account for normatively incorrect responses to comparisons of relative likelihood.

16.2.1.3 Equiprobability Bias

The research of Lecoutre led to the recognition of yet another bias: the equiprobability bias. This research was based on interpreting comparative likelihood task responses that involved a relationship between randomness and equiprobability. As Lecoutre (1992) found, some individuals seem to believe that random events are equiprobable "by nature" (p. 557). For example, when a piece of candy is drawn from a box containing two orange flavoured pieces and one lemon flavoured piece, the events "draw a lemon-flavoured piece" and "draw an orange-flavoured piece" may be considered equally likely by some individuals because the process is random, and thus the two results are equiprobable because "it is a matter of chance" (Lecoutre 1992, p. 561). Worthy of note, this research involved participants comparing events (i.e., *sets* of sequences, where the order of outcomes is not significant) rather than sequences, unlike the aforementioned work of Tversky and Kahneman, Shaughnessy, and Konold et al. This model has been evoked in accounting for certain normatively incorrect responses in more recent studies (e.g., Chernoff 2012a).

16.3 Theoretical Framework

Participants' response justifications for comparing the relative probabilities of events can be framed in terms of heuristic and informal reasoning. In this study, however, a third tool of analysis has emerged, that of logical fallacies, and in particular, the *fallacy of composition*.

16.3.1 Logically Fallacious Reasoning

More recently, a burgeoning area of research suggests that fallacious reasoning, more specifically, the use of informal logical fallacies, can account for certain normatively incorrect responses to probabilistic tasks. Chernoff and Russell (2012) utilized the fallacy of composition to analyse responses when participants (prospective mathematics teachers) were asked to determine which of five possible coin flip sequences were least likely to occur. Simply put, the fallacy of composition occurs when an individual infers something to be true about the *whole* based upon truths associated with *parts* of the *whole*. For example: Bricks (i.e., the parts) are sturdy; buildings (i.e., the whole) are made of bricks; therefore, buildings are sturdy (which is not necessarily true). The fallacy of composition accounted for normatively incorrect responses to the task. More specifically, the researchers demonstrated that participants tended to reference the equiprobability of the coin, note that the sequence is comprised of flips of a fair coin and, as such, fallaciously determine that the sequence of coin flips should also have a heads-to-tails ratio of one to one. In other words, the properties associated with the fair coin (the parts), which make up the sequence (the whole), were also expected in the sequence. As such, the fallacy of composition, in addition to the traditional theories, models, and frameworks associated with heuristic and informal reasoning, accounted for certain incorrect responses.

In breaking from the tradition of presenting sequences, which is found in all previous research on relative probability comparisons, Chernoff (2012a) and Chernoff and Russell (2012) demonstrated that certain individuals (prospective mathematics teachers), when asked to identify which sequence of five flips of a fair coin was least likely to occur, did not use the representativeness heuristic (proposed by Kahneman and Tversky 1972) or the outcome approach (proposed by Konold et al. 1993). Rather, they too were observed to utilize the fallacy of composition. In this instance, the coin flips (the parts) are equiprobable, and the events (the whole) are comprised of coin flips; therefore, the events are deemed to be equiprobable (which, again, is not necessarily true). Worthy of note, the fallacy of composition accounted for both normatively correct and incorrect responses to the new relative likelihood comparison task.

As Chernoff and Russell (2012) note, an advantage of this new interpretation of certain normatively incorrect responses is that it avoids certain assumptions about individuals' understanding of the law of large numbers. While the work of Tversky and Kahneman suggested that the law of small numbers—i.e., an application to small numbers of the law of large numbers, which "guarantees that very large samples will indeed be highly representative of the population from which they are drawn" (Tversky and Kahneman 1971, p. 106)—was a root cause associated with the representativeness heuristic, subsequent research (e.g., Evans and Pollard 1981; Kunda and Nisbett 1986; Stohl 2005) has demonstrated that individuals do not always have the expected understanding of the law of large numbers. Consequently, as Chernoff and Russell (2012) recognized, the assumption that when individuals appear to be employing the law of small numbers, they have an appropriate understanding of the law of large numbers to make such an application, is problematic. On the other hand, the fallacy of composition framework, which accounted for normatively incorrect responses to relative probability comparisons that may have previously been accounted for by the representativeness heuristic, avoids said issues.

In this chapter, we will utilize the fallacy of composition as a theoretical framework to account for certain normatively incorrect responses given by prospective mathematics teachers.

16.3.2 Abductive Reasoning

As hinted in the introduction, we will now focus on the nature of the reasoning behind the aforementioned theories. In particular, the reasoning associated with relative probability comparison theories (e.g., the representativeness heuristic, the outcome approach, the equiprobability bias, sample space partitions, and others) is based on Peirce's (1931) notion of abductive reasoning. Upon hearing or reading a response Q given by a participant, a researcher infers a rule P that implies Q (e.g., if the participant is using the representativeness heuristic, then she will answer Q) and a consequence P (e.g., that the participant is using the representativeness heuristic). Peirce describes abductive inferences as inferences to the *best* explanation. In a more general sense, facts are used as a starting point, a particular hypothesis—derived from inferences and used to best explain the facts observed—is then presented, and, if it is the case that if the hypothesis were true then it would best or most likely explain the observed facts, there exists reason to suspect that the theory hypothesized is true (Lipton 1991).

We contend that relative probability comparison theories were created in a similar fashion. For example, Chernoff (2009) observed through participants' (unconventional) written descriptions of the sample space that certain individuals' relative probability comparison responses were in accord with a *subjective sample space*: that is, a sample space partitioned according to the participant's interpretation of the sequence of coin flips presented (e.g., switches, runs, longest run, switches, and longest run). For example, when asked which sequence is least likely to result from flipping a fair coin five times, some participants responded that HTHTH was least likely, citing the improbability of heads and tails alternating so frequently. When considering a partition of the sample space according to number of switches, there are, in fact, only two ways to get four back-and-forth switches, (HTHTH, THTHT), 8 ways to get 3 switches, 12 ways to get 2 switches, (HHTHH, TTHTT, HHTTH, TTHHT, HTTHH, etc.), and so on. As Chernoff (2009) argued, the participants' interpretation of the sequence based on its characteristic (number of switches) suggests an alternative (i.e., non-traditional) partition of the sample space.

In this case, as with all relative probability comparison theories, the observed facts—the written responses—acted as the starting point. It was then hypothesized that certain individuals answering the question were employing a subjective sample space in answering the task. Then, it was demonstrated that if the newly hypothesized model were true, the results observed would follow accordingly (which occurred through the use of Chernoff's (2009) notions of a *meta-sample space* and *event description alignment*). As such, there existed reason to suspect that the subjective sample space hypothesis is true—that is, to accept the claims of the newly hypothesized model. Subsequently, the notion of subjective sample spaces (as was the case with the representativeness heuristic, the outcome approach, the equiprobability bias, and other theories) exists in the research literature as a theory or model (specific terminology has not yet been adopted) to explain how certain individuals respond to particular probability tasks.

As evidence mounts for a theory or model, there is—on behalf of respective research communities—more and more reason to suspect that the theory is true, or, at least, that it is the best possible explanation for an observed phenomenon: that is, that the participants are employing a particular theory in their answering of a question (see, for example, Chernoff (2012b) for a discussion of the reification of the representativeness heuristic in the field of mathematics education). However, it should be noted that one cannot claim with *certainty* that if an individual declares that, in this instance, one sequence is more or less likely to occur, they have indeed done so according to a particular theory or model. The introduction of new theories and models naturally questions the notion of the 'best possible explanation' associated with previous research, while remaining susceptible to such questioning themselves in the event that still newer and more robust theories are introduced.

16.4 Task and Research Questions

The purpose of this chapter is to extend the recent investigations utilizing logical fallacies, specifically the fallacy of composition, as a theoretical framework to account for certain normatively incorrect responses given by prospective mathematics teachers. In addition, the present study eschews the more traditional approach of presenting sequences and, instead, continues with the recent investigation, where participants are presented with events and asked to determine which is least likely to occur. As is the case with all prior research in this domain, the abductive approach, presented in detail above, was applied throughout the study.

Thus, the purpose of this study is to establish whether the fallacy of composition best (again and hereafter, abductively speaking) accounts for certain incorrect responses given by prospective mathematics teachers when comparing the relative probabilities of given events. To achieve this goal, three different versions of a task familiar to those in the research community were presented to 363 prospective teachers of mathematics enrolled in methodology courses for teaching mathematics. The tasks are presented in Sect. 5. The majority of the prospective teachers were enrolled in their first few years of the Bachelor of Education program, and the topic of probability had yet to be discussed in class at the time of the research. The participants responded on paper to a single question concerning the relative probability of occurrence of given events, and were permitted to work on the task until completion.

16.5 Results and Analysis

The current study, which involves 363 prospective mathematics teachers, is comprised of three separate, yet similar studies consisting of 106, 162, and 95 participants, which have been denoted as *Study I, Study II, and Study III,* respectively. The task given to participants in the different studies varies only slightly in nature and parallels the developmental history of the sequence version of the relative probability comparison task found in the field of mathematics education research (see, e.g., Chernoff and Russell 2012).

16.5.1 Study I

As presented in Table 16.1, of the 106 participants that took part in Study I, 73 (approximately 69% of) respondents incorrectly established that the event 'five heads and one tail' was equally likely to the event 'four heads and two tails.'

Among the 73 participants that incorrectly established that the two events were equiprobable, 14 responses evidenced use of the fallacy of composition. However, given pagination limitations, only five of the fourteen responses are now presented in detail.

Sam: Both results a and b are equally likely to occur because the probability of being heads or tails on every flip is equal (50/50).

Diane: Both are equal in likeliness to occur. There is a 50/50 chance that each flip will get a head or tail, so the probability is equal for every option.

Coach: Each individual flip has the same odds/probability of being either heads or tails (50/50). Therefore all combinations of heads and tails are equally likely to occur.

Cliff: The probability of a coin being heads or tails when flipped is always 50/50. This means that all combinations should be equally probable.

Norm: Both results are equally likely to occur because the probability of flipping a tail or a head is equal

Table 16.1 Task and results from Study I

Which of the following is the least likely result of six flips of a fair coin? Please justify your response	N = 106
a. Five heads and one tail	23
b. Four heads and two tails	10
c. Both results (a and b) are equally likely to occur	73

The above five responses evidence utilization of the fallacy of composition when determining the relative probability of the events presented, although the responses differ in presentation. For example, Sam, Diane, Coach, and Cliff all reference the 50–50 (or 50/50) "probability" (Sam, Coach and Cliff), "odds" (Coach) and "chance" (Diane), whereas Norm references the "probability" as "equal," of the coin being heads or tails. All five respondents, however, are clear in establishing that it is "because" (Sam and Norm) ("means" [Cliff], "therefore," [Coach] or "so" [Diane]) a coin has an equal probability of heads or tails that the "results" (Sam and Norm), "combinations" (Cliff and Coach), or "option[s]" (Diane) should be "equally probable" (Cliff), "equally likely/likeliness to occur" (Cliff, Coach, Sam and Diane). Despite these slight differences in terminology, the core of the responses is strikingly similar: that is, the coin is equiprobable, and as a result, the events are equiprobable. Stated in fallacy of composition parlance, all five individuals infer that equiprobability is true of the events (the whole) from the fact that is true of the coin (the part). Similar findings associated from a similar task were found in Study II, and are presented below.

16.5.2 Study II

As is shown in Table 16.2, of the 162 participants that took part in Study II, 114 (approximately 70% of) respondents incorrectly established that both events, 'three heads and two tails,' and 'four heads and one tail,' are equally likely to occur.

Of the 114 normatively incorrect responses, 5 evidenced use of the fallacy of composition. These five responses are presented in detail below.

Carla: There is a 50/50 chance of flipping either Heads or Tails. Therefore there is a chance that it could easily be either a or b.

Woody: [...] Flipping 3 heads and two tails is just as likely as flipping 4 heads and one tail because there's a 50/50 chance of flipping heads or tails [...]

Rebecca: You have an equal chance (50%) of getting heads as you do tails; therefore, you have an equal likelihood of getting any combination of heads and tails.

Frasier: When flipping a coin there are only 2 outcomes: heads or tails. Therefore, there is a 50% chance of getting heads for one flip and 50% chance of getting tails for the same toss. It is just as likely to get 4 heads and one tail as it is to get 3 heads and 2 tails because looking at each individual toss the coin has equal outcomes of going heads or tails.

Lilith: Both results are equally likely to occur because you have flipped the coin five times, there is a chance each time that you can get either heads or tails, so there is an equal chance of coming out with the outcome of (a) or (b)

Table 16.2 Task and results from Study II

Which of the following (events) is the least likely to result from five flips of a fair coin? Justify your response. In other words, explain your answer…	N = 162
a. Three heads and two tails	13
b. Four heads and one tail	35
c. Both outcomes are equally likely to occur	114

The above responses evidence utilization of the fallacy of composition when determining the relative probability of the events presented. Both Carla and Woody referenced the "50/50 chance" of getting heads or tails, while Rebecca and Frasier stated the probability as a percentage. Lilith, however, does not mention the exact probability of a single coin flip, but simply references the "chance each time that you can get either heads or tails." Each of the five respondents referred to the equiprobability of the coin, and concluded ("therefore" [Carla, Rebecca, Frasier], "because" [Woody] or "so there" [Lilith]) that "a or b" (Carla and Lilith), or "4 heads and one tail [and] 3 heads and 2 tails" (Frasier and Woody) or even "any combination of heads or tails," (Rebecca) could "easily be," (Carla), are just as/ equally "likely" (Woody, Frasier and Lilith), or have "equal likelihood" (Rebecca). As in Study I, although the respondents showed slight variations in their language, each portrays a common belief that since the coin is equiprobable, the events are also equiprobable. The results of Study III, shown below, provide further evidence for the fallacy of composition as the best explanation of certain normatively incorrect responses.

16.5.3 Study III

As Table 16.3 shows, of the 95 participants that took part in Study III, 70 (approximately 73% of) respondents incorrectly established that all of the given events ('four heads and two tails,' 'five heads and one tail,' 'three heads and three tails,' and 'two heads and four tails') were equally likely.

Table 16.3 Task and results from Study III

Which of the following events is least likely to result from flipping a fair coin six times? Give a reason for (i.e., justify) your answer…	N = 95
a. Four heads and two tails	1
b. Five heads and one tail	18
c. Three heads and three tails	6
d. Two heads and four tails	0
e. All four events are equally likely	70

Among the 70 participants who incorrectly established that the two events were equiprobable, 7 responses were flagged as evidencing use of the fallacy of composition. Six of these responses are presented in detail below.

Paul: All four events are equally likely to occur because it is a fair coin meaning 50% chance it will be heads and 50% chance it will be tails leading to any of the outcomes.

Robin: Each time you flip the coin there is a fifty/fifty chance so any of those events are equally likely.

Evan: All equal because there will always be a 50/50 chance of flipping H or T on each flip.

Kelly: Because when you flip a coin you have a 50% chance of it being a head or a tail so each combination could be equally likely to occur? I'm not sure…

Eddie: All four events are equally like because flipping a head or a tail is 1/2 or 50%.

Gary: All four events are equally likely because each time you flip a coin there is a 50% chance of a heads and 50% of getting a tails

As in the previous two studies, the above responses evidence utilization of the fallacy of composition when determining the relative probability of the events presented. Five responses refer explicitly to the 'part': "a fair coin" (Paul), "each time you flip" (Robin and Eddie), "each flip" (Evan), and "flip a coin" (Kelly). All six responses included the characteristic of the part: the equiprobability of a single coin flip stated as a "50% chance" of heads/tails (Paul, Kelly and Gary), a "50/50 chance" (Robin and Evan), or the somewhat more ambiguous "a head or a tail is 1/2 or 50%" (Eddie). Again, the conclusions that the 'wholes', here referred to as "all four events" (Paul, Eddie and Gary), "any of those events" (Robin), "all" (Evan), or "each combination" (Kelly), are "equally likely" (Paul, Robin, Kelly, Eddie and Gary) or "equal" (Evan) is reached "because" (Paul, Evan, Kelly, Eddie and Gary) the part (a single flip) holds the property of equiprobability.

16.6 Discussion

As discussed earlier, the representativeness heuristic has, historically, been widely used to interpret normatively incorrect responses to comparisons of sequences of relative likelihood. Applied to events, Tversky and Kahneman's representativeness heuristic would suggest that an event with a roughly equal number of heads and tails would be considered to be more likely, because the ratio of heads to tails is roughly one. According to Tversky and Kahneman (1974), this conclusion is reached because the same is true of an arbitrarily large event (the parent population) and of an individual coin flip (the process by which the event was generated). Thus, representativeness may have indeed been used by participants who chose the correct response (in Study I, five heads and one tail; in Study II, four heads and one tail; and in Study III, five heads and one tail), because these events are characterized by a ratio of heads to tails furthest from one, and are therefore (according to the representativeness heuristic) least likely. We suggest the investigation of the use of the representativeness heuristic in correct answers to relative probability tasks as a potential area of future research.

In addition, we contend that the above responses are not sufficiently explained by reasoning that participants were using the outcome approach (Konold et al. 1993). This approach, identified by Konold et al., in participants' responses to relative likelihood comparisons involving sequences, is characterized by responses including references to an inability to predict the future [e.g., "anything is really likely" (p. 406)], or to the reality that any of the any of the events listed are, in fact, possible ["it could be anything"; "just because it hasn't happened to me doesn't mean that it doesn't happen" (p. 406)]. In the responses above, however, participants seem confident in their claim that each event is equally likely, not because each event is possible, but rather because the event is made up of individual coin flips that have an equal likelihood of turning up heads or tails.

As discussed earlier, Lecoutre (1992) found that some individuals seem to believe that random events are equiprobable "by nature" (p. 557), or, conversely, misconstrue equiprobability as the notion that "anything can happen" (Chernoff 2012a, p. 80). Although some participants in the studies discussed above do refer to the results being up to chance (as could be argued in the case of Carla), the reasoning of these responses is still clearly connected to the outcome of an individual coin flip. If participants were utilizing the equiprobability bias, as opposed to the fallacy of composition, reference to an individual coin flip would not be necessary. One could simply state the events as equiprobable because the process is a random matter of chance. However, as we have seen, participants are explicit in their reasoning that the equiprobability of the events is based on the equiprobability of the individual coin flips. None of the above responses include any reference to equiprobability based solely on "randomness," "nature," or "chance," and therefore, they cannot be as reasonably accounted for by the equiprobability bias.

It must be noted that the questions presented to the participants differed in wording: While Studies II and III both explicitly use the term "event," Study I asks for the least likely "result." Further investigation would be necessary to determine whether this difference influences participants' reasoning and responses. However, we note that the number of responses evidencing use of the fallacy of composition is strikingly similar across the three studies. By way of specific examples, in Study I, Cliff responds that "the probability of a coin being heads or tails when flipped is always 50/50. This means that all combinations should be equally probable." Similarly, in Study II, Rebecca responds that "you have an equal chance (50%) of getting heads as you do tails; therefore, you have an equal likelihood of getting any combination of heads and tails. Likewise, Kelly responds in Study III that "when you flip a coin you have a 50% chance of it being a head or a tail so each combination could be equally likely to occur." Similar comparisons may be made with other responses, suggesting that this difference in wording did not result in a meaningful difference in these participants' reasoning.

16.7 Concluding Remarks

As discussed in the prelude of this article, in the last decade, there has been a systematic investigation into prospective mathematics teachers' relative probability comparisons, specifically as related to coin flip sequences. More generally, relative probability comparisons have sustained research for the last 50 years, and the investigation is continuing to expand. Much of the research is dedicated to accounting for normatively incorrect relative probability comparisons. As such, over the last 50 years, an impressive array of theories and models have been developed to account for normatively incorrect responses to relative probability comparisons, including the representativeness heuristic (see Shaughnessy 1977; Tversky and Kahneman 1974), the outcome approach (Konold 1989; Konold et al. 1993), the equiprobability bias (Lecoutre 1992), sample space partitions (Chernoff 2009), attribute substitution (Chernoff 2012b; Kahneman and Frederick 2002), inadvertent metonymy (Abrahamson 2009), and others. More recently, Chernoff and Russell (2012) proposed an additional theory or model: the fallacy of composition, which occurs when an individual infers something to be true about the whole based upon truths associated with parts of the whole. Although the theory has been useful in accounting for certain normatively incorrect responses when participants have been asked to determine which sequences of coin flips were least likely to occur among given possibilities, some have dismissed the results on the grounds that the participants were, in fact, comparing events instead of sequences (and doing so correctly). The results presented herein provide a response to this speculation, and suggest that it cannot account for all normatively incorrect responses to relative likelihood tasks.

Rather than being asked to compare sequences of coin flips, prospective mathematics teachers in three separate studies (363 participants in total) were asked to compare events related to coin flips and to determine which were less likely to occur or whether all events were equally likely to occur. In these studies, 69%, 70%, and 73% of participants, respectively, incorrectly responded that all events were equally likely to occur. Moreover, 26 responses in total (of the 363 participants and the 257 participants that deemed events equally likely) were identified as evidencing use of the fallacy of composition. In general, these participants evidenced use of the following reasoning: the coin flips are equiprobable, and the events are comprised of coin flips; therefore, the events are equiprobable. In other words, the properties associated with the fair coin (the part), which makes up the events (the whole), are also expected to be observed in the event. As discussed in the previous section, the presented responses are not well accounted for by the representativeness heuristic, the outcome approach, or the equiprobability bias—theories that have traditionally and widely been used to account for normatively incorrect responses to relative probability comparisons. We have suggested that the fallacy of composition, as a new interpretation of certain normatively incorrect responses to relative likelihood tasks, may account for certain incorrect responses to questions involving not only coin sequences, but also events related to coin flips. While the results herein present a qualitative confirmation of this inference, further research in this area is needed.

Although we have highlighted the differences in the various theories and models accounting for normatively incorrect relative probability comparisons, we have also argued that they all have something in common: Namely, the reasoning associated with these theories is based on Peirce's (1931) notion of abductive reasoning, which involves inferences to the best explanation of observations. We have described how probability comparison theories may have been developed according to this reasoning, and have utilized a similar approach in reaching our own conclusions about responses given by prospective mathematics teachers when tasked with comparing the relative probabilities of events. However, although there is growing evidence to suggest that the use of the fallacy of composition may best explain certain responses to relative probability comparisons, we can only use the term "best" in an abductive sense: In other words, we cannot claim with certainty that if an individual declares that given events are equally likely to occur, they have indeed done so by employing the fallacy of composition. The same applies, of course, for other theories and models related to probability comparisons, such as the representativeness heuristic, the outcome approach, and the equiprobability bias.

References

Abrahamson, D. (2009). Orchestrating semiotic leaps from tacit to cultural quantitative reasoning: The case of anticipating experimental outcomes of a quasi-binomial random generator. *Cognition and Instruction, 27*(3), 175–224.

Batanero, C., & Serrano, L. (1999). The meaning of randomness for secondary school students. *Journal for Research in Mathematics Education, 30*(5), 558–567.

Batanero, C., Chernoff, E., Engel, J., Lee, H., & Sánchez, E. (2016). *Research on teaching and learning probability*. ICME-13 Topical Surveys. New York: Springer.

Borovcnik, M., & Bentz, H. (1991). Empirical research in understanding probability. In R. Kapadia & M. Borovcnik (Eds.), *Chance encounters: Probability in education* (pp. 73–106). Dordrecht, The Netherlands: Kluwer.

Chernoff, E. J. (2009). Sample space partitions: An investigative lens. *Journal of Mathematical Behavior, 28*(1), 19–29.

Chernoff, E. J. (2012a). Logically fallacious relative likelihood comparisons: The fallacy of composition [Special issue: National year of mathematics]. *Experiments in Education, 40*(4), 77–84.

Chernoff, E. J. (2012b). Recognizing revisitation of the representativeness heuristic: An analysis of answer key attributes [Themed issue: Probability in reasoning about data and risk]. *ZDM Mathematics Education, 44*(7), 941–952.

Chernoff, E. J., & Mamolo, A. (2015). Unasked but answered: Comparing the relative probabilities of coin flip sequence attributes. *Canadian Journal of Science, Mathematics and Technology Education, 15*(2), 186–202.

Chernoff, E. J., & Russell, G. L. (2012). The fallacy of composition: Prospective mathematics teachers' use of logical fallacies. *Canadian Journal of Science, Mathematics and Technology Education, 12*(3), 259–271.

Evans, J. S. B. T., & Pollard, P. (1981). Statistical judgment: A further test of the representativeness heuristic. *Acta Psychologica, 51,* 91–103.

Falk, R. (1981). The perception of randomness. In C. Laborde (Ed.), *Proceedings of the Fifth International Conference for the Psychology of Mathematics Education* (pp. 222–229). University of Grenoble.

Garfield, J., & Ahlgren, A. (1988). Difficulties in learning basic concepts in probability and statistics: Implications for research. *Journal for Research in Mathematics Education, 19*(1), 44–63.

Jones, G. A., & Thornton, C. A. (2005). An overview of research into the teaching and learning of probability. In G. A. Jones (Ed.), *Exploring probability in school* (pp. 65–92). New York, NY: Springer.

Jones, G. A., Langrall, C. W., & Mooney, E. S. (2007). Research in probability: Responding to classroom realities. In F. Lester (Ed.), *Handbook of research on mathematics teaching and learning* (2nd ed., Vol. 2, pp. 909–955). Greenwich, CT: Information Age Publishing and National Council of Teachers of Mathematics.

Kahnemann, D., & Frederick, S. (2002). Representativeness revisited: Attribute substitution in intuitive judgment. In T. Gilovich, D. Griffin, & D. Kahnemann (Eds.), *Heuristics and biases: The psychology of intuitive judgment* (pp. 49–81). New York: Cambridge University Press.

Kahneman, D., & Tversky, A. (1972). Subjective probability: A judgement of representativeness. *Cognitive Psychology, 3*(3), 430–454.

Kahneman, D., Slovic, P., & Tversky, A. (1982). *Judgment under uncertainty: Heuristics and biases*. Cambridge, England: Cambridge University Press.

Konold, C. (1989). Informal conceptions of probability. *Cognition and Instruction, 6*(1), 59–98.

Konold, C., Pollatsek, A., Well, A., Lohmeier, J., & Lipson, A. (1993). Inconsistencies in students' reasoning about probability. *Journal for Research in Mathematics Education, 24*(5), 392–414.

Kunda, Z., & Nisbett, R. E. (1986). Prediction and partial understanding of the law of large numbers. *Journal of Experimental Social Psychology, 22*(4), 339–354.

Lecoutre, M. P. (1992). Cognitive models and problem spaces in "purely random" situations. *Educational Studies in Mathematics, 23*(6), 557–568.

Lipton, P. (1991). *Inference to best explanation*. New York, NY: Routledge.

Peirce, C. S. (1931). Principles of philosophy. In C. Hartshorne & P. Weiss (Eds.), *Collected papers of Charles Sanders Peirce* (Vol. 1). Cambridge, MA: Harvard University Press.

Shaughnessy, J. M. (1977). Misconceptions of probability: An experiment with a small-group, activity-based, model building approach to introductory probability at the college level. *Educational Studies in Mathematics, 8*(3), 295–316.

Shaughnessy, J. M. (1992). Research in probability and statistics. In D. A. Grouws (Ed.), *Handbook of research on mathematics teaching and learning* (pp. 465–494). New York: Macmillan.

Stohl, H. (2005). Probability in teacher education and development. In G. A. Jones (Ed.), *Exploring probability in school: Challenges for teaching and learning* (pp. 345–366). New York: Springer.

Tversky, A., & Kahneman, D. (1971). Belief in the law of small numbers. *Psychological Bulletin, 76,* 105–110.

Tversky, A., & Kahneman, D. (1974). Judgement under uncertainty: Heuristics and biases. *Science, 185*(4157), 1124–1131.

Chapter 17
Preparing Teachers for Teaching Probability Through Problem Solving

Pedro M. Huerta

Abstract In this paper we introduce the idea of solving problems by simulation with didactical intention. Based on this idea, we analyse the way we prepare the teachers to teach solving probability problems by simulation, in distinguishing whether the teaching is offered to prospective elementary teachers or secondary school teachers. The process of solving probability problems is seen as an inquiry process, which uses simulation as a resolution method with heuristic content, so that the preparation of teacher is then based on this approach. Therefore, prospective teachers should have the opportunities to learn about probability problems, about the process of solving probability problems, and about students' resolution process by observing the work of other classmates.

Keywords Preparing teachers · Primary school teachers · Secondary school teachers · Probability problem solving · Simulation

17.1 Introduction

The 21st century society needs well-prepared teachers willing to tackle such a complex task as teaching probability at school from a plural perspective of probabilistic reasoning (Chernoff and Siraman 2014). Without well-prepared teachers the teaching of probability and statistics at school will improve with difficulty.

In effect, for years, many researchers have repeatedly suggested to teachers' educators to carry out a better preparation of prospective teachers in statistics and probability in order to improve their competency in these topics and, as a

---

P. M. Huerta (✉)
Departament de Didàctica de la Matemàtica, Facultat de Magisteri,
University of Valencia, Avda. Tarongers, 4, 46022 Valencia, Spain
e-mail: manuel.p.huerta@uv.es

C. Batanero and E. J. Chernoff (eds.), *Teaching and Learning Stochastics*,
ICME-13 Monographs, https://doi.org/10.1007/978-3-319-72871-1_17

consequence, to improve those of their future primary and secondary school students. This is not an easy task, of course, since how this preparation should be still is, in our opinion, an unsolved question for the community.

Indeed, teacher educators agree that training teachers must combine two types of knowledge, the content knowledge and the pedagogical content knowledge (Shulman 1986), or according to Ball et al. (2008) the content knowledge for teaching. Some variations of this dichotomise model include terms such as mathematical proficiency for teaching (MPT in Wilson and Heid 2010). But, it is not easy to design useful tasks for prospective teachers that cover both content knowledge and pedagogical content knowledge. Of course, prospective teachers generally have many difficulties in learning and understanding basic concepts of statistics and probability, still not learnt or understood when they were secondary school students. So, how can they learn about pedagogical content knowledge if they have so many difficulties with content knowledge? An answer to this question could be found in the idea we want to develop here, in preparing teachers by solving probability problems by simulation with didactical intention (Huerta 2015).

Eichler and Vogel (2014) point out that simulation is used in teaching probability as a tool to: (a) explore a model that already exists, (b) develop an unknown model approximately, and (c) represent data generation (p. 93), but they do not mention the role of simulation as a method for solving probability problems with heuristic content, just as we already proposed in Huerta (2015). This method allows working with more than one meaning of the concept of probability during the process of solving problems, in the sense suggested by Borovcnik and Kapadia (2014a, b).

Moreover, Pfannkuch and Ziedins (2014) suggest that “Probability is the only area in the school curriculum where students are afforded the opportunity to view the world non-deterministically and to learn about randomness and decision under uncertainty” (p. 102). We should add that prospective teachers should also have this opportunity at university, for several reasons: firstly, because they need to view the world from the same point of view and secondly because they could improve their professional noticing (Jacobs et al. 2010) about the reasoning of their future students at schools in situations like those lived by them.

Finally, many researchers support a simulation and modelling-based approach to teaching. In particular, Shaughnessy in an interview to Rossman (2013) analyses its benefits in the learning of the probability in prospective teachers and also in helping to the change of students’ thinking about probability and statistics; and Chaput et al. (2011) describe three stages in the teaching modelling-based process: pseudo-concrete model, mathematization—formalization and validation—and interpretation in context. But, the question is how such a perspective can be put into practice and what would be desired learning outcomes across the curricular levels (Chaput et al. 2011, p. 102). Maybe in probability problem solving by simulation we could find a possible answer.

17.2 Probability Problem Solving and Preparing Teachers

It is well known that mathematics teachers, both in primary and secondary school, have many difficulties in understanding basic concepts on statistics and probability in a meaningful way. They find great difficulties when they are tackling statistics and probability tasks, including, of course, problem-solving tasks. Therefore, probability problem solving should be included in the education of teachers, not only considering them as learners of mathematics but also as a future teacher that should understand which aspects relate the process of solving probability problems to the construction of statistics and probabilistic thinking in students, as well as learn to become a monitor of the process of solving probability problems in primary or secondary classrooms. Shaughnessy (1992) already claimed that teaching probability is teaching problem solving.

Santos (2012) formulates himself some questions about what problem solving activities should be considered in a program intended for preparing mathematics teachers, both as learners and future professionals that we can extend to statistics and probability:

- What mathematics problem solving activities help prospective teachers to construct and develop the probabilistic and statistic content knowledge for teaching?
- What kind of problems facilitates the construction of students' mathematics knowledge?
- How can we distinguish and characterize the mathematics thinking that prospective teachers have to display in their experiences in solving probability problems and in the design of activities for teaching?

Moreover, Kilpatrick (2016) proposes looking at the problem solving process as an inquiry/investigation process. A new related approach to the teaching of mathematics is the so called inquiry-based learning (IBL) (Maaβ and Doorman 2013), which is understood as a "teaching culture and classroom practices in which students inquire and pose questions, explore and evaluate" (p. 887). In this framework, both problem formulation and the teachers training become fundamental aspects of the teaching and learning process, these aspects being even more crucial and surely more appropriate in probability than in other curricular areas. Therefore, training teachers for this IBL approach should start at the beginning of their training process and not only during their professional development: "teachers need to acquire experience using IBL in their instruction … (although) an especial opportunity to gain experience with IBL is when they are still being prepared to teach" (Kilpatrick 2016, p. 74).

Having this in mind, we have considered the idea of solving probability problems with *didactical intention* in order to prepare (primary and secondary school) teachers to teach statistics and probability. That is to say, we propose probability problems that have to be solved from an IBL point of view and, at the same time, to pay attention to the didactical potential of problems. This means that prospective teachers explore the possibility that the problems they are solving become useful to be solved by their students in the future and, if so, decide what transformations these problems need to suffer in order to be appropriated for teaching in a given school level.

Hence, in our approach, we place future teachers in a double role: first, as a solver of probability problems, getting the content knowledge needed for solving them, and also as a *proposer of problems* he or she would like to teach, requiring the need to get pedagogical content knowledge in order to infer the problem potentialities and difficulties. Thus, for example, prospective teachers may become conscious of what difficulties they and their classmates have in finding the solution of the problems. Lastly, solving problems with didactical intention—such as we show in this paper, may allow prospective teachers to get basic content knowledge, teaching knowledge and probability problem solving knowledge. Moreover, they also have the opportunity to learn about the uses of problem solving in teaching probability (in the sense of Leonng et al. 2016).

17.3 Solving Probability Problems with Didactical Intention in Preparing Teachers. The Prior Theoretical Analysis

In this section, we contextualize the problem we use in Sect. 17.5 to introduce the method of solving problem by simulation with didactical intention. More specifically, we analyse the notion of simulation, the family of problems to which the problem belongs to and the theoretical framework in which the problem can be analyzed.

17.3.1 Simulation

Many researchers (see, for example, the modelling point of view in Chaput et al. 2011) and also some programs for preparing teachers (see Rossman's (2013) interview of Mike Shaughnessy) suggest that simulation should be present in the teaching of probability. We will use this suggestion to show what we understand by solving problems with didactical intention, in this case, the way of solving probability problems by simulation. We use the notion of simulation in a different, but complementary, role to that reported by Eichler and Vogel (2014), as a method of solving probability problems, which clearly has connexions with modelling processes in probabilistic situations.

But, what do we mean with the noun "simulation" and the verb "to simulate"? As we know, for some random processes their probability distribution are unknown or they are very difficult to obtain. About 1950 mathematicians, such as

Von Neumann (1951) or Ulam (1951), proposed numerical methods for studying stochastic systems without going to analytical calculations in a theoretical model. Basically, their proposal consisted of building a dynamic, numerical model with which the behaviour of a real system is simulated. Questions formulated into the real system are answered from data in the numerical system, by means of sampling statistical procedures. In this proposal, to simulate means to translate a probability problem into a statistic problem, so that an answer to this second problem could be used to approximate an answer to the former. A simulation will consist in acting into the numerical system with random numbers or random devices that generate data related to the numerical system behaviour.

17.3.2 A Family of Problems to Be Solved by Simulation

Here we are going to use an old problem to illustrate the prior steps to approach the method we develop in the next sections:

> Problem: A cereal company has put plastic models of 5 Star Wars characters in boxes of SPACY-0's. The company puts one character in each box. How many boxes would you expect to have to buy in order to get all five characters? (Shaughnessy 1983, p. 340)

We have only changed the context and the number of pieces of the collection (6), in order to facilitate prospective teachers in the searching and choosing of appropriated random devices. However, we only consider three pieces when the problem is put forward to students at the Kindergarten (Martínez and Huerta 2015). Both decisions are not casual, but as a consequence of the prior analysis of the problem. This problem is an instance of a more general situation denominated *the collection problem*. In Borrás (1996) we can find other instances of this family of problems, formulated in other contexts, and modelled with no didactical but mathematical intention.

In preparing teachers, when a problem has to be solved by simulation, it should be analysed and compared to the different situations where the corresponding simulated problems could be formulated. The relationship between reality and models (Pfannkuch and Ziedins 2014) is difficult and provokes confusion and misunderstandings in the students who are solving the problems by simulation.

17.3.3 The Prior Theoretical Analysis

Theoretically, the collection problem may be modelled by an absorbent Markov's Chain (Gordon 2012). Let n be the number of pieces of a given collection. The graph presented in Fig. 17.1 represents the different states of the collection, from 0 (the start) to n (the collection is completed) and one of the two possible types of transition probabilities between states (Gordon 2012). Particularly, in Fig. 17.1 the

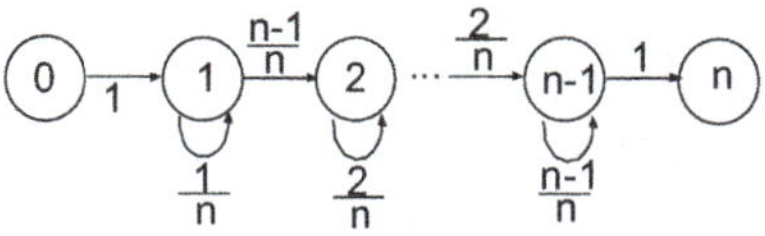

Fig. 17.1 Graph representing the random process

conditional probability of reaching a neighbour state k or $k + 1$ given that we are at the state k, (k: 1 … n) is represented. The conditional probability to reach the state k given that we are at the state j (being $j \neq k$) is not represented. The process always ends at the state n, so this event is certain, but the time, or number of attempts we must wait to reach this state is, indeed, uncertain.

In terms of the collection, the waiting time indicates the expected or average number m of cereal boxes that have to be bought to complete the collection. The following transition probabilities between neighbour states define this stochastic process under, at least, the following two fundamental hypotheses: the number of cereal boxes is unlimited and the pieces are uniformly distributed among the cereal boxes.

$$p_{k,k+1} = \frac{n-k}{n}, \quad p_{k,k} = \frac{k}{n}; \quad \mathrm{k} = 0, 1, 2, \ldots, \mathrm{n} - 1.$$

The waiting time m before the process reaches the state n, starting at the state 0, is given by the following expression (Borrás 1996):

$$m = n \times \left(\frac{1}{n} + \frac{1}{n-1} + \frac{1}{n-2} + \cdots + \frac{1}{2} + 1 \right).$$

In Table 17.1 we provide information on the average number of cereal boxes required to get the whole collection, dependant on the number of pieces in it. Being aware of this information may allow educators of teachers and prospective teachers to make thoughtful methodological decisions before using the problems for teaching.

Table 17.1 Theoretical expected number of cereal boxes needed to complete a collection for different number of pieces

Number of pieces in the collection (n)	Theoretical expected number of boxes	School level where the problem was used
3	5.5	Kindergarten (Martínez and Huerta 2015)
5	11.42	Secondary and college students (Shaughnessy 1983)
6	14.7	Training elementary teachers (this paper)
7	18.5	Training secondary school teachers (this paper)

Of course, theoretically speaking, Markov chains are far away from the basic content knowledge that we can expect in primary and secondary mathematics teachers. This knowledge would only be desirable from a mathematical proficiency for teaching point of view (Wilson and Heid 2010) in preparing secondary school teachers. However, other simpler and more concrete theoretical models can also be available in Engel (1975). We, therefore, need to consider this problem in two differentiate teaching contexts: preparing elementary teachers and preparing secondary mathematics teachers. In both contexts we share the strategy to solve the problem by simulation, but differ in the didactical intentions. In the following we describe the way to solve probability problems and the associated didactical intentions in each context.

17.4 Solving Probability Problems by Simulation

In the Sect. 17.3, our main interest has been the theoretical analysis of a particular family of mathematics problems we propose to our students to be solved by simulation. They are school word-problems that are formulated into an uncertainty situation and ask for a probability or the expected value of a random variable, in the way it is usual at schools or in textbooks. However, to analyse the process of problem solving by simulation, this time from a heuristic point of view, we need take some notions from the heuristic style of problem solving (Puig 1996), such as heuristic tools, heuristic suggestion and solving method with heuristic content, that we will use later.

17.4.1 Some Notions from the Heuristic Style of Problem Solving

Thus, we assume that *heuristic tool* is "a procedure, independent from the content of the problem, which transforms a problem into another" (Puig 1996, p. 41). Indeed, although a heuristic tool transforms a problem into another, it does neither solve it nor guarantees its solution. So, the simulation of the throwing of an ideal coin by means of an Euro coin is a heuristic tool, because a more abstract situation problem (probability to get a tail in throwing a coin) could be transformed into a problem in the most concrete situation generated by the throwing of an Euro (probability to get a tail in this coin). Simulating a problem with random numbers is another heuristic tool because we can transform every probability problem into another problem which is referring, this time, to random numbers.

When, in the context of solving probability problems at schools we suggest the students to simulate the problem, then this suggestion requires that they use heuristic tools.

Siñériz and Puig (2006) define *solving method with heuristic content* as a set of rules that generates new problems from the one to be solved, until getting a problem that either is an elementary problem, and the way to solve it is already known, or we have its solution. We find coincidences between this notion and that of reformulating problems by Kilpatrick (2016). Simulation, then, could be seen as a method of solving probability problem with heuristic content.

17.4.2 A Way of Solving Probability Problems by Simulation. Didactical Implications

With the expression *way of solving probability problems by simulation* we will describe the method consisting in a set of steps needed to generate new problems from the original problem, in such a way that we can tackle it and even solve every new generated problem. We will develop this idea in the Sect. 17.5, by describing simulation as a solving method with heuristic content in a teaching context.

We will assume that a problem has been solved by simulation if during the solving process the *original problem* has been translated into another problem (*the simulated problem*) by means of random devices, in such a way that the new simulated problem is probabilistically equivalent to the original one. So, the new problem has been tackled and an answer to the posed question has been given. Furthermore, from this solution the solvers can infer an answer to the question posed in the original problem.

From a didactical point of view, what is really interesting about simulation is its heuristic character and its potential of inquiry, exploration and discovery and its capacity to be used in many other different problems. The method allows users to get a great amount of numerical information that has to be later treated in order to give an answer to the simulated problem, which could be finally exported to the original problem. This requires, of course, taking into account statistical instruments and methods. The translation of the answer to the simulated problem into an answer to the original problem requires from solvers to consider necessarily the empirical nature of probability, besides its stability in the presence of a large number of empirical essays.

As the process of transforming a problem depends on the available transformation tools, many questions arise if we pay attention to the students while they are solving problems, or to the teaching of solving probability problems by simulation. Paraphrasing Puig (1996, p. 41) some examples are the following: With which intention is this tool being used? How is the original problem related to the simulated problem by means of the considered tool? Is the original problem solution depending on the tool considered? What can we export from the answer to the simulated problem to the original one, and what cannot be exported? Once an answer has been given to the original problem, are new circumstances in which we should look at the problem again? Do these circumstances require the original problem to be reformulated?

These questions do not have general answers. On the contrary, they should be formulated and answered for every susceptible tool: urns, dices, roulettes, o random numbers tables, random numbers generated in spreadsheets, R or Matlab.

Again, let us consider the collection problem analysed in Sect. 17.3.3, for n = 6. Let us assume that the prospective teachers know no theoretical model to solve the problem, but they want to solve it. That is to say, to find an answer as reasonable as possible to the posed question using the simulation as a solving method. So, while solvers are solving the problem, they get in touch with and uncertain situation and learn to manage themselves into it. Our teaching objective is not modelling the problem, as we did in Sect. 17.3, but to solve this particular problem and learning to use probabilistic reasoning for finding an answer that will involve variability and uncertainty.

17.4.3 Scopes in the Process of Solving Problems by Simulation

In the process of solving problems by simulation we can distinguish four scopes in which the work has to be developed: Working in the original problem scope, in the simulated problem scope, in the simulation problem solution scope and, finally, in the original problem solution scope, and of course working with the corresponding relationships/translations between scopes, as we can appreciate in the Fig. 17.2. However, when we consider the teaching of solving probability problems by simulation, this work can then be divided into a number of steps in each of the scopes, described in the next section.

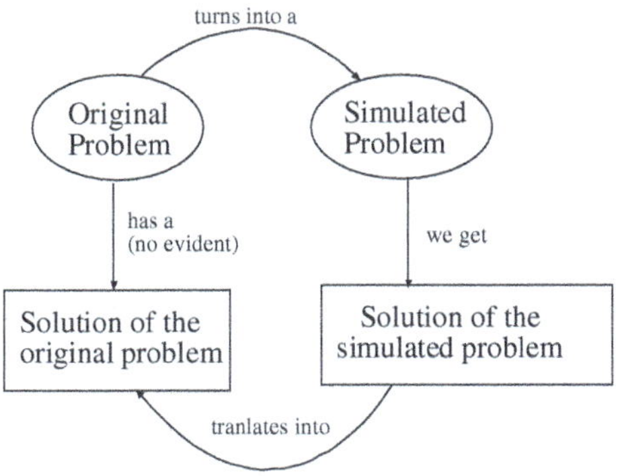

Fig. 17.2 Basic diagram with the scopes in the process of resolution of a probability problem by simulation

Every time I buy a cereal box I get one of the six pieces of the collection. It may happen that I already got this piece or not and, in this case, the state in which the collection is does not change. But, every time I buy a cereal box, until the box is opened, I do not know what piece I am going to get. By hypothesis (hypothesis 1) I can suppose that the pieces are uniformly distributed among the boxes and these as well among the shelves of the supermarket where I usually buy food (hypothesis 2). So, I can suppose that the probability to get piece X is the same that to get piece Y (hypothesis 3). We will finish the buying of cereal boxes just in the moment I have the whole collection. Will it be possible to get the whole collection in any time? Will we have to buy many or few boxes? How many do we expect to buy? These are questions that reasonably solvers can make to themselves and teachers only expect some conjectures from the students. In the approach we are proposing, the formulation of initial hypothesis and conjectures form a crucial part of the probability problem solving process and the understanding of this gives sense to the solvers' work with the method.

The hypothesis of equiprobability is usually accepted when a dice is thrown; therefore the process of throwing a dice is probabilistically equivalent to buying a cereal box under the aforementioned hypotheses 1 and 2, getting one of the six pieces of the collection. Moreover, the random process of repeatedly throwing a dice simulates the uncertainty process in buying the boxes under the hypothesis 1–3. Finally, the whole collection is got throughout some buying of boxes. The resulting piece does not depend on the pieces we already got; that is to say, the uncertainty process is made of independent trials, as does the successively throwing of a dice.

Therefore, if we consider a dice as the tool with which we simulate the original problem, then we will throw a particular dice several times noting down, each time, the resulting face, until all the 6 faces appear at least one time. Thus, the original problem has been replaced by the following: How many times (in average) would you expect to throw a dice in order to get all six numbers at least one time? This is now the *simulated problem*. The original questions are now reformulated for the simulated problem: It is likely that all faces of the dice appear at least one time? Should we make many trials until we have all faces at least one time?

These questions have not obvious answers, neither in simulated nor in the original problem, but something is different: now we can experiment with a dice in order to get enough information that allows us to give a reasonable answer to the simulated problem, throwing the dice so many times as needed. We will consider that a simulation has been carried out when all the six faces have appeared at least one time. Several simulations later, we have information about what has happened in our experimentation with the dice. This information can be translated into numerical information, and thus, we create a (new) statistical problem to answer questions formulated for the simulated problem. The definition of statistical variables and the sample size becomes something consubstantial to the process.

A relevant question is: How many simulations are needed in order to answer the simulated problem? The simulated problem has, in real situations in class, as many possible answers as solvers, and as many accuracy levels as required from the students for giving an answer. We can argue that we need at least two simulations;

otherwise we cannot get the average. But this (two simulations)-based answer involve higher uncertainty than an answer based on a higher number of simulations. But, how many? In many research reports, a number of simulations such as 50, 100 or 1000 is suggested. In textbooks, there is some ambiguity and only a great amount of simulations, the more the better is suggested.

Didactically speaking, what is relevant in this case is to justify this specific number. The conclusive answer is mathematical, although some times the number of simulations depends on the solver and therefore is subjective. The answer given by the students will depend on the reliability, given the number of simulations they decided to carry out. Or perhaps, teacher decides in which conditions an answer will be acceptable. For instance, let us suppose that teachers demand an estimation of probability that differs less than $\varepsilon = 1\%$ from a searched theoretical probability (which may be known or not) with a confidence level of $(1 - \delta) = 95\%$, $\delta = 5\%$.

Under these conditions, the number of required simulations by teachers can be estimated by the following expression, based on the Chebyschev inequality (De Groot and Schervish 2012): $n_0 \geq \frac{p(1-p)}{\varepsilon^2(1-\delta)}$. But, this formula is useful when students know the value of p, although it is possible to estimate a value for n_0 by maximizing the function: $kp(1-p)$, where $k = \frac{1}{\varepsilon^2(1-\delta)}$ is a prefixed constant. This function gets a maximum at p = 1/2, then the number of simulations can be estimated by the expression: $n_0 = \frac{1}{4\varepsilon(1-\delta)}, \forall\varepsilon > 0, \forall\delta > 0$.

From the teaching point of view, this formula help both educators of teachers and prospective teachers to think about the transition from a paper and pencil simulation to a conscious use of software supported simulation in solving the problem. But it does not help the students, whose satisfaction level and requirements could be achieved with a number of simulations smaller than the theoretical.

In any case, the answer to the simulated problem should be translated into an answer of the original problem, or being used to give an answer to the original problem. Moreover, in the terms of the question,—how many boxes would you expect to buy to complete the collection?—it also requires that students provide some reliability to the simulated problem answer—how many times (in average) would you need to throw a dice to get all six faces at least one time?—because this question demands from students the level of reliability they concede to the given average value.

17.5 Teaching to Solve Probability Problems by Simulation: An Example

Based on the analysis in Sects. 17.3 and 17.4, in this section we describe how we approach the collection problem with prospective primary teachers (n = 6 pieces) and secondary school mathematics teachers (n = 7 pieces).

Before introducing this task, prospective primary school teachers complete some other tasks in order to break the ice and to initiate them in probability problem solving from a heuristic point of view. However, with all prospective secondary school teachers having a high mathematical proficiency, they are also introduced to random walks as both absorbent and non-absorbent Markov's Chains.

In both types of prospective teachers, apart from improving their probability and statistics content knowledge, the task was intended to accomplish three main objectives:

- To explore the different meanings of the concept of probability (in the sense of Borovnick and Kapadia 2014b) during the process of solving probability problems by simulation.
- To explore the didactical potential of the simulation, both as a method for solving probability problems and as a context for teaching statistics and probability at schools.
- To analyse the extent to which solving probability problem by simulation facilitates the change in the students' subjective judgement about random phenomena in an uncertain situation in which the problem is put forward.

Another additional aim is increasing the professional noticing of prospective teachers (Jacobs et al. 2010). In particular, their perception about probability problems and their resolution processes, about simulation as a method for solving probability problems and as a teaching context, and about students' reasoning in solving problems by simulation (by observing their classmate's resolutions of problems).

At this point we are going to distinguish two cases: preparing prospective primary school teachers and secondary school teachers, with some differences in the teaching approach.

17.5.1 Preparing Primary School Teachers

When introducing the task to prospective primary school teachers we first formulate the problem as it was described in Sect. 17.4; then we follow the eight steps each of which pose some relevant questions that should be answered. The number and type of heuristic suggestions is obviously variable and depends on the problems and the students.

In Tables 17.2 and 17.3 we describe the steps and heuristics suggestions we set in our experience with prospective teachers. The aim was that these teachers will use this approach with their students at schools, and will approach teaching statistics and probability by solving probability problems as an inquiry process, involving simulation as the resolution method. These questions are formulated as heuristic suggestions (in the sense of Polya (1957), also considered in Kilpatrick 2016) that could help prospective teachers to progress in the problem resolution. The number of steps is related to the scopes in the resolution process we described in Sect. 17.4, and suggest the work that should be done in each of them.

Table 17.2 Steps 1–4 and heuristic suggestions in the process of solving the problem

Steps in the process	Heuristic questions
Step 1. Exploring the real situation: uncertainty; what is and is not already known in the situation.	– Can you say with complete certainty that sooner or later you will complete the collection? Why? – How many boxes do you need to buy to get the whole collection? Why? Please, give the exact number.
Step 2. Prior subjective judgements as a consequence of the analysis of the real situation.	– Do you think it is easier to complete the collection or not? Why? – If you do not know exactly how many cakes you should buy, make a conjecture[a] by considering the hypotheses formulated. Explain.
Step 3. Reliability and credibility of conjectures.	– How would you explore the extent to which the conjectures you made previously are reliable and credible?
Step 4. Simulation: Need and use of heuristic tools. The simulated problem.	– It is possible to experience the problematic situation? Or it is only possible to simulate it? – How are you going to simulate (or to experience)? Would you need to formulate some hypotheses and conjectures? – Can you explain what simulation and experimentation are[b]? – How would you reformulate the original problem in terms of the simulation you are considering? Please, formulate a new problem using only words, symbols and relationships corresponding to the uncertainty situation created by the simulation. – Do you think that there is another way to simulate the problem[c]? Which one? Why is it different? – In what sense do you think that answering the simulated problems could help you in giving an answer to the original problem? Why? Please explain your answer.

[a]The word conjecture is given here a probabilistic meaning, in the sense of Bernoulli (1713/1987, p. 213)

[b]Although at this point we are still distinguishing experimentation and simulation, in general, the prospective teachers have already made a decision about which of them they are going to consider

[c]Here we try to evaluate other simulation models different from manipulative materials with coins, dices, roulettes, urns, etc., such as random digits tables, random digits generators, or software like Excel. This is an inquiring task for the prospective teachers

Table 17.3 Steps 5–8 and heuristic suggestions in the process of solving the problem

Step 5. Information produced by the simulation, and its dependence on the considered tool. Related statistical problem. Data analysis in relation to the sample size (number of simulations).	– Solve each of the simulated problems you formulated before. – Explain, as best as possible, how you would organize the information you get from simulations. Which variables are you considering and how are you going to analyse them? Formulate the statistical problems that allow you to answer the simulated problems.
Step 6. Probabilistic equivalence between simulated problems.	– Please compare the answer you obtained for each of the questions you formulated in each simulated problem. What can you conclude from your comparison? Did you expect that this would happen? Why? – Which of your answers are you considering to use to solve the original problem? Why?
Step 7. From the simulated problem solution to the original problem solution.	– Now, try to give an answer to the question posed in the original problem. Please, explain your answer. – If your answer is a number, what does this answer mean to you?
Step 8. Translating the simulated problem's answers to answers of the original problem.	– To what extent do you think that your answers to the original problem are reliable or credible for you and others? – Can this reliability and credibility be improved? Please justify your answer. – To what extent the answers given to the problem are useful for the real situation you have experienced?

Finally, in order to develop the professional noticing with which prospective teachers should look at probability problem solving by simulation as a teaching and learning context, we add an extra step, in which we want prospective teachers to explore the potential of simulation for teaching. Thus, once the problem has been solved, we want them to revisit the problem and its resolution process as teachers rather than as students, in paying attention to the probability and statistics concepts that are involved, the procedures required to simulate or experience the problem and the role of interaction between the students and teachers with the problem during its resolution.

We formulate this last step by requiring the prospective teachers to imagine that they are going to propose this problem in primary school and then ask the following question, depending on the school grade:

- What influential considerations should be taken into account on your proposal from a cognitive point of view? And, from a curricular point of view?
- What probabilistic and statistical concepts do your students face while they solve the problem by simulation? Could you identify these concepts in each step you follow to approach the resolution of the problem by simulation?

- Methodologically speaking, which functions would you assign to the following elements: the problem, the students and the teacher, during the process of solving probability problem by simulation in a primary school classroom? Please discuss your opinion with other classmates.

17.5.2 Preparing Secondary School Teachers

In this context, we agree with many other authors (for example with Maxara and Biehler 2006), that, unlike prospective elementary teachers, prospective secondary school teachers have an adequate mathematics proficiency to understand what a theoretical model means, and can therefore learn about it and learn how to use it in applications, such as the collection problem. Therefore, they can understand and learn the theoretical model that solves the problem: Markov's Chains, but, perhaps, they do not know how to approach the teaching of solving problems at schools without first introducing the theoretical model and therefore, some didactical preparation is required.

In preparing primary school teachers we moved from a particular situation to a more general one by modelling the original situation with the help of problem solving and the simulation as a method. However, we approach from the opposite direction in preparing secondary school teachers, from the most general situation, described by the theoretical model, to the particular school task given to the students and simulation as a teaching method of probability problem solving.

In preparing future secondary school teachers, another additional aim is reducing the difficulties faced with problems considered as elementary. These difficulties are related to the way they conceive mathematics, where there is only one way to solve the problem and this means applying a theoretical model expressed by a formula. This is also evident in probability where many prospective teachers argue that they cannot solve a basic conditional probability problem: "I know this is a Bayes problem, but I don't remember the formula" (Huerta et al. 2011).

Taking into account the aforementioned hypotheses, in preparing prospective teachers for teaching probability at secondary school, we suggest that probability problem solving by simulation is the most appropriate approach. The approach to introduce the steps described in the previous section is a bit different from that used with the prospective elementary teachers; however, in both types of prospective teachers the work to be done when they are solving probability problems by simulation is similar. Our approach with prospective secondary school teachers can be summarized in Table 17.4.

Table 17.4 Steps in the process of solving problems with prospective secondary school teachers

Steps in the process with prospective mathematics secondary school teachers.	– Connections with steps in Tables 17.2 and 17.3.
Step 1. Theoretical resolution of the original problem.	– Prior theoretical analysis of the original problem. – Theoretical hypotheses and the solution as a theoretical conjecture.
Step 2. Translation of the original problem into a simulated problem. Formulation of the simulated problem for a given random device.	– Steps 1 to 4 in Table 17.2. – Describing possible random devices appropriate for simulating the original problem. – Proving the stochastic equivalence between the original and simulated problem. – Proving the stochastic equivalence among these random devices.
Step 3. Designing the simulation with the random device considered. Collecting information and data with the simulation.	– Steps 5 and 6 in Table 17.3. – Data analysis and solution of the simulated problem (or problems).
Step 4. Translating the simulated problem solution to a solution for the original problem. Inter-dependence of both solutions.	– Steps 7 and 8 in Table 17.3. – Applying the Law of Large Numbers and the Chebyschev's theorem to find the number of simulations needed to get a reliable and credible answer to the simulated problem and to the original problem.

Once the problem has been solved, the prospective secondary school teachers are ready to tackle the 9th step, just as the primary school did, except that both types of teacher should consider the different students that they will have to teach in the future. In summary, the prospective secondary school teachers should discuss;

- The teaching methodology derived from this way of solving the problems (Steps 2–4 related to the Step 1, Table 17.4).
- The advantages of using manipulative (paper and pencil) simulations, using software, or both (derived from the analysis in the Step 4, Table 17.4).
- The different functions that should be assigned to the problems, the students and the teachers who propose these problems. They should explain what to think about and expect from students' learning while they are solving problems by simulation.

17.6 Final Remarks

Throughout this paper, simulation is seen as a numerical and statistical inquiry process with heuristic content for solving probability problems. At the same time, it is also viewed as an appropriate context for preparing teachers to teach probability and statistics at schools. Even more, simulation is conceived as a method for solving problems that allows the teachers and the students to deal with routine and non-routine probability problems at any school level.

Indeed, together with this approach of preparing teachers to teach probability, we are offering them an eight-step method of resolution with heuristic potential for tackling uncertain situations in which probabilities and random variable are involved, independent of their probability and statistics background. Moreover, on the one hand they are learning mathematics problem solving from a heuristic point of view, and on the other hand, due to the strong conceptual character of problems, they are also learning probabilistic and statistics concepts and their meanings, when being used in the resolution process. As Leong et al. (2016) suggest, with this approach we try to infuse the probability problem solving in preparing teachers, in three ways, preparing them for teaching probability and statistics problem solving, for teaching about (probability and statistics) problem solving and finally, for teaching probability and statistics trough problem solving.

At the same time, we are offering a new context to develop new research towards future teachers reasoning on simulation as a probability problem solving method. For example, we need to know how future teachers will think in each of the steps in which we have divided the resolution process. Even more, we also need to know how they notice this process from a professional point of view.

References

Ball, D. L., Thames, M. H., & Phelps, G. (2008). Content knowledge for teaching: What makes it special? *Journal of Teacher Education, 59*(5), 389–407.

Bernoulli, J. (1987). *Ars conjectandi- 4ème partie.* Rouen: IREM. (Original work published in 1713).

Borrás, E. (1996). *Algunos modelos de simulación aleatoria y su aplicación a la enseñanza del azar* (Some random simulation models and their application to the teaching of chance). Unpublished Ph.D. Universitat Politècnica de Catalunya. Barcelona, Spain.

Borovcnik, M., & Kapadia, R. (2014a). A historical and philosophical perspective on probability. In E. J. Chernoff & B. Sriraman (Eds.), *Probabilistic thinking: Presenting plural perspectives.* (pp. 7–34). Berlin: Springer.

Borovcnik, M., & Kapadia, R. (2014b). From puzzles and paradoxes to concepts in probability. In E. J. Chernoff & B. Sriraman (Eds.), *Probabilistic thinking: Presenting plural perspectives* (pp. 35–73). Berlin: Springer.

Chaput, B., Girard, J. C., & Henry, M. (2011). Frequentist approach: modelling and simulation in statistics and probability teaching. In C. Batanero, G. Burril, & C. Reading (Eds.), *Teaching statistics in school mathematics-challenges for teaching and teacher education. A Joint ICMI/ IASE study* (pp. 85–95). New York: Springer.

Chernoff, E., & Sriraman, B. (2014). *Probabilistic thinking: Presenting plural perspectives.* Berlin: Springer.

DeGroot, M. H., & Schervish, M. J. (2012). *Probability and statistics* (4th ed.). Menlo Park, CA: Addison Wesley.

Eichler, A., & Vogel, M. (2014). Three approaches for modelling situations with randomness. In E. J. Chernoff, & B. Sriraman, *Probabilistic thinking: Presenting plural perspectives.* (pp. 75–99). Berlin: Springer.

Engel, A (1975). The probabilistic abacus. *Educational Studies in Mathematics*, *6*(1), 1–22.

Gordon, H. (2012). *Discrete probability* (2nd ed.). New York: Springer.

Huerta, M. P. (2015). La resolución de problemas de probabilidad *con intención didáctica* en la formación de maestros y profesores de matemáticas. (The probability problem solving *with didactical intention* in preparing teachers and mathematics teachers). In C. Fernández, M. Molina, & N. Planas (Eds.), *Investigación en Educación Matemática XIX* (pp. 105–119). Alicante, Spain: Spanish Society for Research in Mathematics Education.

Huerta, M. P., Cerdán, F., Lonjedo, M. A., & Edo, P. (2011). Assessing difficulties of conditional probability problems. In M. Pytlak, T. Rowland, & E. Swoboda (Eds.), *Proceedings of the Seventh Congress of the European Society for Research in Mathematics Education* (pp. 807–817). Rzeszów, Poland: European Society for Research in Mathematics Education.

Jacobs, V. L., Lamb, L. L. C., & Philipp, R. A. (2010). Professional noticing of children's mathematical thinking. *Journal for Research in Mathematics Education 41*(2), 169–202.

Kilpatrick, J. (2016). Reformulating: Approaching mathematical problem solving as inquiry. In P. Felmer; E. Pehkonen, & J. Kilpatrick (2016), *Posing and solving mathematical problems* (pp. 69–81). New York: Springer.

Leonng, Y. H., Tay E. G., Toh, T. L., Quek, K. S., Toh, P. C., & Dindyal, J. (2016). Infusing mathematical problem solving in the mathematics curriculum: replacement units. In P. Felmer; E. Pehkonen, & J. Kilpatrick (2016), *Posing and solving mathematical problems* (pp. 309–325). New York: Springer.

Maaβ, K., & Doorman, M. (2013). A model for a widespread implementation of inquiry-based learning. *ZDM—International Journal on Mathematics Education, 45*(6), 887–899.

Martínez, M. L., & Huerta, M. P. (2015). Diseño e implementación de una situación de incertidumbre en una clase de educación infantil (Design and implementation of an uncertainty situation in a Kindergarten class). *Edma 0-6: Educación matemática en la Infancia* 4(1), 24–36.

Maxara, C., & Biehler, R. (2006). Students' simulation and modelling competence after a computer-intensive elementary course in statistics and probability. In A. Rossman, & B. Chance (Eds.), *Proceedings of the Seventh International Conference on Teaching Statistics*. Salvador de Bahia, Brazil: International Statistical Institute. https://iase-web.org/documents/papers/icots7/7C1_MAXA.pdf Accessed April 14, 2017.

Pfannkuch, M., & Ziedins, I. (2014). A modelling perspective on probability. In E. J. Chernoff, & B. Sriraman, *Probabilistic thinking: Presenting plural perspectives* (pp. 101–115). Berlin: Springer.

Polya, G. (1957). *How to solve it* (2nd Edition). New Jersey: Princeton.

Puig, L. (1996). *Elementos de resolución de problemas* (Elements of problem solving). Granada, Spain: Comares.

Rossman, A. (2013). Interview with Mike Shaughnessy. *Journal of Statistics Education*, *21*(1), 1–27.

Santos, M. (2012). El papel de la resolución de problemas en el desarrollo del conocimiento matemático de los profesores para la enseñanza (The role of problem solving in developing teachers' mathematics knowledge for teaching). *Cuadernos de Investigación y Formación en Educación Matemática 7*(10), 151–163.

Shaughnessy, M. (1983). The psychology of inference and the teaching of probability and statistics: Two sides of the same coin? In R. W. Scholz (Ed.), *Decision making under uncertainty* (pp. 325–350). Amsterdam: Elsevier.

Shaughnessy, J. M. (1992). Research in probability and statistics: Reflections and directions. In D. Grouws (Ed.), *Handbook of research on mathematics teaching and learning* (pp. 465–494). New York: MacMillan.

Shulman, L. S. (1986). Those who understand. Knowledge growth in teaching. *Educational Researcher, 15*(2), 4–14.

Siñeriz, L., & Puig, L. (2006). Un modelo de competencia para la resolución de problemas de construcción con regla y compás (A model of competence for solving problems of constructions with rule and compass). In J. Aymerich, & S. Macario, (Eds.), *Matemáticas para el siglo* XXI (pp. 323–332). Castellón de la Plana, España: Publicacions de la Universitat Jaume I.

Ulam, S. (1951). On the Monte Carlo method. *Proceedings of a Second Symposium on Large-scale digital Calculating Machinery* (pp. 207–212). Cambridge: Harvard Academic Press.

Von Newmann, J. (1951). Various techniques used in connection with random digits, Monte Carlo method. *National Bureau of Standards Applied Math. Series 12*, 36–38.

Wilson, P. S., & Heid, M. K. (Eds.). (2010). *Framework for mathematical proficiency for teaching*. Athens, GA: Mid-Atlantic Center for Mathematics Teaching and Learning. http://jwilson.coe.uga.edu/Situations/Framework%20Folder/111021MPT%20Framework.pdf. Accessed 15 April, 2017.

Chapter 18
Exploring Teachers' Attitudes Towards Probability and Its Teaching

Assumpta Estrada, Carmen Batanero and Carmen Díaz

Abstract Since probability is a new topic at the primary school level in many countries, many primary school teachers may consider themselves not well prepared, or may not value the formative role of probability as a topic for their students to learn. Consequently, the assessment of teachers' attitudes using valid and reliable instruments is a preliminary step in organizing formative actions for these teachers. In this paper, we describe the development of a scale aimed at measuring primary school teachers' attitudes towards probability and the teaching of probability. The components considered in building the instrument, the process of selecting the items, as well as the results of a pilot trial of the instrument are presented. The validity and reliability of the instrument are also discussed, and the preliminary results of a second sample of 232 prospective teachers in Spain is analysed. Our results suggest that future teachers' attitudes towards probability and its teaching are generally positive in all of the components considered.

Keywords Attitudes · Probability · Teaching probability · Prospective teachers

---

A. Estrada (✉)
Facultat de Ciències de l'Educació, Universitat de Lleida, Campus Cappont, Avinguda Estudi General, 4, 25001 Lleida, Spain
e-mail: aestrada@matematica.udl.cat

C. Batanero
Facultad de Ciencias de la Educación, University of Granada, Granada, Spain
e-mail: batanero@ugr.es

C. Díaz
University of Huelva, Huelva, Spain
e-mail: carmen.diaz@dpsi.uhu.es

C. Batanero and E. J. Chernoff (eds.), *Teaching and Learning Stochastics*, ICME-13 Monographs, https://doi.org/10.1007/978-3-319-72871-1_18

18.1 Introduction

Probability is a new topic at the primary school level in many countries (Australian Curriculum, Assessment and Reporting Authority [ACARA] 2010; Ministerio de Educación Pública in Costa Rica [MEP] 2012; New Zealand Ministry of Education [ME] 2007), while in other countries the curricular guidelines the study of probability has been delayed to either Grade 6 or to secondary education (Common Core State Standards Initiative [CCSSI] 2010). In Spain (MECD 2014), the topic is included in the curriculum guidelines since the first grade (6-year-old students) that include content from the *Standards* developed by the National Council of Teachers of Mathematics (NCTM 2000).

The introduction of this topic in the education of all citizens reflects the need to overcome deterministic thinking and to accept the existence of fundamental chance in nature (Batanero et al. 2016). At the same time, we need to equip citizens with strategies and ways of reasoning that can help them in making adequate decisions in everyday and professional situations in which chance is present. We also observe a change in the teaching approach towards the frequentist view, which is grounded in simulation and experiments with the aim of providing children with a stochastic experience from their childhood. An advantage of this approach is the close link it establishes between statistics and probability (Batanero 2013).

However, the success of the new curricular proposals about probability directly depends on teachers' willingness to and interest in teaching probability. Due to the novelty of this topic, many prospective and in-service primary school teachers have not received specific education in probability or teaching probability, and may therefore feel uncomfortable when confronted with the necessity of teaching it. These teachers may consider themselves not well prepared, or may not value the formative role of probability as a topic for their students to learn (Batanero and Díaz 2012).

This chapter focuses on attitudes towards probability and its teaching, given that improving the attitudes of teachers is a goal of the statistics education reform (Tishkovskaya and Lancaster 2012). The assessment of attitudes is particularly relevant in the preparation of teachers, since positive attitudes towards a topic (in this case, probability) help teachers understand that their students can be trained to understand and use the topic themselves (Schau 2003). In addition, teachers with a positive attitude towards a topic are more likely to transmit this attitude to their students, as well as to foster an appreciation for its potential uses in future personal and professional life (Gal and Ginsburg 1994). Consequently, the assessment of teachers' attitudes using valid and reliable instruments is a preliminary step in organizing formative actions for these teachers that take into account not only the teachers' knowledge of probability, but also the components in their attitudes.

While there exist various instruments to measure attitudes towards statistics, we have only found one that measures attitude towards learning probability: the Probability Attitude Inventory (PAI), developed for use with undergraduate students (Tan et al. 2011). Consequently, there is no available instrument aimed at measuring teachers' attitudes towards probability, and even the few instruments for measuring attitudes towards statistics developed specifically for use with teachers do not consider their attitudes towards teaching statistics.

To fulfil this need, we started developing a valid and reliable questionnaire that can be used to assess the different components of attitudes towards probability and its teaching. In this paper, we describe the first steps in building this questionnaire, which include the definition of the instrument components, the development of an item bank, the selection of the items in the scale using expert judgment, a preliminary pilot trial of the instrument in a sample of 71 prospective teachers, and the results from a second sample of 232 prospective teachers.

18.2 Theoretical Background

18.2.1 Attitudes and Their Relevance in Education

Theoretical issues related to attitudes have received some attention over the years, and different perspectives and conceptualizations have emerged. In their analysis of the affective domain in mathematics education several authors (Gómez-Chacón 2000; Phillipp 2007) discriminated between emotions, attitudes, and beliefs that differ in the stability of the affective responses that they represent, the degree to which cognition plays a role and the time they take to develop.

Due to the complexity, attitudes have been conceptualized in different ways. For example, Ajzen (1989, p. 241) define attitudes as an "Individual's disposition to respond favourably or unfavourably to an object, person, institution, or event, or to any other discriminable aspect of the individual's world," and McLeod (1992 p. 581) described attitudes as "Affective responses that involve negative or positive feelings of moderate intensity." Gal et al. (1997, p. 40) suggest that they are "A summation of emotions and feelings experienced over time in the context of learning mathematics or statistics." Aiken (2002, p. 3) considers attitudes as "A relatively enduring organization of beliefs around an object or situation predisposing one to respond in some preferential manner." Philipp (2007) describes attitudes as ways of acting, feeling, or thinking that show a person's disposition or opinion towards a topic, and Kislenko (2009) as an affective response including negative or positive feelings of moderate stability and strength.

According to Di Martino and Zan (2015), attitudes act as a bridge between beliefs and emotions, since attitudes take into account both beliefs (about oneself and about mathematics) and emotions, as well as the relationships between beliefs and emotions. Hannula (2002) distinguishes four aspects of attitudes: emotions aroused in a situation, emotions associated with particular stimuli, expected consequences, and relating the situation to personal values.

Attitudes towards a topic derive from positive or negative experiences with the topic in school or in informal learning settings outside of school; students may also transfer their attitudes about mathematics in general towards probability (Estrada et al. 2011). Attitudes are relatively stable, involve a greater cognitive component and less emotional intensity than emotions, develop as a result of repeated positive or negative emotional responses, and are automatized over time.

Attitudes might influence a person's behaviour in relation to the topic inside and outside the classroom, as well as their willingness to attend courses on the topic in the future (Gal et al. 1997). Moreover, according to Veloo and Chairhany (2013), a student's attitude is the main factor in his or her orientation to learning a topic. It is therefore important for teacher educators to identify prospective teachers' attitudes towards the topics they are to teach, especially those with which they are less familiar, and to use this knowledge in taking appropriate measures to improve these attitudes (Veloo and Chairhany 2013).

18.2.2 Training Teachers to Teach Probability

Including a topic in the curriculum does not automatically ensure that it will be successfully taught and learned. Therefore, successful implementation of the curriculum will depend on the adequate education of teachers. This education should take into account the characteristics of probability, its multifaceted views (classical, frequentist, subjective or axiomatic probability), common misconceptions, and wrong intuitions, which are less frequent in the study of other mathematical topics and which present special challenges for teachers (Batanero et al. 2016).

Although school textbooks provide examples and teaching resources, some texts present too narrow a view of probabilistic concepts or only one approach to probability. The applications of probability in textbooks may be restricted to games of chance, and definitions of concepts may be incorrect or incomplete (Gómez-Torres et al. 2014). Consequently, teachers need specific preparation in teaching probability. However, while there are different models that describe the components of teachers' knowledge needed to teach probability, research that adapts or describes these components for the specific purpose of teaching probability is very scarce.

There is general consensus in the mathematics education community that teachers need to develop adequate probabilistic knowledge and reasoning. However, many teachers, in particular primary school teachers, have only studied theoretical probability and lack experience in designing investigations or simulations in the classroom context (Kvatinsky and Even 2002; Stohl 2005). Moreover, recent research suggests that many prospective teachers hold the same biases about probabilistic reasoning as their students (Batanero et al. 2014; Prodromou 2012).

One of these components is the pedagogical knowledge needed to teach probability, where general principles that are valid for other areas of mathematics are not always appropriate. For example, teachers should not only present different probabilistic concepts and their applications, they should also be aware of the different meanings of probability and philosophical controversies around them (Batanero et al. 2004). Teachers should also be acquainted with research results that describe children's reasoning and beliefs in uncertain situations and with didactic materials and technological resources that can help their students develop correct intuitions in this area. Finally, it is important to consider teachers' beliefs and attitudes, since according to Veloo and Chairhany (2013), positive attitudes about learning mathematics and probability have a positive effect on developing understanding. In fact, positive attitudes are also a major contributor in the learning of probability (Tan et al. 2011).

18.3 Previous Research

In our review of the literature, we only found two papers dealing with attitudes towards probability, both of which measured students' attitudes and involved small samples. Tan et al. (2011) examined the effectiveness of interactive learning involving graphing calculators (GC) on the attitudes towards probability of 65 students in Malaysia. The authors reported significantly more positive attitudes towards probability (particularly in terms of interest and self-concept in probability, as well as the perceived usefulness of probability) in the experimental group as compared to a control group. A very similar study was carried out by Veloo and Chairhany (2013) in Indonesia with 64 students, where again it was found that interactive learning supported by a technological tool improved attitudes towards probability of the experimental group. Both studies used a scale for assessing attitudes towards mathematics in which they only changed the word "mathematics" to "probability" in all of the items. While this scale assessed attitudes about learning probability, it did not measure attitudes towards the *teaching* of probability.

Early research on attitudes towards statistics was aimed at building scales to measure these attitudes, such as the Statistics Attitude Survey (SAS; Roberts and Bilderback 1980), the Attitudes Toward Statistics survey (ATS; Wise 1985), and the Survey of Attitudes Toward Statistics (SATS; Schau et al. 1995). An extensive review of these and other attitude scales and of studies analysing the influence of different variables on attitudes towards statistics is given in Carmona (2004).

Some research has focused on attitudes towards statistics among teachers in particular; these studies offer varied results (see Estrada et al. 2011 for a survey). For example, Begg and Edwards (1999) reported negative attitudes in the affective component, with teachers expressing feelings like "fear," "horror," and "lack of interest," even though they considered statistics to be an important topic. Despite the teachers in the study were concerned about possible manipulation of statistics, most of them thought that a good understanding of mathematics was not needed to grasp basic statistical concepts. They did not consider themselves good at statistics, although they were confident in their ability to read and understand statistical information in the media.

Onwuegbuzie (1998) used the ATS to compare prospective teachers' attitudes towards statistics with those of graduate students enrolled in other courses, and reported that teachers in his sample had less positive attitudes towards statistics than did the graduate students. He also examined attitudes and knowledge among teachers and concluded that attitudes towards statistics are strongly correlated to achievement in statistics.

Nasser (2004) used the SATS to examine the relationships between attitudes towards statistics, anxiety about statistics, mathematical aptitude, and achievement in statistics in a study involving prospective teachers in Egypt taking part in an introductory statistics course. The author found a small positive effect of attitudes towards statistics on achievement in statistics.

Chick and Pierce (2008) gave 10 items taken from the SATS to 27 prospective primary school teachers and reported that the teachers had neutral attitudes towards statistics. With regards to attitudes about the value of statistics, even when teachers agreed with the item "To be an intelligent consumer, it is necessary to know something about statistics," a majority also agreed that "When buying a new car, asking a few friends is preferable to consulting an owner satisfaction survey in a consumer magazine."

Estrada et al. (2005) administered the SATS instrument to a sample of 367 prospective primary school teachers in Spain. Results showed moderately positive attitudes on cognitive competence and value items. Moreover, liking or disliking statistics was related to the teachers' perception of their capacity to learn statistics and to the value they attributed to statistics. A qualitative analysis of the ten SATS open items that had yielded lower scores (Estrada and Batanero 2008) was used to classify the main reasons for positive and negative scorings. This research suggests that two main influences on teachers' attitudes are understanding of statistics and positive learning experiences.

Estrada et al. (2010) compared the attitudes towards statistics of Spanish and Peruvian primary education teachers and found positive attitudes in both countries with a slight positive difference among Spanish teachers, who found statistics easier than their Peruvians colleagues who suggested that statistics is a more adequate topic of study for science students. In Portugal, Martins et al. (2011, 2012) assessed attitudes towards statistics of in-service primary education teachers. There, teachers were reported to generally like learning and teaching statistics, and considered statistics to be a tool in facing real-world problems and connecting the cognitive and social components of education. There were, however, significant differences in the overall scores according to the teaching cycle, time of service, area of initial training, and statistics study level.

Considered together, these results indicate the need to strengthen teachers' attitudes towards statistics, since these attitudes have a significant effect on teachers' statistical education, their teaching of statistics, and the attitudes of their students. Moreover, our literature review revealed a lack of research related to teachers' attitudes towards probability and its teaching; the aim of the present study is to fulfil this need.

18.4 Developing an Attitude Scale

As is the case with many educational or psychological variables, attitudes towards probability and its teaching are not directly observable. We must therefore infer attitudes from other empirical indicators (e.g., subjects' responses to items on a scale; McDonald 2013). In developing a scale aimed at measuring teachers' attitudes towards probability and the teaching of probability, we proceeded as follows.

18.4.1 Semantic Definition: Components of Teachers' Attitudes Towards Probability

The first step in building such a scale involved developing a detailed definition of the construct, elaborating upon its semantic definition. The variable to be measured (attitude) is not one-dimensional, but is rather made up of different components that should be understood, analysed, and specified in advance. This definition will be used later to analyse the content validity of the instrument, that is, the extent to which the questionnaire represents all of the facets that were considered in our definition of attitudes towards probability and its teaching.

Taking into account research on attitudes by different authors, we conceptualized attitudes as a multidimensional concept. In research discussions about the concept of attitude towards any particular topic, it is very common to identify the following three components of attitudes: affect, behavior, and cognition (Breckler 1984; Wyer and Albarracín 2014). Consequently, after a revision of the research on attitudes towards statistics, we settled upon three generally accepted components of attitudes (Estrada et al. 2011; Gal and Ginsburg 1994; Gómez-Chacón 2000; Ramirez et al. 2012; Schau et al. 1995), which are defined as follows:

- *Affective component towards probability* (AP): Feelings about probability (for example, pleasure or displeasure towards the topic, interest or disinterest in probability, rejection of or anxiety towards the topic).
- *Cognitive competence towards probability* (CCP): Self-perception with regards to self-competence, knowledge, and intellectual skills when applied to probability. This component is very important in the case of teachers, who need a favourable perception of their own abilities in order to teach a particular subject.
- *Behavioural component towards probability* (BP): Inclinations to act towards the attitude object in a particular way, to make decisions in situations involving the attitude object, to help colleagues to learn and use probability.

Since this scale is aimed at teachers, we complemented the three classical attitude components with three additional components of attitude that are related specifically to teaching probability:

- *Affective component towards teaching probability* (AT): Personal feelings about teaching probability, which may differ from (although may be associated with) feelings towards the topic itself: pleasure-displeasure, fear-confidence, interest-disinterest in teaching probability.
- *Teaching probability competence component* (CT): This component assesses the teacher's perception of his/her ability to teach probability, to help students, to pose effective tasks to seek relevant resources, etc.
- *Behavioural component towards teaching probability* (BT): This component is related to didactic action: whether the teacher has or has not taught (or whether he/she is willing or not willing to teach) probability, whether he/she gives priority to probability over other topics, and whether he/she thinks the topic should be postponed or given emphasis.

Finally, we have included a *Value component towards probability and its teaching* (VPT): appreciation of the usefulness, relevance, and importance of probability and its teaching in personal and professional life. Although value has been included as a component of attitude, for example, in Schau et al. (1995), none of the scales analysed above assess the value attributed to teaching a particular topic.

18.4.2 Item Selection

To identify suitable items for the questionnaire that covered the intended components, a preliminary pool of items was created by translating and adapting items from different research on attitudes towards statistics, since we expect some relationship between attitudes towards statistics and probability. We examined the following scales:

- *Statistics Attitude Survey* (SAS) (Roberts and Bilderback 1980). This was the first instrument developed to measure attitudes towards statistics. This scale was designed to be one-dimensional, with 33 Likert-type items.
- *Attitudes Towards Statistics* (ATS) (Wise 1985). The ATS is a 29-item, Likert-type scale with five response possibilities consisting of two subscales. The *Course* (9 items) and *Field* (20 items) subscales respectively aim to measure attitudes towards the particular statistics course in which students are enrolled and the use of statistics in their fields of study.
- *Survey of Attitudes Towards Statistics* (SATS) (Schau et al. 1995). This scale consists of 28 items measuring four components: *Affect* (feelings about statistics), *Cognitive competence* (perception of self-competence, knowledge, and intellectual skills when applied to statistics), *Value* (appreciation of the usefulness and value of statistics in personal and professional life), and *Difficulty* (the perceived difficulty of statistics as a subject). We also analysed a later version (Schau 2003) that also includes effort and interest as components.

- *Multidimensional Auzmendi's Scale* (Auzmendi 1992). This scale considers attitudes towards both mathematics and statistics. It includes five factors (usefulness, anxiety, confidence, pleasantness, and motivation).
- *Scale of Attitudes Towards Statistics* (EAEE) (Estrada 2002). This scale was developed specifically for use with teachers, and considers didactic aspects of attitudes. Its 25 items are distributed across three classical components (affective, cognitive, and behavioural dimensions), in addition to three additional components (social, educational, and instrumental dimensions).

All of the aforementioned instruments are Likert scales with 5 or 7 graduations in the responses, have been validated with university students and have been found to possess good reliability. After listing all the items in the different scales, we identified and adapted (to the case of probability and its teaching) those which a priori could be useful in assessing the components of interest. Each resultant item was analysed several times and compared with the intended content with the help of researchers who volunteered to provide feedback on the adequacy of potential items.

18.4.3 Item Format and Wording

Once a set of 56 items was fixed (eight items per component), their format and wording was revised. All of the items were in the format of statements, in response to which respondents scored their level of agreement or disagreement on a five-point Likert scale (from 1: strongly disagree to 5: strongly agree). We decided to use negative and positive sentences to avoid the problem of acquiescence and to take into account different aspects of the pedagogical and didactical components as described before. For example, the sentence "I enjoy the lessons in which probability is explained" is related to the affective component towards probability, and strong agreement suggests a positive attitude in this component. At the same time, the sentence "I never use probability outside mathematics" is related to the behavioural component towards probability, and strong agreement suggests a negative attitude in this component. Scores for items that are presented in negative form should be reversed before the scale and component scores are computed by adding the scores for different items.

18.4.4 Expert Judgement and Content Validity

In order to assure content validity, the 56 items in the preliminary questionnaire were submitted to a panel of 14 experts (statisticians, mathematicians, psychologists, and statistics educators), all of whom were at the time university lecturers with a doctoral degree and who possessed a great deal of experience in either teaching probability or in training teachers (or both). Each of these experts provided a numerical value (ranging from 1 to 5) for each item, rating its adequacy for assessing the intended component. For each component, the four items with the highest average scores assigned by the expert judges were selected; the judges also provided suggestions for improving item wording. After performing a statistical analysis of the scores provided by the experts, we selected the four items with the highest average scores and lowest standard deviation for each component so as to assure high adequacy and agreement of experts. All of the items that were selected had an average score of 4 points or greater. This process assured the content validity of the instrument.

The final scale includes 28 items (Appendix), 14 of which are expressed affirmatively (1, 2, 4, 5, 6, 8, 9, 10, 14, 15, 19, 20, 24, and 28) and 14 of which are expressed in a negative manner (3, 7, 11, 12, 13, 16, 17, 18, 21, 22, 23, 25, 26, and 27). The distribution of items according to the components is presented in Table 18.1.

As Table 18.1 shows, the same number of items was included for each of the seven components, with about the same number of negative and positive wordings of items in each component.

18.5 Preliminary Results

18.5.1 Pilot Trial

The instrument was administered to a pilot sample of 71 prospective primary school teachers in Spain who were enrolled in a statistics education course. 83% of the participants were women, and participants' ages ranged between 20 and 24 years. The participants had studied probability during their secondary education. The questionnaire was administered at the beginning of the second semester of the 2014–2015 academic year. The average value per item ranged from 2.84 to 4.44, with a mean of 3.7 and a standard deviation of .42 (while the expected average value was 3). The instrument reliability was quite high (Cronbach's Alpha = .934).

Box plots for the scores in the different components are displayed in Fig. 18.1, where the theoretical mean for a neutral attitude in each component is 12. These results suggest that the prospective teachers' attitudes were generally positive across the different components, in particular with regards to their willingness to teach probability (BT) and the value attributed to probability and the teaching of probability (VPT).

Table 18.1 Components and corresponding items

Component	Scale items
Affective component towards probability (AP)	1, 5, 16, 27
Cognitive competence towards probability (CCP)	6, 8, 17, 22
Behavioural component towards probability (BP)	2, 7, 15, 18
Affective component towards teaching probability (AT)	9, 21, 26, 28
Teaching probability competence component (CT)	3, 10, 14, 23,
Behavioural component towards teaching probability (BT)	11, 20, 24, 25
Value component towards probability and its teaching (VPT)	4, 12, 13, 19

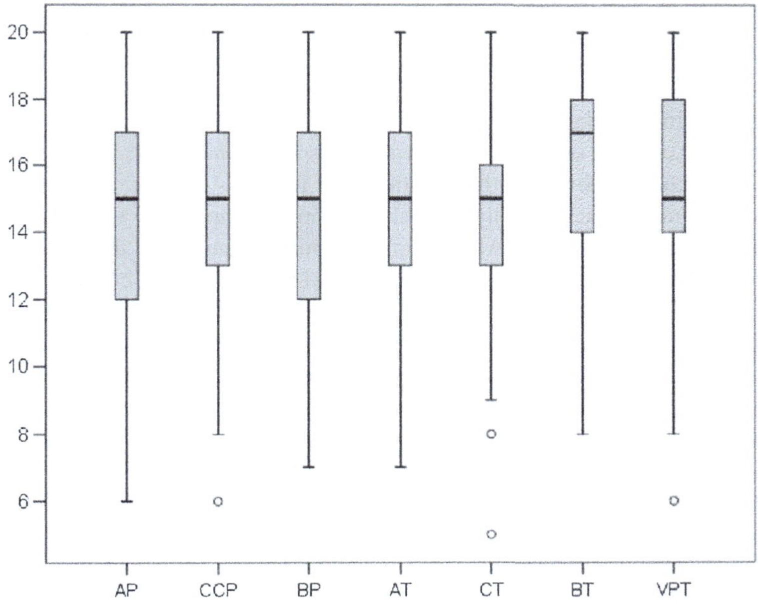

Fig. 18.1 Box plots of scores for attitude components in the pilot trial

18.5.2 Detailed Analysis in a New Sample

A second assessment was carried out with a sample of 232 prospective teachers (67 males and 165 females, aged between 20 and 25 years), 79 of whom had studied probability for the last time in high school, 124 in middle school, and 23 in primary school. Six of the prospective teachers had studied probability during professional training. They belonged to different groups in the same university in Catalonia and were in their 3rd year of university, preparing to become primary school teachers. They had not studied probability in previous courses at university before they completed the questionnaire.

Results for each item are included in the Appendix. The mean total score for the scale was $\bar{x} = 102.6$ with a standard deviation of $\sigma = 13.5$. Since the theoretical "neutral" score (when a participant answers with a score of 3 to all the items) is 84, the results suggest a generally positive attitude towards probability and its teaching in this second sample of students. The instrument reliability was again quite high (Cronbach's Alpha = .892) and lower[1] in the different components (AP = .759, CCP = .637, BP = .537, AT = .713, CT = .612, BT = .584, V = .599) but still acceptable, according to Nunnally (1978), in the first steps of exploratory research. These values of reliability are similar to those reported for the components of some studies of attitudes towards statistics, as reviewed in Carmona (2004).

We also carried out a study of the discriminative power of the items. We ordered the subjects by their total score on the questionnaire and compared the scores for each item for the upper percentile (over the 70% percentile in the total score) and the lower percentile (below the 30% percentile in the total score). Both of these groups were made up of 70 students; the results in the t-test of difference of averages in independent groups was statistically significant for all of the items. Moreover, except for one item, the *p*-value was lower than .001, which suggests a high discriminative power of the questionnaire items.

An exploratory factor analysis with principal components as the extraction method yielded seven factors with eigenvalues higher than 1 and thus provisionally confirmed the suggested structure of the questionnaire and its construct validity. These seven factors accounted for the 67.99% of the explained variance in the model. Since the sample size was still moderate, we were unable to carry out a confirmatory factor analysis and thus recognize the need to collect more data to continue the process of validation of the instrument.

The items with the highest scores (4 or more points on average, and very similar scores in the pilot sample) were item 12 (probability is useful in situations besides games of chance; $\bar{x} = 4.41$), items 25 and 11 (intention to teach probability; $\bar{x} = 4.40$ and $\bar{x} = 4.13$), item 17 (probability is understandable; $\bar{x} = 4.26$), item 26 (interest in teaching probability; $\bar{x} = 4.19$, and item 18 (non-avoidance of reading information that refers to probability; $\bar{x} = 3.99$).

The items with the lowest scores (scores below 3 points) were item 14 (facility for preparing teaching materials, $\bar{x} = 2.68$), item 6 (probability is easy; $\bar{x} = 2.87$), and item 8 (I have mastered probability content; $\bar{x} = 2.91$), with scores similar to those in the pilot sample. Thus, according to these results, the prospective teachers in the study perceive the usefulness of probability and are willing to teach it, but do not feel well prepared enough in either the mathematical or the pedagogical content to do so. In Fig. 18.2, we observe that the majority of students are distributed above

[1]This is reasonable, given than Alpha value increases with the number of items and we only included four items per component in order to have a scale of reasonable length.

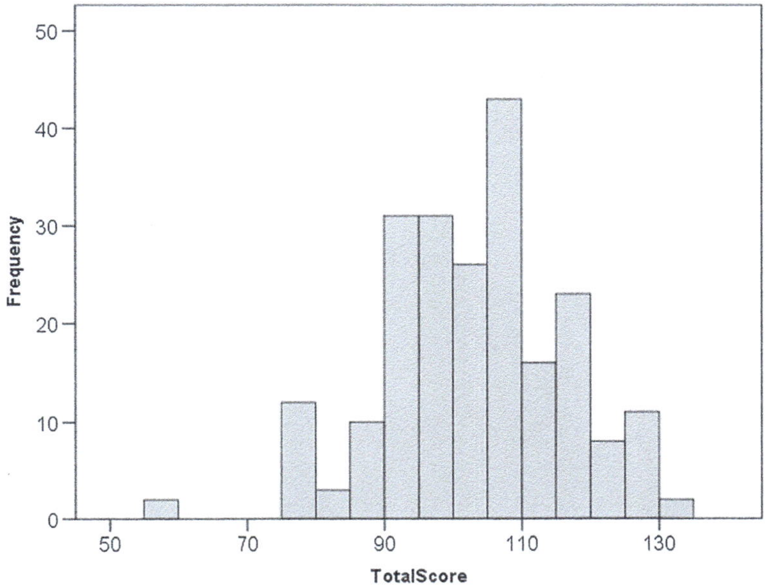

Fig. 18.2 Distribution of total scores in the second sample

the theoretical mean value of 84, suggesting that these prospective teachers have a generally positive attitude towards probability; this finding coincides with the results of the pilot trial.

Statistics summaries for the total score and the scores in the different components are displayed in Table 18.2 and in Fig. 18.3, which suggest that attitudes of these teachers were generally positive across the seven components. The components with the highest scores correspond to the participants' willingness to teach probability (BT) and the value attributed to probability and its teaching (VPT), with the lowest (but still positive) scores corresponding to cognitive competence for teaching (CT), cognitive competence towards probability (CCP), and affect (AT). The small size of the standard deviation values as compared to the mean suggest that the sample was very homogeneous with regards to attitudes across the components.

Table 18.2 Mean and standard deviation of the items, total score, and components in the second sample

Variable	Mean	St. deviation	Variable	Mean	St. deviation
Scale	102.6	13.5	AT	14.0	2.6
AP	13.2	2.9	CT	12.9	2.2
CCP	13.1	2.6	BT	15.5	2.4
BP	14	2.6	VPT	15.7	2.3

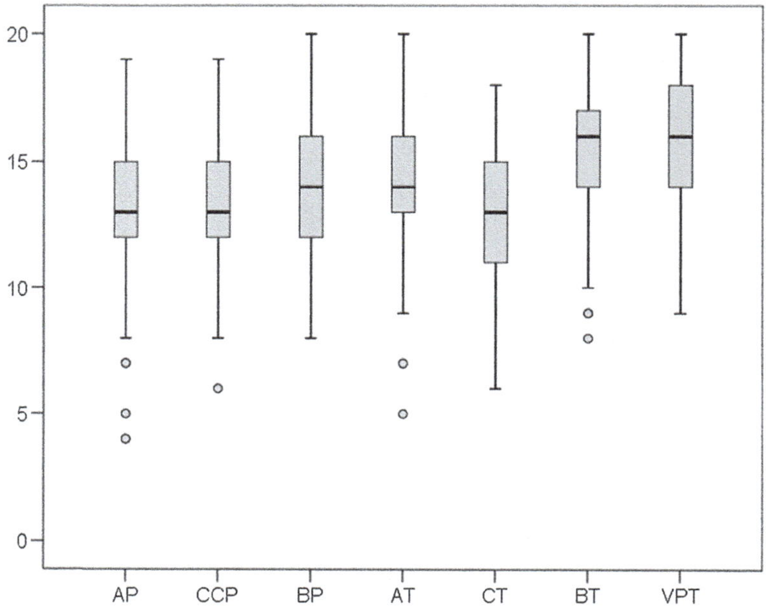

Fig. 18.3 Box plots of scores in attitude components in the second sample

Table 18.3 Pearson's correlation between components of attitudes

	Total score	AP	CCP	BP	AT	CT	BT	VPT
Total score	1	.804**	.782**	.634**	.850**	.722**	.784**	.539**
AP	.804**	1	.555**	.410**	.712**	.536**	.549**	.286**
CCP	.782**	.555**	1	.380**	.682**	.623**	.397**	.303**
BP	.634**	.410**	.380**	1	.350**	.260**	.467**	.441**
AT	.850**	.712**	.682**	.350**	1	.655**	.623**	.270**
CT	.722**	.536**	.623**	.260**	.655**	1	.437**	.106
BT	.784**	.549**	.397**	.467**	.623**	.437**	1	.530**
VPT	.539**	.286**	.303**	.441**	.270**	.106	.530**	1

**Significant ($p < 0.01$)

The Pearson correlation coefficient between scores in the scale and the different components are given in Table 18.3, which shows that the most influential on global attitude (Total score) are attitudes towards teaching (AT) and affect towards probability (AP), while the less influential component is the value given to the teaching of probability (VP). Moreover, the components correlate significantly: in particular, affect towards probability (AP) is highly correlated with affect towards teaching (AT), cognitive competence (CCP) is highly correlated with affect towards teaching

(AT), and affect towards teaching (AT) is highly correlated with competence for teaching (CT). Our interpretation is that affect towards probability influences affect towards teaching and perceived competence to teach the topic, with less influence of other factors such as cognitive competence or value attributed to probability.

18.6 Discussion and Research Perspectives

Our results suggest a positive attitude towards probability and its teaching in the pilot trial as well as in this second sample of students, where results from the pilot study were reproduced. Not only were the average scores positive (higher than 3) for almost all the items, but the average scores for all of the components were higher than the theoretical mean value ($\bar{x} = 12$). These results are likely to be replicated with other prospective teachers from different contexts, as the instrument reliability was quite high for both samples.

These results are in contrast with those of studies measuring prospective teachers' attitudes towards statistics. For example, Begg and Edwards (1999) found negative attitudes in the affective component related to statistics, with participants in the study considering themselves to be poor at statistics. Estrada et al. (2005) found only moderately positive attitudes about cognitive competence and the value of statistics, and neutral or negative scores for the affect and difficulty components in a study that involved prospective teachers in Spain of similar age and preparation. Consequently, our results suggest that prospective teachers' attitudes towards probability and its teaching may be more positive than attitudes towards statistics, and will favour the introduction of probability in schools in Spain. Indirectly, these results also point to these prospective teachers' interest in professional development in probability. Our research also showed a lower (but still positive) score for the items of cognitive competence for teaching (CT), cognitive competence towards probability (CCP), and affect (AT). Our suggestion to improve attitudes for these components is to have teachers explore different methods for teaching probability and gain confidence in their ability to learn and to teach probability.

In agreement with Goldin et al. (2016), we consider two main directions to expand this research. On the one hand, more research aimed at describing the different profiles of attitudes is required, as well as research investigating their origin; in particular, it is important to describe the role of didactical, social, and cultural factors in the development of recurrent profiles of attitudes, which can be achieved by comparing future teachers' attitudes in different countries, cultures, and school systems.

To achieve this aim, we plan to continue to study the usefulness of the scale in comparing attitudes of pre-service and in-service teachers at both the primary and secondary school levels (the scale is not specific to a particular grade level). We also plan to collect new samples from teachers in Latin American countries to carry out some transnational studies and analyse the effect of different educational and sociological variables on these attitudes.

A second line of research is taking into account the role of non-cognitive factors such as attitudes and motivations in the education of teachers to teach probability, given the relevance of attitudes in teaching and learning (Gal and Ginsburg 1994). A first step is making teachers conscious of the relevance of their attitudes and preparing them to use scales such as the one developed in our research to assess their own attitudes and those of their students.

Acknowledgements Research supported by the projects: EDU 2013-41141-P (MEC) and EDU2016-74848-P (AEI, FEDER).

Appendix: Attitudes' Scale Towards Probability and Its Teaching

Note: Each item below was followed by a 5-point scale ranging from 1 (strongly disagree) to 5 (strongly agree). Scores in items with negative wordings were rescaled before computing the mean.

Item	Mean	Std. Dev.
I1. I enjoy the lessons in which probability is explained	3.04	0.80
I2. I use probability information when making decisions	3.17	0.95
I3. It will be hard for me to teach probability	3.29	0.98
I4. Probability helps you understand today's world	3.55	0.94
I5. I like probability; it is a subject that always interested me	2.91	0.92
I6. Probability is easy	2.87	0.94
I7. I never used probability outside mathematics	3.73	1.04
I8. I have mastered the main contents of probability	2.69	0.96
I9. I am sure I will enjoy teaching probability in school	3.31	0.90
I10. I think I can notice and correct students' errors and difficulties with probability	3.32	0.84
I11. I will only teach probability if there is time available after teaching the other topics	4.13	0.90
I12. Probability is only useful for games of chance	4.41	0.75
I13. Probability is not as valuable as other areas of mathematics	4.09	0.90
I14. It will be easy for me to design assessment tasks related to probability	2.68	0.70
I15. I use probability in everyday life	3.36	0.93
I16. I feel scared when faced with probability information	3.53	0.99
I17. Probability is understandable only to "science people."	4.26	0.90
I18. I avoid reading information that contains probability terms (e.g. in drugs prospects)	3.99	1.00
I19. Probability knowledge helps students to reason critically	3.73	0.76
I20. Probability should be taught since the first teaching levels	3.48	0.95

(continued)

(continued)

Item	Mean	Std. Dev.
I21. I feel worried about being able to reply to my students' questions about probability	3.07	0.96
I22. I do not feel well enough prepared to solve any basic probability problem	3.35	0.99
I23. I would not be able to prepare suitable didactic resources for the probability lesson	3.67	0.91
I24. I will use probability when needed in other topics I teach	3.48	0.83
I25. If I could skip a topic, I would skip probability	4.40	0.90
I26. I am not interested in teaching probability, even if it appears in the curriculum	4.19	0.89
I27. I do not enjoy solving probability problems	3.75	1.04
I28. As a teacher, I would feel comfortable when teaching probability	3.47	0.87

References

Aiken, L. (2002). *Attitudes and related psychosocial constructs—Theories, assessement, and research*. London: Sage.

Ajzen, I. (1989). Attitude structure and behavior. In A. Pratkanis, S. Breckler, & A. Greenwald (Eds.), *Attitude structure and function* (pp. 241–274). Hillsdale, NJ: Erlbaum.

Australian Curriculum, Assessment and Reporting Authority. (2010). *The Australian curriculum: Mathematics*. Sidney, NSW: Author. www.australiancurriculum.edu.au/mathematics/curriculum/f-10?layout=1. Accessed: 25 January 2016.

Auzmendi, E. (1992). *Las actitudes hacia la matemática estadística en las enseñanzas medias y universitarias* (Attitudes towards teaching mathematics in high school and university). Bilbao: Mensajero.

Batanero, C. (2013). Teaching and learning probability. In S. Lerman (Ed.), *Encyclopedia of Mathematics Education* (pp. 491–496). Heidelberg: Springer.

Batanero, C., Arteaga, P., Serrano, L., & Ruiz, B. (2014). Prospective primary school teachers' perception of randomness. In E. Chernoff & B. Sriraman (Eds.), *Probabilistic thinking: Presenting plural perspectives* (pp. 345–366). New York: Springer.

Batanero, C., Chernoff, E., Engel, J. Lee, H, & Sánchez, E. (2016). *Research on teaching and learning probability*. ICME-13. Topical Survey series. New York: Springer.

Batanero, C., & Díaz, C. (2012). Training teachers to teach probability: Reflections and challenges. *Chilean Journal of Statistics 3*(1), 3–13.

Batanero, C., Godino, J. D., & Roa, R. (2004). Training teachers to teach probability. *Journal of Statistics Education, 12*. www.amstat.org/publications/jse/v12n1/batanero.html. Accessed: 21 May 2016.

Begg, A., & Edwards, R. (1999). Teachers' ideas about teaching statistics. *Proceedings of the 1999 combined conference of the Australian Association for Research in Education and the New Zealand Association for Research in Education*. Melbourne: AARE & NZARE. http://eric.ed.gov/?id=ED455195. Accessed: 23 February 2016.

Breckler, S. J. (1984). Empirical validation of affect, behavior, and cognition as distinct components of attitude. *Journal of Personality and Social Psychology*, *47*(6), 1191.

Carmona, J. (2004). Una revisión de las evidencias de fiabilidad y validez de los cuestionarios de actitudes y ansiedad hacia la estadística (Revising the reliability and validity evidence of attitudes and anxiety towards statistics questionnaires). *Statistics Education Research Journal, 1*(1), 5–28. www.stat.auckland.ac.nz/~iase/serj/. Accessed: 12 February 2016.

Chick, H. L., & Pierce, R. U. (2008). Teaching statistics at the primary school level: Beliefs, affordances, and pedagogical content knowledge. In C. Batanero, G. Burrill, C. Reading, & A. Rossman (2008). (Eds.). *Proceedings of the Joint ICMI/IASE Study: Teaching statistics in school mathematics - Challenges for teaching and teacher education. Proceedings of the ICMI Study 18 and 2008 IASE Round Table Conference.* Monterrey, Mexico: International Commission on Mathematical Instruction e International Association for Statistical Education.

Common Core State Standards Initiative (CCSSI). (2010). *Common Core State Standards for Mathematics.* Washington, DC: National Governors Association for Best Practices and the Council of Chief State School Officers. www.corestandards.org/Math/. Accessed: 30 June 2016.

Di Martino, P., & Zan, R. (2015). The construct of attitude in mathematics education. In B. Pepin & B. Roesken-Winter (Eds.), *From beliefs to dynamic affect systems in mathematics education: Exploring a mosaic of relationships and interactions* (pp. 51–72). New York: Springer.

Estrada, A. (2002). *Análisis de las actitudes y conocimientos estadísticos elementales en la formación del profesorado* (Analysis of attitudes and statistical knowledge in teacher' education). Ph.D. dissertation. Universidad Autónoma de Barcelona, Spain. www.tesisenxarxa.net/TDX-0502103-191818/. Accessed: 18 April 2016.

Estrada, A., & Batanero, C. (2008). Explaining teachers' attitudes towards statistics. In C. Batanero, G. Burrill, C. Reading, & A. Rossman, (Eds.). *Proceedings of the Joint ICMI/IASE Study: Teaching statistics in school mathematics - Challenges for teaching and teacher education. Proceedings of the ICMI Study 18 and 2008 IASE Round Table Conference.* Monterrey, México: International Commission on Mathematical Instruction e International Association for Statistical Education.

Estrada, A., Batanero, C., Fortuny, J. M., & Diaz, C. (2005). A structural study of future teachers' attitudes towards statistics. In M. Bosch (Ed.), *Proceedings of the Fourth Congress of the European Society for Research in Mathematics Education* (pp. 508–517). Barcelona: IQS Fundemi. ermeweb.free.fr/CERME4/. Accessed: 25 March 2016.

Estrada, A., Batanero, C., & Lancaster, S. (2011). Teachers' attitudes towards statistics. In C. Batanero, G. Burrill, & C. Reading (Eds.), *Teaching statistics in school mathematics. Challenges for teaching and teacher education* (pp. 173–174). Springer.

Estrada, A., Bazán, J., & Aparício, A. (2010). A cross-cultural psychometric evaluation of the attitude toward statistics scale Estrada's in teachers. In C. Reading (Ed.), *Proceedings of the Eight International Conference on Teaching Statistics.* Ljubljana, Slovenia: International Statistical Institute. iase-web.org/documents/papers/icots8/ICOTS8_7C3_ESTRADA.pdf. Accessed: 13 June 2016.

Gal, I., & Ginsburg, L. (1994). The role of beliefs and attitudes in learning statistics: Towards an assessment framework. *Journal of Statistics Education, 2*(2). www.amstat.org/publications/jse/v2n2/gal.html. Accessed: 24 May 2016.

Gal, I., Ginsburg, L., & Schau, C. (1997). Monitoring attitudes and beliefs in statistics education. In I. Gal & J. B. Garfield (Eds.), *The assessment challenge in statistics education* (pp. 37–51). Voorburg, Netherlands: IOS Press.

Goldin, G., Hannula, M., Heyd, E., Jansen, A., Kaasila, R., Lutovac, S., Di Martino, P., Morselli, F., Middeleton, J., Pantziara, M., & Zhang, Q. (2016). *Attitudes, beliefs, motivation and identity in mathematics education.* ICME-13 Topical Surveys. New York: Springer.

Gómez-Chacón, I. (2000) Affective influences in the knowledge of mathematics, *Educational Studies in Mathematics, 43*(2), 149–168.

Gómez-Torres, E., Ortiz, J. J., & Gea, M. M. (2014). Conceptos y propiedades de probabilidad en libros de texto españoles de educación primaria (Probability concepts and properties in Spanish primary education textbooks). *Avances de Investigación en Educación Matemática, 5*, 49–71.

Hannula, M. S. (2002). Attitude towards mathematics: Emotions, expectations and values. *Educational Studies in Mathematics, 49*(1), 25–46.

Kislenko, K. (2009). Mathematics is a bit difficult but you need it a lot: Estonian pupils' beliefs about Mathematics. In J. Maaβ & W. Schlöglmann (Eds.), *Beliefs and attitudes in mathematics education: New research results* (pp. 143–163). Rotterdam: Sense Publishers.

Kvatinsky, T., & Even, R. (2002). Framework for teacher knowledge and understanding of probability. In B. Phillips (Ed.), *Proceedings of the Sixth International Conference on Teaching Statistics*. Voorburg, The Netherlands: International Statistical Institute. icots6.haifa.ac.il/PAPERS/6A4_KVAT.PDF. Accessed: 2 May 2016.

Martins, J., Nascimento, M., & Estrada, A. (2011). Attitudes of teachers towards statistics: a preliminary study with Portuguese teachers. *Proceedings of CERME 7*. Rzeszow, Poland: ERME. www.cerme7.univ.rzeszow.pl/WG/5/CERME_Martins-Nascimento-Estrada.pdf. Accessed: 4 May 2016.

Martins, J. A., Nascimento, M. M., & Estrada, A. (2012). Looking back over their shoulders: A qualitative analysis of Portuguese teachers' attitudes towards statistics. *Statistics Education Research Journal, 11*(2), 26–44. www.stat.auckland.ac.nz/~iase/serj/SERJ11(2)_Martins.pdf. Accessed: 27 March 2016.

McDonald, R. P. (2013). *Test theory: A unified treatment*. Sussex, UK: Psychology Press.

McLeod, D. (1992). Research on affect in mathematics education: A reconceptualization. In D.A. Grouws (Ed.), *Handbook of research on mathematics teaching and learning* (pp. 575–596). New York, NY: Macmillan.

Ministerio de Educación, Cultura y Deporte. (2014). *Real Decreto 126/2014, de 28 de febrero, por el que se establece el currículo básico de la Educación Primaria* (Royal Decree establishing the minimum content for primary education). Madrid: Author.

Ministerio de Educación Pública. (2012). *Programas de estudio de matemáticas* (Study programs for mathematics). San José: Costa Rica: Ministry of Education.

Ministry of Education. (2007). *The New Zealand curriculum*. Wellington, New Zealand: Learning Media.

Nasser, F. M. (2004). Structural Model of the Effects of Cognitive and Affective Factors on the Achievement of Arabic-Speaking Pre-service Teachers in Introductory Statistics *Journal of Statistics Education, 12*(1). https://ww2.amstat.org/publications/jse/v12n1/nasser.html. Accessed: 6 March 2016.

National Council of Teachers of Mathematics. (2000). *Principles and standards for school mathematics*. Reston, VA: Author.

Nunnally, J. C. (1978). *Psychometric theory* (2nd ed.). New York: McGraw-Hill.

Onwuegbuzie, A. J. (1998). Teachers' attitudes toward statistics. *Psychological Reports, 83*, 1008–1010.

Philipp, R. A. (2007). Mathematics teachers' beliefs and affects. In F. Lester (Ed.), *Second handbook of research on mathematics teaching and learning* (pp. 257–315). Charlotte, NC: Information Age Publishing & National Council of Teachers of Mathematics.

Prodromou, T. (2012). Connecting experimental probability and theoretical probability. *ZDM-The International Journal on Mathematics Education, 44*(7), 855–868.

Ramirez, C., Schau, C., & Emmioğlu, E. S. (2012). The importance of attitudes in statistics education. *Statistics Education Research Journal, 11*(2), 57–71.

Roberts, D. M., & Bilderback, E. W. (1980). Reliability and validity of a statistics attitude survey. *Educational and Psychological Measurement, 40*(1), 235–238.

Schau, C. (2003, August). Students' attitudes: The other important outcome in statistics education. Paper presented at the *Joint Statistical Meeting of the American Statistical Association*, San Francisco, CA: American Statistical Association.

Schau, C., Stevens, J., Dauphine, T., & Del Vecchio, A. (1995). The development and validation of the survey of attitudes towards statistics. *Educational and Psychological Measurement, 55* (5), 868–875.

Stohl, H. (2005). Probability in teacher education and development. In G. Jones (Ed.), *Exploring probability in schools: Challenges for teaching and learning* (pp. 345–366). New York: Springer.

Tan, C. K., Harji, M .B., & Lau, S. H. (2011). Fostering positive attitude in probability learning using graphing calculator. *Computers & Education*, *57*(3), 2011–2024.

Tishkovskaya, S., & Lancaster, G. (2012). Statistical education in the 21st century: a Review of challenges, teaching innovations and strategies for reform. *Journal of Statistics Education*, *20* (2), 1–24. www.amstat.org/publications/jse/v20n2/tishkovskaya.pdf. Accessed: 9 May 2016.

Veloo, A., & Chairhany, S. (2013). Fostering students' attitudes and achievement in probability using teams-games-tournaments. *Procedia—Social and Behavioral Sciences, 93,* 59–64.

Wise, S. L. (1985). The development and validation of a scale measuring attitudes toward statistics. *Educational and Psychological Measurement, 45*, 401–405.

Wyer, R. S., & Albarracín, D. (2014). Belief formation, organisation and change: Cognitive and motivational influences. In D. Albarracìn, T. Johnson, & M. P. Zanna (Eds.). *The handbook of attitudes* (pp. 273–322). New York: Psychology Press.

Chapter 19
Students' Reflections About a Course for Learning Inferential Reasoning Via Simulations

Susanne Podworny

Abstract This chapter discusses the development of a one semester course for preservice teachers at the university level based on intensive use of simulation software. The course is designed to deepen the understanding of inference. Some design ideas for the course are presented. One major aspect of students' work was to reflect in written form on their own statistical learning process for each of the five topics. An analysis of these reflections shows that the statistical learning process is closely related with the practiced handling of the software. Many positive statements in the reflections show where the design of the course worked well and which aspects have been understood by students. Statements concerning difficulties identify aspects for a future redesign of the course.

Keywords Probability course · Students' reflections · Simulation technology Inferential reasoning

19.1 Introduction

In Germany, simulations are included in the national educational standards for mathematics, and the *German Special Working Group of Stochastics* strongly recommends the use of computers for data analysis and probability (GDM 2003), so it is necessary to prepare future teachers to teach these topics. Moreover, research shows some benefits of a simulation approach in teaching probability at the school and university level (e.g. Batanero et al. 2005; Konold et al. 2007).

In this chapter we are analysing a one-semester elective course on simulation, titled "Applied stochastics—Understanding and solving complex problems via simulations" which was developed as a follow-up to an introductory statistics course. In the introductory course, the preservice teachers learn the basics of

S. Podworny (✉)
Institut für Mathematik, Fakultät Elektrotechnik, Informatik und Mathematik, Universität Paderborn, Warburger Str. 100, 33098 Paderborn, Germany
e-mail: podworny@math.upb.de

C. Batanero and E. J. Chernoff (eds.), *Teaching and Learning Stochastics*, ICME-13 Monographs, https://doi.org/10.1007/978-3-319-72871-1_19

statistics and probability. Learning to simulate chance experiments and deepening the understanding of probability and inferential reasoning (Cobb 2007) are the main goals of the new elective course.

19.2 Principles of Course Design

The course design was based on ideas of two existing courses that were developed at Paderborn University, Germany, and with ideas from the CATALST curriculum (Zieffler and Catalysts 2013). Following a design-based research approach (Cobb et al. 2003), five hypothetical learning trajectories on different topics were prepared (see next section). The previous analysis of the software used as an instrument as defined by Rabardel (2002) regarding his instrumental genesis was an important part for designing the learning trajectories (see Podworny and Biehler 2014).

The design of the course followed the six principles of the Statistical Reasoning Learning Environment (SRLE) by Garfield and Ben-Zvi (2008). For instance we have focused on central statistical ideas (inferential reasoning), used real and motivating data, used classroom activities (cooperative learning environments) and also have integrated the use of appropriate technological tools (TinkerPlots). One major goal was to enable participants to view simulations as one solution to solving probability problems and to be aware of the limitations that come along with simulations. The final aim was to introduce students to inferential reasoning. Monitoring of the learning process happened with weekly homework and a cumulative portfolio with regular written reflections.

TinkerPlots (Konold and Miller 2011) was used as simulation software, because it is adequate software for preservice primary and secondary teachers, it is quite easy to use and, according to Pratt and Ainley (2014), there is much potential in this software. As a tool for "expressive modelling" (Doerr and Pratt 2008), the simulation process in TinkerPlots is visual and not based on commands, which simplifies the procedure. New learning goals related to the software were presented briefly at the beginning of a lesson, and afterwards, students worked in groups of two or three on selected statistical topics. Each topic dealt with one major statistical concept.

One main aspect was to construct models appropriate to a problem with the help of the software (c.f. Maxara and Biehler 2006; Garfield et al. 2012). These models should reproduce and simplify reality and enable users to make predictions or identify trends or patterns. The context of the activity played an important role, as this is what makes problems meaningful and relevant to students (Pfannkuch 2011). Self-assessment should be conducted during the working process so that students were able to select, refine, and elaborate their models meaningfully.

Another important aspect was documentation of the problem-solving process. A simulation scheme was developed as supporting material (Podworny and Biehler 2014) and was used throughout the course. The solution to a probability problem via simulation should integrate the thinking process, beginning with assumptions, goals, and a solution path (Konold et al. 2007). A general way of thinking should be represented by the model (Lesh and Doerr 2003), and this ensured that students' models could be used by others too. A model should be as simple as possible, so that it represented a kind of prototype for other problems with the same structure. A problem-solving iteration was like a cycle with several runs. This cycle started with a description and modelling of a real-world problem, a manipulation of the model, translation of relevant results back to the real world, and finally verification of the usefulness of solutions completed the cycle. Revising a model was an important part of all the activities used, and TinkerPlots offered very good possibilities for changing models [this is according to some ideas of model-eliciting activities (MEAs) as proposed by Lesh et al. (2000)].

For the design of the course, small-group and large-group discussions were very important. They were used as an assessment "to monitor the development of their statistical learning as well as to evaluate instructional plans and progress" (Garfield and Ben-Zvi 2008, p. 48). Many learning settings took place via the instrumental orchestration approach by Drijvers et al. (2010, p. 219) with short periods of "technical demo", "explain the screen" and longer periods of "discuss the screen", "Sherpa at work" and "work-and-walk-by" (Drijvers 2012, p. 278). The first four configurations are situated in a whole class setting with one central screen. In "technical demo" the teacher explains technical aspects on tool use. "Explain the screen" is enriched by the teacher with mathematical content. A whole class discussion takes place in the "discuss the screen" orchestration, and a student is explaining his work on a central screen in a "Sherpa at work" orchestration. In the "work-and-walk-by" orchestration the teacher walks among the working students and guides their process when necessary. All tasks and additional reflections on each topic were collected in portfolios by every participant to enable participants keeping track of their knowledge (Stratmann et al. 2009). The portfolios were used to monitor students' understanding and to evaluate material and tasks, and were the final output of students' work during the course.

19.3 Realization of the Course

The course was designed for preservice teachers of primary and lower secondary school at Paderborn University. The prerequisite for taking the "Applied stochastics" course was to pass the exam of the introductory statistics course. This meant that students entering the new course should be experienced with Fathom and with basic ideas of probability and simulation. The course covered five topics of different lengths for a total of 13 sessions (90 min each). The topics were: (1) Data analysis; (2) Basic simulations; (3) Accuracy of simulations; (4) Independence and dependence; (5) Hypothesis testing with p-values and randomization tests. The English version of TinkerPlots was used, because a German version was not available at that time.

19.3.1 Topic 1, Data Analysis

In the first topic (one session) the software was introduced and connected with the participants' previous knowledge from the introductory course. As a first learning goal, students should refresh their knowledge on data analysis. The second learning goal was to get used to the new software for data analysis on a basic level (use data cards, add new cases, use formulas, plot data, use mean and median, use absolute and relative frequencies in plots). In this topic, students were mainly asked to "discuss the screen" with their peers in teams of two.

19.3.2 Topic 2, Basic Simulations

The second topic, covering four sessions, consisted of two learning goals. First, participants should learn technical aspects of simulations and, more importantly, they should be introduced to first ideas of inferential reasoning via modelling and simulation like demanded by Cobb (2007). Then, the following six steps of a simulation process adapted from Maxara and Biehler (2006) were discussed and implemented: (1) Modelling a real situation, (2) Choosing the model of the chance experiment, (3) Defining the interesting events or random variables, (4) Realization and repetition of the simulation, (5) Data analysis, (6) Interpretation of the data. Step 2–5 are software related and represented in the TinkerPlots simulation scheme (Podworny and Biehler 2014), which was used since the third session. To face the question of how many repetitions of a simulation are required, a "rule of thumb" was established concerning the accuracy of simulations in the second session. This was the distance between the probability and the relative frequency for several numbers of repetitions (for n = 100, 1000, 5000, 10,000). Several tasks from different contexts were included in this topic, each of them with the focus on a special feature of the software and the different modelling possibilities offered by the software.

At the end of this topic, participants should know different options for setting up a model with the software (Garfield et al. 2012; Maxara and Biehler 2006; Noll and Kirin 2016). This included knowing the devices box and spinner and their options, how to arrange more than one device in the sampler, and how to sample without replacement. This also included knowing the meaning of "draw" and "repeat", how to define random variables or chance events by a formula in the table or by plotting the results, how to analyse the data, and knowing about collecting history. The first task for the autonomous work of the participants was the following.

> Some pupils want to build a wheel of fortune booth on Christmas market. They want to earn 50 cents per game on average. The wheel allows the following outcomes: There is a 1% chance of getting 50 Euro, a 45% chance of getting 1 Euro, and a 54% chance of getting nothing. Rule: The visitor may turn the wheel three times in one game and receives the sum as benefit. Question 1: What should be the price for one game in order for the pupils to make 50 cents on average? Question 2: What is the probability for a random visitor to win nothing?

With some homework and some more subtasks, this problem was used to introduce many aspects of different routes through the software. After short introductions with "technical demo," there were long periods for participants to work on their own with the help of the teacher while walking around the class ("work and walk by" as proposed by Drijvers 2012).

19.3.3 Topic 3, Accuracy of Simulations

A hands-on activity about throwing coins (two sessions) was used for the empirical law of large numbers and dealt with both directions of the so called *$1/\sqrt{n}$-law*. At first, the preservice teachers learned that $p \pm 1/\sqrt{n}$ is predicting the interval where the relative frequency of an event can be expected with 95% probability, if the probability p is known. The second direction of the *$1/\sqrt{n}$-law* is to make inferences about the unknown probability p, when a relative frequency of h_n of an event was observed. This was communicated as a rule that one can estimate p in the interval $h_n \pm 1/\sqrt{n}$ and specify the uncertainty of this estimation as 95%. This ended the level of teaching on informal confidence intervals and did not go into more details.

The whole topic was more formal and focussed mainly on throwing coins. Only the last task of this topic was situated in the context of overbooked planes and the probability of overbooking depending on the number of seats in different sizes of planes. The learning goal for this topic was to enable students to give more accurate answers on problems solved via simulations with regard to the *$1/\sqrt{n}$-law* as described above. A second learning goal was to deepen their knowledge on the law of large numbers and on the development of absolute and relative frequencies when increasing the number n of repetitions. Figure 19.1 shows one plot of a learning environment for the *$1/\sqrt{n}$-law*. Tasks on this plot were (*A*) *Make an interpretation of the hat plots getting smaller.* (*B*) *What is the relation between the width of the hat plot and the number n of repetitions?* Mainly "discuss the screen" and "explain the screen" (Drijvers 2012) were orchestrations of this topic, because there was much explanation and group discussion needed in this topic.

19.3.4 Topic 4, Independence and Dependence

During the four sessions of this topic, two paths were discussed in two sessions: from model to data and from data to model. Reasoning for both directions was a learning goal. In the approach from model to data, the software was used to make predictions in a conditional probability task design. The second direction was led by the question "which model fits best to given data?" (c.f. Konold and Kazak 2008). There was room for plenty of small-group and whole-class discussion during this topic.

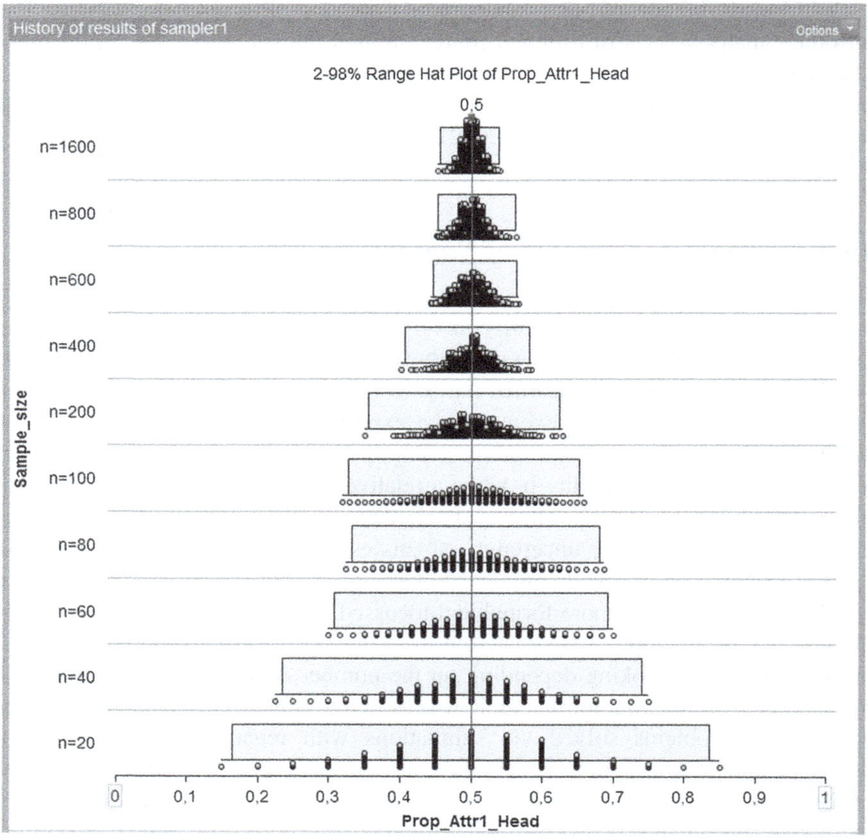

Fig. 19.1 Simulated proportions of 'heads' when throwing a coin n times, with the 2–98 percentile hat plot

19.3.5 Topic 5, Hypothesis Testing and Randomization Tests

First ideas of testing hypotheses were introduced in two sessions and resulted in randomization tests in another two sessions. Together with the second topic, this was the main focus of the course. The starting point was a model-eliciting activity (Lesh et al. 2000) about a fictitious person who identified the quality of music (low mp3 or CD quality) and correctly identified the quality of 10 out of 12 music tracks. The question was whether this person has the ability of hearing differences in quality of music or if she was just guessing. This task lasted for a whole session with several subtasks and discussions, and resulted in deepening questions in a second session. For supporting the inferential reasoning the following steps were suggested: note an observation, choose a null hypothesis, simulate under this null hypothesis, identify the test statistics, estimate the p-value, interpret the results with regard to Cobb's three 'Rs': randomize, repeat, reject (Cobb 2007). For setting up the null hypothesis we focused on a "just guessing" or "randomly chosen" vocabulary. The learning goal of this topic was on testing hypotheses reasoning and to enable students to identify and interpret a p-value. All types of orchestrations were used during the four sessions in several tasks.

19.4 Students' Reflections in Their Portfolios

In this section, research based on students' reflection is presented for redesigning the course in the next cycle. Two research questions are posed:

R1: How do the participants reflect on their own statistical learning process along several topics?

R2: To what extent are students' reflections suitable for doing research on redesign in a design-based environment?

19.4.1 Participants and Data Collection

There were twenty-five students participating in the course, but only twenty-three of them decided to finalize the course by writing a portfolio. Twenty of them were female and three were male. Seventeen students are at the university to become primary school teachers and six want to become teachers at lower secondary schools, which in Germany range from grade five to grade ten. This is a typical audience for a course like this, because there are many preservice primary teachers at Paderborn University and a very large percentage of them are female. All of the preservice primary teachers get the teacher license up to grade ten and therefore should be introduced to probability too, especially since every teacher should know more mathematical background than just what she or he will be teaching.

All of the participants passed the compulsory introductory course, but this had occurred one or two years ago, if not earlier. The prior knowledge of the participants was rather mixed, if the written test at the end of the introductory course is taken into account. Four of them needed to take the test a second time, because they failed the first time. All but one took the introductory course when Fathom was used, so all but one had prior software knowledge with Fathom for data analysis as well as for simulation.

The course was divided into five topics, as written above, and for each topic the participants had to write a reflection on their individual learning successes, their difficulties and suggestions for improvement. There were two questions guiding the reflections:

1. Summarize what you have learned statistically in this topic. Identify issues you did not understand or about which you would have liked to get more information.
2. Report your difficulties with the software, if you had any. Additionally, you may report tasks you liked or did not like and why.

Research on the second question is not addressed in this chapter. There are twenty-three reflections for each topic, which results in 115 reflections in total. A reflection had to be at least one page long. All reflections were in written form (pdf) and were collected after the course. The reflections were sometimes enriched by TinkerPlots screenshots, tasks of the course, and everything else on which the participants wanted to reflect, so they are rather mixed in content.

19.4.2 Methodology for Data Analysis

The main purpose of the data analysis was to categorize students' reflections related to statistical learning process (research question 1), and to qualify whether the statements were positive, neutral or reported difficulties. A look at the statements coded as positive or as containing difficulties will hopefully give valuable hints for a redesign of the topics. Due to the large amount of data (115 reflections of at least one page each) the analysis method was to reduce data by categorization. Qualitative content analysis (Mayring 2015) was used for the categorization of the data.

In the same way the software use (SU) of the participants was coded, but this is not reported in this chapter. Within the first step of this approach, the research questions were stated, and two dimensions of the analysis were determined in the second step. These dimensions are (1) students' reflections on their own statistical learning process (SLP) and (2) students' reflections on their software use (SU). This second dimension is not part of the analysis in this chapter. The third step of this analysis method is to define the minimum and maximum coding unit. As a coding unit, a unit of meaning was chosen. A minimum coding unit was a sentence, and a maximum coding unit were a bundle of sentences as long as they referred to the same meaning. For step four, the definition of categories, a scaling-structuring content analysis was used with the codings presented in Table 19.1. Codings were subcoded in relation to the topic to which they referred. Afterwards, a scaling coding was done by rating student' statements as positive, neutral or afflicted with difficulties. All topics (1–5) were coded in the same manner.

19.5 Results

Each reflection of the twenty-three participants was coded with the codings reported above. For each one-page reflection, between one and ten different codings were given related to either statistical learning process (SLP), or software use (SU).

Table 19.1 Examples of codings of statistical learning process (SLP)

Coding	Definition	Key example
SLP positive	Either the student evaluates a statistical issue as "positive", "interesting", "clear", or "easy" or the student reports that "I learned…" or "I got more knowledge about a statistical issue …"	"All these findings have been crucial for me for understanding the $1/\sqrt{n}$-*law*." (Lilly[a], reflection on topic 3)
SLP neutral	A statistical issue is reported as neither positive nor negative. It is a description of statistical teaching content	"All this was about multi-level random experiments, so we had to think about whether they were dependent or independent." (Edward, reflection on topic 4)
SLP difficulties	A statistical issue is reported and evaluated as "not understood" or at least with some difficulties. This often goes along with proposals for redesign	"This topic was very extensive and I think it was the hardest topic. Thereby working on the tasks with TinkerPlots was less hard, but to determine the null hypothesis as well as to find the observed value for the hypothesis or randomization test." (Camilla, reflection on topic 5)

[a]All names are pseudonyms

The contents of students' reflections were rather mixed. Some only reported in a neutral way about the statistical content of the topic, some responses were related very much to software use. Some mixed it up all through, starting with SLP, going on with SU, coming back to SLP, mentioning a task, writing again about SLP and so on.

Altogether, there are 252 codings related to SLP within all 115 reflections. The number of codings for SLP is mainly increasing from topic 1 with 33 to topic 5 with 68 codings. This mirrors the content of the course in some way, because the first two topics were mainly on how to use the software and starting from topic 3 the content was more about statistical ideas, so there have been more responses to the statistical learning process.

For each topic there will be examples on the different ratings of students' statements, due to the amount of space the leading principle for the selection is the idea of "interesting or representative" episodes (Voigt 1984).

19.5.1 Results for Topic 1 on Data Analysis

Only eight codings (24%) were given to a positive statistical learning process, and they came from eight different participants. For example, participants stated that they liked to link their previous knowledge gained in the introductory course with the new software. They especially liked to discuss several different options to plot data and how to reason with different plots. This was a main feature of the first topic, because every plot had to be 'created' by the user from the beginning. Some examples follow:

Anita: "I have learned to understand and interpret graphs. I knew this before, but the course [topic 1] helped me to develop a deeper understanding."

Anastasia: "I found the discussion of plots with the same content, but different ways of graphing very good and very important. A freshman like me did not know from the beginning to pay attention to which aspects."

The 25 (76%) codings of neutral SLP are divided among 19 participants. Mainly these reflections described the things that were done in the course like the conscious selection of a plot. Often these statements were quite similar to the statements rated as "positive", but without key words like "I have learned to…", "interesting", "easy" and so on. No hints for redesign (or keeping the actual design) could be identified from these neutral ratings.

None of the participants reported difficulties on a statistical level for this topic. This may be due to the fact that most statistical content of this topic was known from the introductory course and the focus was on the software use.

19.5.2 Results for Topic 2 on Basic Simulations

The number of reflections on SLP in topic 2 were with a total of 36 the fewest between all topics. This is surprising, because it was a major topic of the course with four sessions and introduced first ideas on inferential reasoning, but it may be due to the fact that software aspects seemed to be the focus of these sessions from studetns' point of view and not about modelling different probability situations. But at least 26 (76%) codings were given to a positive or neutral statement in students' reflections and only eight difficulties (24%) were reported. The positive remarks were mainly superficial, about how a simulation could help to solve a task in everyday life. The same was reported by many more in a neutral way, such as:

Lucy: "In this topic we started to work elaborately on complex problems of everyday life. By simulating different random experiments we translated these problems into mathematics and analysed them. Typically, we used realistic themes."

Additionally, a theoretical solution for two tasks was part of this topic and was reported in contrast to the simulated solution by participants, but without an evaluation. This mirrors one problem of the reflections quite early. There are lots of neutral statemants that do not give hints for a researcher what was gained or not understood by students.

Some participants mentioned the simulation scheme as helpful like:

Ivonne: "The simulation scheme can be used for structuring a simulation and is very helpful for each step."

One difficulty reported by two participants mentioned that using the simulation scheme for planning a simulation before using the computer was difficult.

Modelling a situation was another difficulty, as reported by:

Rebecca: "Here I had some difficulties. I understood the situation of many tasks and had a notion of how to answer the question. But I missed a trigger for conducting a simulation."

This problem of how to model a simulation was typical for many students. Rebecca called this a "trigger", and after getting an idea of how to start with the sampler, most students worked well on their tasks, so this may be a difficulty on a software level, but hardly on a statistical level. Very few participants mentioned details about their gains in knowledge or even in their problems with these sessions, so from a broader point of view it was merely useful for a researcher's perspective.

19.5.3 Results for Topic 3 on Accuracy of Simulations

Topic 3 was the most theoretical topic between all five. A total of 57 codings relate to SLP here. Few participants reflected on the topic as a whole, such as:

Anton: "As a whole, this topic showed to me how important a large number of repetitions is in matters of the explanatory power of a simulation."

But even though only few participants mentioned this, the following sessions showed, that nearly all of the participants understood the relation of the accuracy of a simulation and the number of repetitions.

Many participants repeated in a neutral way statistical findings from this topic (35 out of 57, 61%). One example is

Melissa: "With $1/\sqrt{n}$ you can compute the deviation of a theoretical probability. You see in this formula, that by decreasing the number n the deviation of a theoretical probability is decreasing. To double the accuracy of the results of a simulation, you have to quadruple the number of repetitions. If you multiply the formula $1/\sqrt{n}$ by two, you can compute the width of the interval where the theoretical probability will be."

Many explanations like this were found in the SLP reflections of this topic, not 100% correct, but not totally incorrect too like the one by Melissa. These statements help to get insights into the statistical learning process, because it is possible to identify what was learnt and what was not understood, even though it is not rated by the students as benefit or not understood.

Because it was much content for two sessions, some complained about the shortness of time and reported difficulties, like

Angelina: "I know now about the formula [$1/\sqrt{n}$] but I would have liked to talk about its derivation for a longer amount of time, because I did not fully understand every step of its derivation."

But no major difficulties were reported about the understanding of the *$1/\sqrt{n}$-law*. At least every participant seemed to understand how to use this in combination with accuracy of simulations. Seen from the end of the course, this was the most challenging topic, but most of the participants learnt to rate the accuracy of a simulation and used this knowledge in further sessions.

19.5.4 Results for Topic 4 Independence and Dependence

Many participants (20 out of 66 codings, 30%) liked parts or all of topic 4 and reported in a positive way things they understood. The 20 codings on SLP rated as positive are divided among 13 participants. One example with a good summary of the knowledge gained is

Anita: "In this topic we looked at dependent and independent attributes. I knew these terms before. The tasks in the course deepened my knowledge and I learned how to analyse data concerning dependence, for example by simulating a dependent or independent model based on real data."

Additionally, all but two participants (36 codings, 55%) reported in a neutral way on this topic too. But there appeared interesting conclusions about the topic, in a similar way like Anita did. This topic on independence and dependence was again short, including only with two sessions. This is what many participants struggled about (10 codings, 15%), such as:

Claudia: "All in all it was a short topic and I would have liked to discuss the difference between dependence and independence in more depth."

The way the software supports the modelling process graphically is mentioned positively by many participants too, like Ingrid:

Ingrid: "As mentioned above, the software shows this dependence very well, because there are labels from the first to both depending spinners. If the event "ill" occurred, it went to the second spinner. If the event "healthy" occurred, it went to the other depending spinner."

In this topic, one of the big advantages of the software was seen by the participants. The modelling activities and its representation in the software helped the understanding and the reasoning process.

19.5.5 Results for Topic 5 Hypothesis Testing and Randomization Tests

The topic of testing hypotheses was totally new to the participants, so they reflected about their SLP a lot, with a total of 68 codings, the most of all topics. Testing with *p*-values was reflected many times, positively or neutral. For example Kerstin wrote:

Kerstin: "In this topic I learned how to argue with several *p*-values. I state, that for *p*-values larger than. 1 I cannot say something about the null hypothesis. I cannot say for sure, for example, if a test person just guessed or did her result by knowledge. For this reason I cannot reject the null hypothesis, but not say it is true either."

This is from a teacher's point of view a good summary of reasoning with *p*-values. What could and could not be reasoned by small or large *p*-values was part of many reflections. Interestingly, there was no wrong statement concerning a *p*-value in the reflections. Only Anton was concerned about what to conclude from a medium *p*-value:

Anton: "At some point, my partner and I were at odds about interpreting *p*-values at medium level. Perhaps the interpretation of medium *p*-values can be discussed in more details. Or state clearly that it is not important how to interpret medium *p*-values as long as the interpretation is comprehensible. For me, I decided on the background of the problem how to decide with a medium *p*-value, sometimes like this, sometimes like that. Perhaps this was correct, perhaps wrong."

Compared with the other topics, the most difficulties (19, 28%) were reported for topic 5. This is not surprising, because testing hypotheses was not known in advance by the participants. Some students mentioned difficulties on how to locate a *p*-value, how to choose a null hypothesis, the sequences of a test. But all in all, the reflections on this topic were satisfying.

The use of software was well known at the end of the course. A good summary of the positive codings is exemplified by

Lucy: "In the end I can say, the complex tasks and simulations we did in this topic were supported by the clearly represented structure of the software. The uncomplicated use of the software was helpful for the display and analysis of the tasks."

The codings of the last topic showed the satisfaction by students of their own learning process concerning their use of TinkerPlots.

19.6 Discussion and Implication for Research

The statistical learning process of students (research question 1) depended and was combined with the software use in this course. In total there were 253 codings for SLP. Many of the codings related to a neutral rating. The number of difficulties reported in students' reflections for SLP were quite low. They started with zero in topic 1 and increased to 19 in topic 5. This was a good result, as the amount of content for the course was rising for each topic. Only few participants reported concrete aspects they did not understand or did not like, there were more statements about too little time or a wish for deepening some aspects like the formula $1/\sqrt{n}$.

One major topic of the course was topic two, because all basics were taught in these four sessions for using the software as an instrument (Rabardel 2002) and for learning about modelling (Maxara and Biehler 2006; Garfield and Ben-Zvi 2008; Garfield et al. 2012). Many students reported that they liked the orchestration of "work and walk-by" of the teacher (Drijvers 2012), so this is one aspect that will be kept in the next iteration. There was plenty time for students to work on tasks with their partners and only few moments of central lecturing, what seemed fine to most students and helped to develop and discuss their reasoning. The selection of problems seemed to be interesting for students; at some point nearly everybody reported a specific task that related to their everyday life. For SLP, the selection of problems seemed to be adequate. The context of the problems seemed interesting, but not disguising the inferential reasoning process (Pfannkuch 2011). The simulation capacity that was analysed by the author in detail before the course for SU was represented well in these four sessions. There were students like Sara, who noticed the software as a good expressive medium (like in the sense of Doerr and Pratt 2008), which was one of the intentions for the course. This is an aspect about the software use that was important for the design and worked well.

The third topic was a more mathematical one and seemed to be quite hard to understand for the students. Like Anastasia, many students expressed a wish for more time with this topic. But the following sessions showed, that the students learned to reason with the *$1/\sqrt{n}$-law* and to specify the accuracy of a simulation, even though this was reported as difficulties. Together with the next topic, a redesign should be to lengthen this topic.

In topic four, students learned to create "factories" (Konold et al. 2007) with TinkerPlots. This worked on both levels, SLP and SU, which were again strongly related. For evaluating their models, students were introduced to p-values; this seemed not to be an appropriate way of starting hypothesis testing. There were some difficulties reported on SLP, so it must be carefully decided which parts of this topic will be kept and which parts will be integrated in the last topic. Because of the great acceptance of this topic it will be redesigned, rather than erased as was intended after first teaching the topic.

Like Anton said in his reflection, topic five was the topic that brought everything together. Small and large p-values were discussed in detail, but as he mentioned, there were few discussions on medium p-values, that are p-values between .05 and .1.

On the SLP view, there should have been more emphasis on these *p*-values. But in general, students' reflections on SLP revealed their understanding of hypothesis testing. In summary, most reflections were quite positive or at least neutral, and students were highly engaged in the course, although it was more weekly work for them than participating in a "normal" course.

With regard to the methodological research question *How far are students' reflections suitable for doing research on redesign in a design-based environment?* it is arguable whether reflections of students are a valuable instrument for research or not. Due to the large amount of neutral ratings, there were only few hints on students' performances. Only about half the codings related to either positive or difficulty ratings. And only these ratings gave insights into students' thinking. But students expressed their own benefit of reflecting each topic, so the reflections were a good type of assessment, but there should rather be an adapted leading question. Perhaps stronger hints to write from an "I" perspective and to report on difficulties and challenges can help to strengthen the value of reflections for research.

For preservice teachers in Germany, there exist only few courses that build on an introductory statistics course. This course is one attempt to deepen the understanding of inferential reasoning and offers a new approach to probability problems via simulation. The simulation method enables students to solve more complex problems than could be solved by calculation only (Engel 2007), and allows them to solve unfamiliar problems. The reflections as being part of the portfolios helped the participants to keep track of their knowledge and knowledge gaps, and appeared to be a valuable instrument for assessment. But it cannot be assumed that the participants of the course fully understood the underlying statistical ideas and sampling that is occurring. In face of much work for the participants, they were highly motivated and worked hard during the whole course. From this first point of view, the design worked well and will be redefined in small details for a future iteration.

Of course, the reflections of this study are only from a sample of students of Paderborn University and only for the newly designed course. There may be the thought of "social accessibility" when writing a reflection as part of passing a course. Students probably withheld some criticisms in their reflections because of this. And another reason is the metacognition of the participants, who might not always be able to formulate difficulties they had. But all in all, the picture given by the reflections seems to represent students' learning processes quite well.

References

Batanero, C., Biehler, R., Maxara, C., Engel, J., & Vogel, M. (2005, June). Using simulation to bridge teachers content and pedagogical knowledge in probability. *Paper presented at the 15th ICMI Study Conference: The Professional Education and Development of Teachers of Mathematics*. Aguas de Lindoia, Brazil: ICMI.

Cobb, G. (2007). The introductory statistics course: A ptolemaic curriculum? *Technology Innovations in Statistics Education, 1*(1), 1–15.

Cobb, P., Confrey, J., deSessa, A., Lehrer, R., & Schauble, L. (2003). Design experiments in educational research. *Educational Researcher, 32*(1), 9–13.

Doerr, H. M., & Pratt, D. (2008). The learning of mathematics and mathematical modeling. In M. K. Heid & G. W. Blume (Eds.), *Research on technology and the teaching and learning of mathematics: Research syntheses* (pp. 259–285). Charlotte: Information Age Publishing.

Drijvers, P. (2012). Teachers transforming resources into orchestrations. In G. Gueudet, B. Pepin & L. Trouche (Eds.), *From text to 'lived' resources* (pp. 265–281). Dordrecht, The Netherlands: Springer.

Drijvers, P., Doorman, M., Boon, P., Reed, H., & Gravemeijer, K. (2010). The teacher and the tool: Instrumental orchestrations in the technology-rich mathematics classroom. *Educational Studies in mathematics, 75*(2), 213–234.

Engel, J. (2007. August). *On teaching the bootstrap.* Paper presented at the International Statistical Institute 56th Session, Lisbon: International Statistical Institute. https://iase-web.org/documents/papers/isi56/IPM40_Engel.pdf. Accessed 1 July, 2017.

Garfield, J., & Ben-Zvi, D. (2008). *Developing students' statistical reasoning*. New York: Springer.

Garfield, J., delMas, J., & Zieffler, A. (2012). Developing statistical modelers and thinkers in an introductory, tertiary-level statistics course. *ZDM Mathematics Education 44*(7), 883–898.

GDM, Working Group of Stochastics (2003). Empfehlung zu Zielen und zur Gestaltung des Stochastikunterrichts (Recommendations for goals and configuration of statistics lessons). *Stochastik in der Schule, 23*(3), 21–26.

Konold, C., Harradine, A. & Kazak, S. (2007). Understanding distributions by modeling them. *International Journal of Computers for Mathematical Learning 12*(3), 217–230.

Konold, C., & Kazak, S. (2008). Reconnecting data and chance. *Technology Innovations in Statistics Education 2*(1). http://escholarship.org/uc/item/38p7c94v. Accessed: 30 June 2017.

Konold, C., & Miller, C. (2011). *TinkerPlots TM Version 2.* Emeryville, CA: KCP.

Lesh, R., & Doerr, H. M. (2003). Foundations of a models and modelling perspective on mathematics teaching, learning and problem solving. In R. Lesh & H. M. Doerr (Eds.), *Beyond constructivism. models and modelling perspectives on mathematics problem solving, learning, and teaching* (pp. 3–33). Mahwah: Lawrence Erlbaum.

Lesh, R., Hoover, M., Hole, B., Kelly, A., & Post, T. (2000). Principles for developing thought-revealing activities for students and teachers. In R. Lesh, & A. Kelly (Eds.), *Handbook of research design in mathematics and science education* (pp. 591–645). Mahwah: Lawrence Erlbaum.

Maxara, C. & Biehler, R. (2006). Students' probabilistic simulation and modeling competence after a computer-intensive elementary course in statistics and probability. In A. Rossman & B. Chance (Eds.), *Proceedings of the Seventh International Conference on Teaching Statistics.* Salvador de Bahia (Brazil); International Statistical Institute. https://www.stat.auckland.ac.nz/~iase/publications/17/7C1_MAXA.pdf. Accessed 1 July, 2017.

Mayring, P. (2015). Qualitative content analysis: Theoretical background and procedures. In A. Bikner-Ahsbahs, C Knipping,. & N. Presmeg, (Eds.), *Approaches to qualitative research in mathematics education* (pp. 365–380). Dordrecht, The Netherlands: Springer.

Noll, J. & Kirin, D. (2016). Student approaches to constructing statistical models using TinkerPlots. *Technology Innovations in Statistics Education 9*(1). http://escholarship.org/uc/item/05b643r9. Accessed: 30 June 2017.

Pfannkuch, M. (2011). The role of context in developing informal statistical inferential reasoning: A classroom study. *Mathematical Thinking and Learning 13*(1–2), 27–46. doi: https://doi.org/10.1080/10986065.2011.538302.

Podworny, S. & Biehler, R. (2014). A learning trajectory on hypothesis testing with TinkerPlots - design and exploratory evaluation. In K. Makar; B. de Sousa; R. Gould (Eds.), *Proceedings of the Ninth International Conference on Teaching Statistics* Flagstaff, AR: International Statistical Institute. http://iase-web.org/icots/9/proceedings/pdfs/ICOTS9_9A2_PODWORNY.pdf. Accessed: 30 June 2017.

Pratt, D., & Ainley, J. (2014). Chance *re*-encounters: 'Computers in probability education' revisited. In T. Wassong, D. Frischemeier, P. R. Fischer, R. Hochmuth, & P. Bender (Eds.), *Mit Werkzeugen Mathematik und Stochastik lernen. Using tools for mathematics and statistics* (pp. 165–177). Wiesbaden: Springer.

Rabardel, P. (2002). *People and technology*. Occasional paper. Université Paris 8. https://hal.archives-ouvertes.fr/hal-01020705. Accessed: 30 June 2017.

Stratmann, J., Preußler, A., & Kerres, U. (2009). Lernerfolg und Kompetenz: Didaktische Potentiale von „Portfolios" im Hochschulstudium (Learning success and competency: Didactical potential of "portfolios" in universities). *Zeitschrift für Hochschulentwicklung 4* (1), 90–103.

Voigt, J. (1984). *Interaktionsmuster und -routinen im Mathematikunterricht: Theoretische Grundlagen u. mikroethnographische Falluntersuchungen* (Behaviour patterns and –routines in mathematics education. Theoretical basics and microethnoraphic analysis). Weinheim: Beltz.

Zieffler, A., & Catalysts. (2013). *Statistical thinking. A simulation approach to uncertainty.* Minneapolis, MN: Catalyst Press.

Chapter 20
Prospective Teachers' Probabilistic Reasoning in the Context of Sampling

Emilse Gómez-Torres, Carmen Díaz, José Miguel Contreras and Juan Jesús Ortiz

Abstract In this paper, we analyse the knowledge of sampling in 157 prospective primary school teachers in Spain. Using two different tasks, and taking into account common and horizon content knowledge (described in the model proposed by Ball et al. in J Teacher Educ 59:389–407, 2008), we assess the teachers' understanding of the following concepts: population and sample, frequency, proportion, estimation, variability of estimates, and the effect of sample size on this variability. Our results suggest that these prospective teachers have correct intuitions when estimating the sample proportion when the population proportion is known. However, they tend to confuse samples and populations, sometimes fail to apply proportional reasoning, misinterpret unpredictability, and show the representativeness heuristic and the equiprobability bias.

Keywords Sampling · Common mathematical knowledge · Assessment Prospective teachers · Primary education

E. Gómez-Torres (✉)
Statistics Department, Universidad Nacional de Colombia, CP: 111321 Bogotá, D.C., Colombia
e-mail: egomezt@unal.edu.co

C. Díaz
Universidad de Huelva, Huelva, Spain
e-mail: carmen.diaz@dpsi.uhu.es

J. M. Contreras
Universidad de Granada, Granada, Spain
e-mail: jmcontreras@ugr.es

J. J. Ortiz
Universidad de Granada, Granada, Spain
e-mail: jortiz@ugr.es

C. Batanero and E. J. Chernoff (eds.), *Teaching and Learning Stochastics*, ICME-13 Monographs, https://doi.org/10.1007/978-3-319-72871-1_20

20.1 Introduction

The omnipresence of randomness implies our need to understand random phenomena in order to make informed decisions when confronted with uncertainty. For this reason, probability is part of the primary school mathematics curriculum in many countries. For example, the Standards of the National Council of Teachers of Mathematics of the United States of America (NCTM 2000, p. 400) include the following recommendations for children in Grades 3–5 (8- to 10-year-old children):

- Describe events as likely or unlikely and discuss the degree of likelihood using such words as *certain*, *equally likely*, and *impossible*;
- Predict the probability of outcomes of simple experiments and test the predictions;
- Understand that the measurement of the likelihood of an event can be represented by a number from 0 to 1.

These recommendations have been reproduced in other countries and specifically in Spain (Ministerio de Educación y Ciencia [MEC] 2006; Ministerio de Educación, Cultura y Deporte [MECD] 2014). The Spanish curricular guidelines include the following probability content (MEC 2006):

- First level (Grades 1–2, 6–7 year olds): Recognising randomness in everyday settings: Distinguishing among impossible, certain and something that is possible but not sure; using probability words or expressions in everyday language (p. 43098).
- Second level (Grades 3–4, 8–9 year olds): Recognising randomness in everyday settings: Assessing outcomes in situations where chance is present. Recognising that there are more and less probable events, and that it is impossible to predict a particular outcome. Introduction to the language of chance. Confidence in their own possibilities, curiosity, interest, and perseverance in communicating conclusions from data displayed in graphical representations (p. 43099).
- Third level (Grades 5–6, 10–11 year olds): Recognising randomness in everyday settings: Recognising chance in daily life. Estimating the level of probability of an event (p. 43101).

These proposals also suggest a change in the teaching approach, moving from an exclusively classical meaning of probability to a frequentist view, which involves children in making predictions about the behaviour of simple experiments, carrying out experiments and simulations, collecting and analysing data, and comparing data with predictions (Batanero et al. 2016).

Given that the topic of probability was only included for the first time in 2006 in Spanish primary school curricula, many current prospective primary school teachers did not themselves study probability in primary school and may have forgotten the relevant ideas that they did learn in secondary school. This situation is common in other countries, as suggested in the ICMI and IASE joint study (Batanero et al. 2011). However, formative actions aimed at empowering these teachers in both their content and pedagogical content knowledge of probability should be based on a previous assessment of their educational needs.

The aim of the research described in this paper was to contribute to this need by assessing prospective teachers' content knowledge of the following concepts related to sampling that are needed for teaching probability with a frequentist approach, and which are included in the curricular guidelines for primary or secondary school level: population, frequency, probability, estimating probability from frequencies in repeated trials, and sampling variability in small and large samples. Our study complements previous research dealing with teachers' knowledge in producing new information about how prospective teachers use the above ideas in the context of sampling.

In the following sections, we summarise the study background and methodology, analyse and discuss the responses by 157 prospective primary school teachers to two different problems, and finally, describe some activities that helped in improving the prospective teachers' knowledge of concepts related to sampling.

20.2 Background

In this section we summarise the model developed by Ball et al. (2008) describing the components of teachers' knowledge, part of which was assessed in our research. We also describe the main research dealing with teachers' understanding of sampling.

20.2.1 Components of Teacher Knowledge

We base our research on the model of "mathematical knowledge for teaching" (MKT) proposed by Ball and her colleagues (Ball et al. 2008) to describe the knowledge required by teachers for effective teaching of mathematics. This model includes *content knowledge* and *pedagogical content knowledge*. In the present study, we only analyse content knowledge, which Ball and her colleagues decompose into three categories:

- *Common content knowledge (CCK)*: according to Ball et al. (2008, p. 399), this is "the mathematical knowledge and skill used in settings other than teaching." In our interpretation, this type of knowledge includes all of the basic skills and general knowledge about the particular topic that are to be taught to students.
- *Specialized content knowledge (SCK)*: "the mathematical knowledge that allows teachers to engage in particular teaching tasks" (Hill et al. 2008, pp. 377–378). This type of knowledge supports the teacher in representing mathematical knowledge, providing explanations, and understanding students' solutions to problems, as well as in planning lessons and assessing students' knowledge.
- *Knowledge in the mathematical horizon*: We interpret this as knowledge that is broader than what the teacher explicitly teaches; It involves, for example, knowledge of the content that students will encounter in future courses:

> Horizon knowledge is an awareness of how mathematical topics are related over the span of mathematics included in the curriculum. First grade teachers, for example, may need to know how the mathematics they teach is related to the mathematics students will learn in third grade to be able to set the mathematical foundation for what is yet to come (Ball et al. 2008, p. 403).

As described in the method section, in our study we selected two complementary tasks that were used to assess common and horizon knowledge of the sampling ideas described in Sect. 20.1.

20.2.2 Teachers' Understanding of Sampling

Understanding sampling requires the assimilation of two apparently antagonistic ideas: sample representativeness and (sampling) variability (Batanero et al. 1994). In these situations, the purpose of drawing a single sample is to quantify that sample's level of unusualness relative to the many other samples that could have been drawn. Saldahna and Thompson (2007) observed that a suitable sense of the variation across many possible samples extends to the notion of the distribution of a statistic. They also suggest that students tend to judge a sample's representativeness only in relation to the variation of the population parameter, but not in relation to the variation of all the possible samples taken from the population.

Research on people's understanding of the idea of sampling started with the heuristics and biases programme, which is summarized in Kahneman et al. (1982). As this research has shown, few people follow the normative mathematical rules to reason about sampling. Instead, they commit systematic errors in decision making. One of the problems proposed in their research is the hospital problem; the original formulation of the problem is as follows:

> A certain town is served by two hospitals. In the larger hospital about 45 babies are born each day, and in the smaller hospital about 15 babies are born each day. As you know, about 50% of all babies are boys. The exact percentage of baby boys, however, varies from day to day. Sometimes it may be higher than 50%, sometimes lower.
>
> For a period of one year, each hospital recorded the days on which (more/less) than 60% of the babies born were boys. Which hospital do you think recorded more such days? (Kahneman et al. 1982, p. 44)

Although the correct response is that it is more likely to get this result in the small hospital (since small samples are more variable), people often respond that both hospitals are equally likely to obtain this result. Kahneman et al. (1982) suggest that people forget the relevance of the sample size in responding to this task. They refer to this phenomenon as the *representativeness heuristic*, by which people only focus on the representativeness of a sample and do not take into account the variability and the effect of sample size on this variability. Another result of this research is the identification of the belief in the*law of small numbers*, whereby people expect that small samples should exactly reflect all of the characteristics in the population distribution.

Very few studies investigating understanding of sampling have been carried out with prospective teachers. Azcárate (1995) analysed responses given by 57 prospective teachers to two versions of the hospital problem and reported that 67% of participants reasoned according to the representativeness heuristic, 56% recognized the difference between probability and observed frequency, and 39% expected exact convergence of frequencies towards theoretical probability.

Similar results were observed by Batanero et al. (1996) in the responses of a sample of 130 prospective teachers to whom they Problem 2 (in Fig. 20.1), which is a variation of the hospital problem by Kahneman et al. (1982). Only 26.9% of the prospective teachers in the study responded correctly to part (a) and only 43.8% responded correctly to part (b). It was also found that 30% of participants considered the different options in part (a) to be equally likely, which the researchers attributed to the equiprobability bias (the belief that all of the events in any random experiment are equally likely; Lecoutre 1992). Moreover, 30% of the correct responses were based on the incorrect argument that anything may happen because the experiment was random. The majority of correct responses to the second part of the problem were also based on equiprobability arguments.

Problem 1. There is a fishing pond on a farm. The owner wants to know how many fish there are in the pond. He took out 200 fish and marked each of them with a coloured sign. He put the marked fish back into the pond and let them mix with the others. On the second day, he took out 250 fish in a random fashion, and found that, among them, 25 were marked.

a. What is the approximate number of fish in the pond?
b. If the owner takes out 100 more fish, approximately how many of these will be marked?

Problem 2. In a town hospital, a record of the number of boys and girls born in the hospital is kept.

a. Which of these cases is more likely:
    - There will be 8 or more boys among the next 10 babies born at this hospital ___
    - There will be 80 or more boys among the next 100 babies born at this hospital ___
    - Both results are equally likely___

    Explain your answer:
b. Which of these cases is more likely among the next 10 babies born at this hospital:
    - There will be 7 or more boys ____
    - There will be 3 or less boys ____
    - The number of boys will be between 4 and 6 ____
    - These three results are equally likely ____

    Explain your answer:

Fig. 20.1 Tasks posed to participants

Watson (2000) carried out a study in which the hospital problem was posed to 33 prospective teachers. 18 of the prospective teachers provided a correct response. Seven of these responses were obtained by intuitive reasoning, eight by mathematical reasoning, and three by a mixture of mathematics and intuition. Few teachers (4 of 15) who responded incorrectly used mathematical reasoning. After administering the questionnaire, the author implemented a simulation activity where the participants were able to discuss the problem and improve their intuitions.

One conclusion of the aforementioned research is that prospective teachers recognise the difference between probability and observed frequency in sampling situations. However, they also experience difficulties related to the notion of sampling, relying on heuristics and biases such as the representativeness heuristic and the equiprobability bias, or expecting exact convergence of frequencies towards theoretical probability, even in small samples. This research has mainly been based on responses to the hospital problem, where participants know the composition of the population and should reason about the features of one or several samples. Research analysing teachers' reasoning in the inverse situation (estimating the characteristics of a population when information from a sample is given) or the relationship between reasoning in both situations has not previously been conducted.

In this paper, we intend to fill this gap and complement previous research with a new analysis of the hospital problem that was first formulated by Kahneman et al. (1982) to which we add a second question to assess understanding of the likelihood of different outcomes in small samples. We also analyse responses to a problem adapted from research with 285 10–13 year old children (Fischbein and Gazit 1984), who assessed the children's ability to infer the characteristics of a finite population when information about the sample is known. Fischbein and Gazit's research described misunderstanding of inverse proportionality and a tendency to use additive strategies in solving this problem. In this paper, we analyse the strategies used by the teachers to justify their solutions to both problems.

20.3 Method

The sample consisted of 157 prospective primary school teachers enrolled in a mathematics education course at the Faculty of Education at the University of Granada in Spain. All of the prospective teachers had studied simple and conditional probability in the previous academic year, as well as random experiments, sample space, sample and population, and Laplace's rule to compute probability throughout secondary school. The data were taken as part of a practical activity during which the participants solved and analysed different probability problems. Three 1-hour sessions (with an interval of a week between sessions) were dedicated to the activity. During the first session, the prospective teachers solved the problems in Fig. 20.1 individually and explained their solutions in detail in a written form. In the second session, the teacher educator organised a debate, during which the different solutions to each problem were discussed until an agreement about the correct solution was reached. In the third session, students carried out simulations of the problems in pairs.

To assess the prospective teachers' knowledge, we use the suggestion by Godino (2009) that common content knowledge should be assessed through problems where the participants use their knowledge of probability concepts included in the primary school curriculum (see Gómez and Contreras 2014). Godino (2009) also suggests that knowledge in the mathematical horizon should be evaluated through problems whose solution require a more advanced knowledge of probability. We expected this knowledge to have been acquired by prospective teachers during their secondary school education. The tasks posed to participants in Spanish (translated into English in Fig. 20.1) include questions related to both categories of knowledge (common knowledge and knowledge in the mathematical horizon).

The first problem was adapted from Fischbein and Gazit (1984). The context of this problem is estimation using the capture-recapture method, which is used to estimate the size of the total population of animals from a sample in moving populations. To estimate the total number of fish in part (a), the prospective teachers first need to find the proportion p of marked fish in the population. This proportion can be estimated by taking into account that p should be close to the observed proportion $\widehat{p}$ in the sample of 250 fish. In this case, $\widehat{p} = \frac{\textit{Number of marked fish in the sample}}{\textit{Number of fish in the sample}} = \frac{25}{250} = 0.1$, which is a reliable estimate of the population proportion, given the large sample size. Since the total number of marked fish in the pond is known, we have $p = \frac{200}{N}$, which gives $N = \frac{200}{p}$, and when substituting p with its estimate, $\widehat{p}$, we obtain $N = \frac{200}{0.1} = 2000$. Due to randomness in the sampling process, this value is only approximate, and some variation around it can be expected.[1] Therefore, although proportional reasoning is important to solve the problem, it is not enough to understand estimation.

To answer part (b), the prospective teacher needs to understand that the expected relative frequency of marked fish in a new sample of 100 fish, f_r, should be close to the probability that a fish is marked in the population, which can also be interpreted as the population proportion p. In particular, $f_r = \frac{y}{100}$, where y is the number of marked fish in the new sample of 100 fish. The expected value of f_r is p, so in a new sample, $f_r \approx p \approx 0.1$, and consequently, the expected number of marked fish in the new sample is $y \approx 0.1 \times 100 = 10$. As with the estimation of p, there is variation around the value of y [2] and, therefore, the prospective teachers need to reason about sampling and variability, in addition to using proportional reasoning.

[1]We can compute a confidence interval for p to better reflect the sampling uncertainty. Since 250 fish is a large sample, we can use the normal approximation. For a 95% confidence, the lower and upper limits of the interval are $\widehat{p} \mp 1.96\sqrt{\frac{\widehat{p}(1-\widehat{p})}{n}}$. Substituting p with its estimate and n with 250, we obtain the interval (0.0628, 0.1372) for the population proportion.

[2]We can also compute a 95% confidence interval for y or for f_r.

In selecting this problem, we followed Godino's (2009) suggestions for assessing common content knowledge, since in the problem participants must use the knowledge of probability included in the mathematical curriculum. First, the variability of estimates (due to sampling variability, as well as due to uncertainty regarding the composition of the population) should be recognised. Second, sampling representativeness means that some potential values of the parameter in the population are more likely than others. The above analysis suggests that to solve the first question in Problem 1, the participants should remember and apply some concepts, properties, and procedures linked to sampling and to the frequentist approach of probability, such as those of frequency (absolute and relative), non-equiprobable events, probability, and estimation, which are studied in Spain during the last year of primary school (and then are considered to be part of common content knowledge, according to Ball et al. 2008 model). The second question requires an understanding of the variability of estimates, as well as of the notions of population, sample, proportion, and inverse proportion. These concepts are included in Spanish curricula for the first years of secondary school (which are included in horizon knowledge in the Ball et al. model).

Problem 2 is a variation of the hospital problem, originally formulated by Tversky and Kahneman (1974) and used in many psychological studies of decision making; our version is taken from Batanero et al. (1996). Question (a) assesses use of the representativeness heuristic, as well as sensitivity to the effect of sample size on sampling variability. In this question, the first option is more likely than the second, because sampling variability is affected by the sample size (this understanding would be part of horizon knowledge in the Ball et al. 2008 model). We know from birth rate data that the percentage of boys among new-born babies varies around 50%; for this reason it is more likely to have 8 boys among 10 new-born babies (the expected value is 5 boys, and the difference is only 3 units) than 80 boys among 100 new-born babies (a difference of 30 boys from the expected value of 50).

Theoretically, let X be the number of boys among 10 new-born babies and Y the number of boys among 100 new-born babies. Then, X has a binomial distribution B (10, 0.5) and Y a binomial distribution $B(100, 0.5)$. In a binomial distribution $B(n, p)$, the standard deviation is equal to $\sqrt{np(1-p)}$. Thus, the standard deviation of X is 1.6 and the difference between 8 and the expected value (5) is only about 2 standard deviations, while the standard deviation of Y is 5 and the difference between 80 and the expected value (50) is 6 standard deviations. More precisely, $P(X \geq 8) = 0.0547$ and $P(Y \geq 80) = 5.6 \times 10^{-10} \approx 0$. As a result, the probability of having 80 boys or more among 100 new-borns in the large hospital is almost zero.

In part (b) of Problem 2, we assess intuition about the behaviour of the same variable in small samples. The correct answer is the third option, because the expected value in a binomial distribution $B(10, 0.5)$ is 5 (the variable is symmetric with higher probability for values around 5). More precisely, let X be the number of boys among 10 new-born babies, $P(X \geq 7) = 0.1719$, $P(X \leq 3) = 0.1719$ and

$P(4 \leq X \leq 6) = 0.6562$. Again this knowledge is part of the horizon knowledge in the Ball et al. (2008) model.

The above analysis suggests that to solve Problem 2, participants should remember and apply some concepts, properties, and procedures linked to sampling and the frequentist approach of probability, such as those of more/less probable events, absolute frequency, repeated trials, sample size, and probability; these concepts are included at the end of the primary school curriculum (and are therefore part of common content knowledge, according to the Ball et al. 2008 model). Solving the problem also requires understanding of the concepts of variability, population and sample, random variables, and the binomial distribution; these concepts are included in Spanish secondary school curricula (which are included in horizon knowledge in Ball et al. model).

These two tasks confront participants with real-life situations involving two binomial random variables. In both tasks we ask the students to consider samples of different sizes and repetitions of the same experiment; as a result, students must reason about samples and populations, parameters and estimates, and samplingvariability. As explained at the end of Sect. 20.2, the tasks are complementary: in the second task, participants know the composition of the population and should reason about features of particular samples, while in the first task, they are asked to estimate the characteristics of a population given information about a particular sample.

In analysing the responses to Problem 1, we considered the estimate provided and the procedure followed to obtain the estimate, and established a priori categories for each of these components of students' responses. An estimate was classified as correct when its value was included in the 95% confidence intervals mentioned in footnotes 1 and 2. To classify the incorrect responses, we took into account results reported in previous research.

To analyse the responses provided to Problem 2, we considered the option selected for each part of the problem and the argument provided to justify the option. Categories for the arguments were selected a priori and took into account results reported in previous research. Some new categories (not reported in previous research) also emerged during the analysis of responses.

20.4 Results and Discussion

Once the written responses to the task were collected, a quantitative and qualitative analysis was carried out to identify categories of responses and to compare with previous research. As our participants responded in Spanish, we translated some responses to show examples of responses when necessary.

20.4.1 Responses to Problem 1

The estimates given by participants in part (a) of Problem 1 were classified into the following three categories:

C1. Correct response ($N = 2000$) (32 out of 157 participants, or 20.4%): These participants computed the proportion of marked fish in the sample ($p = 0.1$) and correctly assumed that the proportion in the population should be close to this value. Consequently, as there were 200 marked fish in the population, they estimated that the most likely number of fish in the pond is 2000 (ten times the number of marked fish). This response was provided by 44% of 12-year-old children who had attended experimental classes on probability in a study by Fischbein and Gazit (1984).

C2. Incorrect response ($N = 425$): 425 was the most common estimate for the number of fish in the pond. The participants who provided this response added the 250 fish in the sample to the additional number of marked fish in the population (200 − 25). They incorrectly assumed that there were only marked fish in the population after the first sample was drawn, and consequently revealed poor intuition about the relationship between the proportion in the population and in the sample. This response was provided by 37% of the children in Fischbein and Gazit's (1984) study.

C3. Other numerical values: Some students confused the notions of sample and population and estimated 225 fish in the total population, a response that was not reported by Fischbein and Gazit (1984). Five students added the fish in the sample to the 200 fish initially marked, assuming that only marked fish remained in the pond and not subtracting the marked fish in the sample, obtaining a value of 450; this response was also documented in Fischbein and Gazit's (1984) study. Other students provided values as high as 2500, multiplying by 10 the sample size to obtain their estimate.

These responses are summarised in Table 20.1 where we observe 14% of the participants did not provide a response. These results are worse in comparison to those observed by Fischbein and Gazit (1984) in the responses of 12-year-old children who had attended experimental classes on probability, 44% of whom provided correct responses to question (a). We attribute these results, which are a matter of concern, to a lack of prior instruction on the idea of sampling and weak proportional reasoning in these prospective teachers.

Table 20.1 Percentages of responses to Part a

Response	Percentage of participants (n = 157)
Correct ($N = 2000$)	20.4
Incorrect ($N = 425$)	49.7
Other numerical values	15.9
No answer	14.0

Participants' estimates in part (b) were classified into the following three categories:

C1. Correct response: 53.6% of participants provided a value close to 10 (10–13), which suggests good reasoning about the relationship between the proportion in the population and in the sample. We also accepted 35 as a correct response, since some participants added the 10 fish (the estimate for the second sample) to the 25 fish of the first sample. As their written responses revealed, these participants assumed that the first sample was not mixed again with population in the pond and that we had asked them to estimate the total number of fish marked in the two samples.

C2. Incorrect response (smaller than the sample size): There are two types of answers within this category:

   (a) Some of the students suggested a number of marked fish in the new sample between 1 and 4. While there might be variability in the expected value of marked fish in the sample, these values are too low when compared to the expected value (10 fish), taking the large sample size into account.

   (b) Other students provided an estimate of the number of fish in the new sample that was between 20 and 100 (10.8% participants[3]). These estimates reveal weak proportional reasoning and too much variability as regards what is expected in a random sample.

C3. Incorrect response (greater than the sample size): Some participants estimated a number of marked fish in the sample comprised between 150 and 525 and therefore greater than the sample size itself. For example, a participant misinterpreted the question and responded 525, having added the size of the second sample (100) with the number of marked fish in the pond (200) and the 225 non-marked fish in the first sample.

These results are summarised in Table 20.2, which shows that 28.8% of participants provided no response to the question. The percentage of correct responses to part (b) was higher than in part (a); however, about half of the participants were unable to solve the problem correctly.

Table 20.2 Percentages of responses to question b

Response	Percentage of students (n = 157)
Correct (10–13 or 35)	54.1
Incorrect (smaller than the sample size)	15.4
Incorrect (greater than the sample size)	1.8
No answer	28.7

Number of marked fish in a second sample (n = 10)

[3]Here, we do not count the 3.2% of the participants who responded 35 (considered to be a correct response, as explained previously).

Table 20.3 Frequencies and row percentage of correct/incorrect responses to both questions in Problem 1

	Question b				
	Correct		Incorrect/no response		Total
Question a	Frequency	Row percentage	Frequency	Row percentage	
Correct	27	84.4	5	15.6	32
Incorrect/no response	58	46.4	67	53.6	125
Total	85	54.1	72	45.9	157

We cross-tabulate the responses to both questions in Problem 1 in Table 20.3, which suggests the relationship between correct responses to both questions. This association was confirmed to be a significant result by the Chi-square test of independence between rows and columns in the table (Chi = 14.8, dg = 1, $p < 0.001$). Our interpretation of this result is that the relationships appear because an understanding of proportional reasoning, as well as the idea of sampling variability, are needed to respond correctly to both questions. Is also worth pointing out from the table how few teachers (only 17.2%) were able to provide correct estimates to both questions.

We additionally performed a qualitative analysis of the strategies participants used in solving the problem. These strategies were identified in the prospective teachers' written justification of their responses, and were classified in the following way:

- *Proportional reasoning*. These strategies involve multiplicative operations and comparing proportions. Participants who solved both questions using these strategies, in general, obtained correct results. Some participants combined proportional reasoning with verbal arguments or used a rote algorithm such as cross multiplication instead of working directly with proportions. In the following example, the subject (S119) first establishes in part (a) the ratio between marked and non-marked fish. Then, he computes the number of non-marked fish in the pond using this ratio, and finally, he adds the number of marked and unmarked fish in the pond to estimate the population size. In part (b), he directly applies the cross product strategy to estimate the number of marked fish in the sample, but confuses the size of the first sample (250) and instead used the number of unmarked fish in that sample (225). However, in spite of this confusion, he obtained a correct estimate, since 11 is very close to the expected number of fish in the sample, taking sample variability into account.

  S119: Part a: If for each 25 marked fish, there are 225 non-marked fish, then $225 \times 8 = 1800$ without marks; there are 200 more marked fish, so that $1800 + 200 = 2000$

$$If \frac{225}{100} \; \frac{-}{-} \; \frac{25}{x} \; x = \frac{100 * 25}{225} = 11.11 \cong 11 \, marked \, fish$$

- *Additive strategies*. These strategies, described by Fischbein and Gazit (1984), involve confusion between the intersection and the union of events; consequently, participants add or subtract data, instead of performing multiplicative operations. The concept of sampling is not applied by most participants using these strategies, as the problem is interpreted in a deterministic way. For example (S122): "250 − 25 = 225 are non-marked. If there were 200 marked fish, then there should be a total of 425". In this example, the total number of fish in the pond is assumed to be made up of the first sample (250 fish) and the fish initially caught and released back into the pond (200 marked fish); consequently, when the number of fish in the intersection is subtracted, the total is determined to be 425 fish (200 marked fish + 250 − 25 non-marked fish in the first sample).
- *Other strategies*. Other strategies included: (a) Reasoning according to the equiprobability bias (Lecoutre 1992), that is, assuming that all of the events have the same probability: "25 fish marked in the first sample + 50 more = 75 possible marked fish. When we take 100 fish, there is a 50% chance that the fish are marked or unmarked" (S33, part b); (b) Providing non-numerical answers based on deterministic reasoning: "There might be any number of marked fish because you never know; it is impossible to count all the fish at once" (S146, part a); (c) Assuming unpredictability: "we only know that there is a minimum of 200 + (250 − 25) = 425 fish, but do not know exactly how many there are" (S143, part a); and (d) Misinterpreting the question: "We do not know how many fish there are, we do not know how the fish are collected, it seems random" (S143, part b).
- *Unable to describe the strategy used.* The questionnaire given to participants asked them to solve the problems and to justify the steps taken to find the solutions in detail. However, some participants limited themselves to only providing numerical values for their estimates, neither justifying the reasons why they had obtained these values nor explaining any of the steps used to reach these estimates.

In Table 20.4, we cross-tabulate the participants' responses and strategies used to solve the questions in Problem 1 and observe that an important percentage of prospective teachers were unable to describe their strategies, which is a matter of concern, since a teacher will need to explain solutions to similar problems to their students. The majority of correct responses were associated with proportional reasoning, while incorrect responses were mainly obtained by using additive strategies or were not accompanied by an explanation of the procedure used. In our sample, the percentage of participants using proportional strategies to obtain correct responses was higher than that reported in the Fischbein and Gazit (1984) study, in which almost all children used additive strategies (only 5% of the older children who provided correct responses used proportional strategies).

Table 20.4 Frequencies and percentages of different strategies in each question

Strategy	(a) Estimate for total: 2000 fish				(b) Estimate for the number of marked fish in second sample: 10 fish			
	Correct		Incorrect/NA		Correct		Incorrect/NA	
	Freq.	Column percentage	Freq.	Column percentage	Freq.	Column percentage	Freq.	Column percentage
Proportional reasoning	24	75.0	2	1.6	55	64.7	14	19.4
Additive strategies			44	36.8				
Other strategies			2	1.6			5	7.0
Unable to describe the strategy	8	25.0	61	48.0	30	35.3	26	36.1
No answer			14	12.0			27	37.5
Total	32	100	125	100	85	100	72	100

20.4.2 Responses to Problem 2

Most participants (77%) reasoned according to the representativeness heuristic (Tversky and Kahneman 1974) in question (a) and assumed that all the results were equiprobable, revealing their insensitivity to the effect of sample size on sampling variability. Only 20% of prospective teachers responded correctly to this part of the problem, while in part (b) 39.5% provided correct responses, and only 29% reasoned according to the equiprobability bias.

In Table 20.4 we summarise the results, where those who reply "equiprobable" to both questions reason according to the equiprobability bias (Batanero et al. 1996). Our results were worse than those reported in previous research in Spain. Azcárate (1995) only used part (a) in a similar item and found that 67% of prospective teachers exhibited use of the representativeness heuristic, while 10% responded correctly and 23% made other errors. Batanero and colleagues (1996) found the representativeness heuristic being used by 62% of their participants in part (a) of Problem 2 and 44% in part (b), with 25% responding correctly to the first question and 39% responding correctly to the second.

We also performed a qualitative analysis of arguments, which were classified in the following way:

- *Sample size*. Some participants focused on both the proportion and the number of new-born babies in each sample. Most of these participants chose the correct option, revealing an intuitive understanding of the law of large numbers, which was also observed by Batanero et al. (1996). For example:

In general, there is the same number of boys and girls among new-born babies (50%). Then, among 100 new-born babies there are 50 girls and 50 boys. But this will only happen in a large group. Thus in small groups it is possible that this number varies, although when the observed group is large, the estimate is more accurate. (S148)

- *Range*. A few participants based their argument on the comparison of ranges of possible values for the number of boys in both hospitals; this is a variation of the previous correct argument:

The range of the second hospital is larger, thus the probability of obtaining eighty boys is smaller. (S97)

- *Representativeness heuristic*. Some participants exclusively focused on the identical proportions in both situations and used numerical reasoning instead of probabilistic reasoning, not taking into account the potential random variation around the expected values (eight or more, eighty or more) and revealing insensitivity to sample size. This type of reasoning, where subjects only take into account the equal expected proportion in two samples of different sizes and do not take into account the effect of sample size on sample variability. We assumed they based on the representativeness heuristic described by Kahneman et al. (1982), which has also been shown in responses to the hospital problem by Azcárate (1995) and Batanero et al. (1996). An example follows:

According to both responses the result is the same on a larger or smaller scale (8/10; 80/100), so that the probability is the same. (S3)

- *Other arguments*: A few participants exhibited reasoning based on the equiprobability bias (Lecoutre 1992), providing arguments that appealed to uncertainty or that did not take the problem data into account (a type response also observed in Azcárate 1995). For example:

Probability is an uncertain science, and so we always have the same probability of observed men as women. (S24)

- Finally, some students provided confused arguments that are difficult to classify:

There is the same number of boys 10→8; 100→80. However, it is not very common for 80 to be male among 100 new born babies. (S83)

In Table 20.5, we observe that most prospective teachers reasoned according to the representativeness heuristic and showed insensitivity to sample size. Correct responses were in general based on an adequate understanding of variability and its relation to sample size, and a few corresponded to a partial understanding of probability distribution. Incorrect responses were based on numeric reasoning and evidenced use of the representativeness heuristic or equiprobability bias.

Table 20.5 Frequencies of responses to Problem 2 and percentages according to Part a

Responses in Part b	Responses to Part a						Total	
	8 or more out of 10 (correct)		Equiprobable		Other			
	Freq.	Column percentage	Freq.	Column percentage	Freq.	Column percentage	Freq.	Global percentage
Between 4/10 and 6/10 (correct)	14	45.2	46	38.0	2	40.0	62	39.5
7/10 or more	5	16.1	17	14.1	1	20.0	23	14.6
Equiprobable	5	16.1	39	32.2	1	20.0	45	28.7
3/10 or less	4	12.9	8	6.6	1	20.0	13	8.3
No answer	3	9.7	11	9.1			14	8.9
Total	31	100	121	100	5	100	157	100

Our results are consistent with those of Azcárate (1995) and Batanero et al. (1996), although we obtained better arguments and probabilistic reasoning and fewer confusing or missing explanations (12% in Azcárate 1995; 19% in Batanero et al. 1996).

In part (b), there is again evidence of use of the representativeness heuristic, as well as the following new arguments:

- *Close to distribution centre*: Some participants chose the third option because its values are closer to the distribution midpoint. This reasoning was also observed by Batanero et al. (1996). who suggest that it is difficult to identify whether this response is due to students' understanding of the binomial distribution or to the representativeness heuristic. For example, "There are only two events, boy/girl, it is nearer to the middle of 10" (S139).
- *Equal proportion of genders*: Some participants demonstrated correct understanding of the binomial distribution, arguing that the third option is more probable because it is the set of values closest to the expected value.

> The 3rd option is more probable because for each baby there is 50% of probability to be a boy or a girl. Thus it is more probable that we observe a similar percentage in both genders. (S100)

- *Misconception of randomness*: Other participants who selected the last option assumed that equiprobability is a property of randomness. For example, "All possibilities are equally likely since the result is random" (S143).
- *Unpredictability*: Some students linked unpredictability to determinism, a response that was observed by Batanero et al. (1996). For example, "There is no right answer, since we cannot determine the result." (S83)

In Table 20.6, we find no prevalent argument; all five were frequently used. Correct answers were mostly based on reasoning about the closeness of the response to the centre of the distribution of possible values. The response that all possibilities are equally likely was linked to a misconception of both randomness and unpredictability, while other incorrect responses were linked to reasoning according to the representativeness heuristic or other arguments. Our participants exhibited better performance than those in Batanero et al. (1996), with better argumentation and a smaller number of non-responses. Their knowledge of the variability in small samples was poorer than that related to large samples. Table 20.7

20.4.3 Relating Responses to Both Problems

To analyse the relationships between responses to both problems, we used 2 × 2 contingency tables to cross-tabulate the responses to each part of Problem 1 with the responses to each part of Problem 2, taking into account only two categories of responses for each question (correct or incorrect/no response). For each of these contingency tables, we performed the Chi-square test of independence for responses to each pair of questions. The only significant result was found in the test of independence between responses to Part 1 in both problems (Chi = 5.42; d.g. = 1, $p = 0.02$). Our interpretation of this result is that only the responses to the first question of both problems were related in our sample, with participants providing correct responses to the estimate of the proportion in the population (number of fish in the pond in Problem 1, part a) being more likely to perceive the effect of sample size on samplingvariability (part a) in Problem 2, and vice versa. Since the responses to other questions were not related, our results suggest the need to educate prospective teachers in all the different concepts assessed in these questions.

20.5 Discussion and Suggestions for Teacher Education

Our results suggest a low level of probability reasoning in the specific context of sampling among the prospective teachers who took part in the study, which might also be common among other prospective teachers in Spain. The participants' responses to Problem 1 indicate weak proportional reasoning skills, as well as a poor understanding of estimation and of the relationship between probability and relative frequency, notions which are part of common content knowledge of probability in the context of sampling (Ball et al. 2008). Our prospective teachers performed worse than the children in the Fischbein and Gazit (1984) study, with only one out of every five participants obtaining a correct estimate of the population size in a finite population given information about a sample. In the second part of this problem, only about 50% of participants provided a correct estimate for the expected frequency of a given result in a new sample.

Table 20.6 Frequencies and percentages of arguments by responses in Part a of Problem 2

Argument	Response to Part a							
	Correct		Equiprobable		Other		Total	
	Freq.	Column percentage	Freq.	Column percentage	Freq.	Colum percentage	Freq.	Global percentage
Sample size	25	80.5	2	1.7			27	17.2
Range	2	6.5					2	1.3
Representativeness heuristic	2	6.5	100	83.3	3	50.0	105	66.9
Other arguments	2	6.5	18	15.0	1	16.7	21	13.3
No explanation					2	33.3	2	1.3
Total	31	100	120	100	6	100	157	100

Table 20.7 Frequencies and percentages of arguments in responses to Part b in Problem 2

Argument	Response to part b							
	Correct		Equiprobable		Other		Total	
	Freq.	Column percentage	Freq.	Column percentage	Freq.	Column percentage	Freq.	Global percentage
Close to the distribution centre	21	33.9	4	8.7			25	15.9
Equal proportion of genders	4	6.5	3	6.5	8	22.2	15	9.6
Misconception of randomness	2	3.2	11	23.9	2	5.6	15	9.6
Unpredictability	5	8.1	16	34.8			21	13.4
Representativeness	10	16.1	2	4.4	8	22.2	20	12.7
Other arguments	12	19.3	6	13.0	15	41.7	33	21.0
No explanation	8	12.9	4	8.7	3	8.3	28	17.8
Total	62	100	46	100	36	100	157	100

In the first part of Problem 2, only 39.5% of participants selected the correct response; these results are worse than those reported in previous research conducted in Spain by Azcárate (1995) and Batanero et al. (1996). Finally, for the second part of the problem, the percentage of correct responses was 45.2%. Many teachers did not recognise the effect of sample size on variability (part a) or identify the more probable events in a binomial distribution (part b), which are in the domain of horizon knowledge of sampling in the model of Ball et al. (2008).

This study provides new information about teachers' probabilistic reasoning when tasked with estimating the characteristics of a population given information about a sample (Problem 1), as well as about the lack of a relationship between reasoning in this problem and in predicting characteristics of a sample when the population is known (Problem 2). This result points to the complexity of the probabilistic concepts underlying sampling.

We also complement previous research in its analysis of the strategies prospective teachers that use to solve the given problems. Once again, the analysis of the strategies used revealed the participants' poor common knowledge about probability in the context of sampling: although many used proportional reasoning to solve the problems, the majority used additive strategies, which are typical of pre-formal operational stages in the theory of Piaget and Inhelder (1951); other participants applied rote algorithms in their solutions. There was also evidence of reliance on heuristics and biases, such as the representativeness heuristic and the equiprobability bias (Lecoutre 1992). Moreover, the arguments participants provided to explain their reasoning were poor, a result that suggests that the specialized content knowledge was also weak among the participants in this study.

Considered together, these results suggest that these prospective teachers need extensive training in probability and sampling, in agreement with previous research by Azcárate (1995) and Batanero et al. (1996). In order to provide the teachers with this training, in our experience, after the assessment, we organised a teaching session, where the prospective teachers presented their solutions to both problems and tried to justify them. This discussion helped to reveal potential conflicts in the teachers' understanding of sampling and probability and in distinguishing the correct and incorrect responses. Moreover, as in Watson's (2000) study, in a third session the prospective teachers performed some simulation activities using freely available applets that simulated the two situations described in the questionnaire (capture-recapture sampling[4] and the hospital problem[5]). During this session, the prospective teachers worked in pairs with the applet and compared their estimates with the applet results. The simulations helped the teachers discover their flawed reasoning on the questionnaire and to successfully solve new, related problems. As a part of the discussion, participants were also asked to list the concepts and

[4]See science.jburroughs.org/mbahe/BioA/starranimations/chapter40/videos_animations/capture_recapture.html.

[5]The number of boys and girls in the small and large hospitals were simulated using an applet from the Rossman and Chance collection (http://www.rossmanchance.com/applets/OneProp/OneProp.htm).

properties they used to solve each problem in order to reinforce their common and horizon knowledge of sampling. Finally, the teachers were asked to write a summary of all of the activities they were involved in during the sessions, including the correct and incorrect answers (with underlying biases) for each problem and a summary of the simulation activities. In agreement with Watson (2000), this practice helped to develop the participants' knowledge about probability and sampling and may be useful in other courses for prospective teachers.

Acknowledgements Research Projects EDU2016-74848-P (AEI, FEDER), and EDU2013-41141-P (MINECO).

References

Azcárate, P. (1995). *El conocimiento profesional de los profesores sobre las nociones de aleatoriedad y probabilidad* (Teachers' professional knowledge of randomness and probability). Unpublished Ph.D. dissertation. University of Cádiz, Spain.

Ball, D. L., Thames, M. H., & Phelps, G. (2008). Content knowledge for teaching. What makes it special? *Journal of Teacher Education, 59*(5), 389–407.

Batanero, C., Burrill, G., & Reading, C. (Eds.) (2011). *Teaching statistics in school mathematics. Challenges for teaching and teacher education. A joint ICMI and IASE study*. New York: Springer.

Batanero, C., Chernoff, E., Engel, J., Lee, H. & Sánchez, E. (2016). *Research on teaching and learning probability*. Berlin: Springer.

Batanero, C., Garfield, J. B., & Serrano, L. (1996). Heuristics and biases in secondary school students' reasoning about probability. In L. Puig & A. Gutiérrez (Eds.). *Proceedings of the Twentieth Conference of the International Group for the Psychology of Mathematics Education* (Vol 2, pp. 51–59). Valencia, Spain: IGPME.

Batanero, C., Godino, J. D., Vallecillos, A., Green, D. R., & Holmes, P. (1994). Errors and difficulties in understanding elementary statistical concepts. *International Journal of Mathematics Education in Science and Technology, 25*(4), 527–547.

Fischbein, E., & Gazit, A. (1984). Does the teaching of probability improve probabilistic intuitions? *Educational Studies in Mathematics. 15*(1). 1–24.

Godino, J. D. (2009). Categorías de análisis de los conocimientos del profesor de matemáticas (Categories in the analysis of teacher's kowledge). *UNIÓN. Revista Iberoamericana de Educación Matemática, 20*, 13–31.

Gómez, E., & Contreras, J. M. (2014). Meanings of probability in Spanish curriculum for primary school. In K. Makar, B. de Sousa, & R. Gould (Eds.). *Proceedings of the Ninth International Conference on Teaching Statistics,* Flagstaff, AR: International Statistical Institute. http://iase-web.org/icots/9/proceedings/pdfs/ICOTS9_C143_GOMEZTORRES.pdf. Accessed 14 April 2017.

Hill, H. C., Ball, D. L., & Schilling, S. (2008). Unpacking pedagogical content knowledge: Conceptualizing and measuring teachers' topic-specific knowledge of students. *Journal for Research in Mathematics Education, 39*(4), 372–400.

Kahneman, D., Slovic, P., & Tversky, A. (Eds.) (1982). *Judgment under uncertainty: Heuristic and biases*. New York: Cambridge University Press.

Lecoutre, M. P. (1992). Cognitive models and problem spaces in "purely random" situations. *Educational Studies in Mathematics, 23*(6), 557–568.

Ministerio de Educación y Ciencia, MEC (2006). *Real Decreto 1513/2006, de 7 de diciembre, por el que se establecen las enseñanzas mínimas de la Educación Primaria* (Royal Decreet establishing the minimum content for Primary Education). Madrid: Author.

Ministerio de Educación, Cultura y Deporte. (2014). *Real Decreto 126/2014, de 28 de febrero, por el que se establece el currículo básico de la Educación Primaria* (Royal Decreet establishing the basic curriculum for primary education). Madrid: Author.

National Council of Teachers of Mathematics. (2000). *Principles and standards for school mathematics*. Reston, Virginia: Author.

Piaget, J., & Inhelder, B. (1951). *La genése de l'idée de hasard chez l'enfant*. Paris: Presses Universitaires de France.

Saldanha. L., & Thompson, P. (2007) Exploring connections between sampling distributions and statistical inference: an analysis of students' engagement and thinking in the context of instruction involving repeated sampling. *International Electronic Journal of Mathematics Education, 2*(3), 270–297.

Tversky, A., & Kahneman, D. (1974). Judgement under uncertainty: Heuristics and biases. *Science, 185*(4157), 1124–1131.

Watson, J. (2000). Preservice mathematics teachers' understanding of sampling: Intuition or mathematics. *Mathematics teacher education and development, 2*(1), 121–135.

A Commentary on *Teaching and Learning Stochastics: Advances in Probability Education Research*

Egan J. Chernoff and Carmen Batanero

There are numerous ways that one could describe this particular monograph. For example, this monograph is comprised of selected papers that represent the current variety of research related to the teaching and learning of probability. More specifically, this monograph houses research concerning (1) teaching of probability, (2) probabilistic reasoning and learning, (3) and teacher education; and has been structured accordingly. Alternatively, the aim of the editors was that *Teaching and Learning Stochastics* would be a continuation—perhaps, culmination—of the work presented during Topic Study Group 14: Teaching and Learning Probability, which took place at the 13th International Congress on Mathematical Education (ICME-13) in Hamburg, Germany. As we mentioned in our Preface, this monograph can be described as aspirational, that is, we hope this book will prove helpful toward improving the teaching and learning of probability in schools, thus informing probabilistic thinking, reasoning and literacy for both teachers and students. A second goal is that the experiences and ideas reported in the book encourage other researchers to investigate the many unsolved problems related to the teaching and learning of probability. Given the options presented above, we will focus the remainder of our commentary on our pronouncement that *Teaching and Learning Stochastics: Advances in Probability Education* can be described as a contribution to probability education, in particular, and mathematics education, in general.

According to Jones and Thornton (2005), research in (what has become to be known as) the field of probability education took place over the last 50 (now the last 60 plus) years. Further, in 2005, the authors declared that investigations into probabilistic thinking and the teaching and learning of probability, those by researchers in the field of mathematics education, could be categorized into three phases: the Piagetian Period, the Post-Piagetian Period and Contemporary Research (pp. 65–66). In addition to the phases, from amongst the numerous, salient declarations found throughout their historical overview, one statement, in particular, stands out: "Although it is premature to evaluate historically the significance of

E. J. Chernoff
College of Education, University of Saskatchewan, Saskatoon, SK, Canada

C. Batanero
Facultad de Ciencias de la Educación, University of Granada, Granada, Spain

C. Batanero and E. J. Chernoff (eds.), *Teaching and Learning Stochastics*, ICME-13 Monographs, https://doi.org/10.1007/978-3-319-72871-1

probability research in this third and contemporary phase, it is clear that the volume and diversity of the research is greater than in the previous two phases" (p. 83). It stands out for us, we contend, because the culmination of our efforts associated with *Teaching and Learning Stochastics* are kindred in nature.

In 2014, based on the many emergent contributions to the field and the predefined length of Jones and Thornton's (2005) research phases, Chernoff and Sriraman, hinted at the dawn of a "new phase of research" (p. xvii) in the field of probability education. They also raised the question of what to name the next period of research. Based on their indicators that probability education, as a field, was moving from the fringe to mainstream mathematics education, they "prematurely" (ibid.) proposed the next phase be denoted the "Assimilation Period" (ibid.). Hedging, though, they also noted that certain developments that occurred or may occur during this particular phase may better, at some point, define the Assimilation Period. Although no particular name, to the best of our knowledge, has been associated with this current phase of research, it is clear that the field of probability education is in a fourth phase. As was the case with Jones and Thornton in 2005, it is a tad premature to evaluate, historically, the significance of probability education research in this fourth phase; however, as the volume and diversity of probability education publication ever grows, this evaluation will eventually take place as it did for the Contemporary Phase.

Sixty years of research has resulted in numerous important publications in probability education (e.g., Batanero et al. 2016; Borovcnik and Kapadia 2009; Borovcnik and Peard 1996; Burrill and Elliot 2006; Fischbein 1975; Garfield and Ahlgren 1988; Hawkins and Kapadia 1984; Jones 2005; Jones et al. 2007; Kapadia and Borovcnik 1991; Piaget and Inhelder 1951/1975; Shaughnessy 1992, 2003; Shulte and Smart 1981). As time marches on and as the field ever grows, we, once again, echo Jones and Thornton (2005) by noting "that the volume and diversity of the research [in this fourth phase] is greater than in the previous [three] phases" (p. 83). Although more is not necessarily better, more does equate to more opportunity for potentially important publications in probability education. Although only time will tell whether or not more recent publications become added to the list of recent books and special journal issues in terms of importance (e.g., Batanero et al. 2016; Biehler and Pratt 2012; Chernoff 2015; Chernoff et al. 2016; Chernoff and Sriraman 2014), the volume and diversity of research in said publications is greater than in previous phases, which brings us back to *Teaching and Learning of Stochastics*.

As presented at the beginning on this commentary, there are various ways to describe this particular monograph. We contended—and as you made your way through this book we hope you agree—the research housed in this monograph can be described as a contribution to the field of probability education. Concurrently, as opposed to alternatively, perhaps the present day ability to culminate the probability education research from dedicated working and topic study groups taking place at major conferences into special journal issues and monographs is yet another indicator that probability education, as a research topic, is further becoming assimilated into mainstream mathematics education.

We thoroughly enjoyed the opportunity to present the research from Topic Study Group 14: Teaching and Learning Probability at the 13th International Congress on

Mathematical Education to a larger audience with this monograph. We hope you enjoyed *Teaching and Learning Stochastics: Advances in Probability Education Research.*

References

Batanero, C., Chernoff, E., Engel, J., Lee, H., & Sánchez, E. (2016). *Research on teaching and learning probability*. Springer. ICME-13. Topical Survey series.

Biehler, R., & Pratt, D. (Eds.) (2012). Probability in reasoning about data and risk [special issue]. *ZDM—The International Journal on Mathematics Education, 44*(7), 819–952.

Borovcnik, M., & Kapadia, R. (Eds.) (2009). Research and developments in probability education [special issue]. *International Electronic Journal of Mathematics Education, 4*(3). Retrieved from http://www.iejme.com/032009/main.htm.

Borovcnik, M., & Peard, R. (1996). Probability. In A. Bishop, K. Clements, C. Keitel, J. Kilpatrick, & C. Laborde (Eds.), *International handbook of mathematics education* (Part 1, pp. 239–288). Dordrecht: Kluwer.

Burrill, G. F., & Elliott, P. C. (Eds.) (2006). *Thinking and reasoning with data and chance: Sixty-eighth yearbook.* Reston: National Council of Teachers of Mathematics.

Chernoff, E. J. (Guest Editor) (2015). Special Issue: Risk—Mathematical or otherwise. *The Mathematics Enthusiast, 12*(1, 2 & 3). 479 pages.

Chernoff, E. J., Paparistodemou, E., Bakogianni, D., & Petocz, P. (Guest Editors) (2016). Special Issue: Research on learning and teaching probability within statistics. *Statistics Education Research Journal, 15*(2). 265 pages.

Chernoff, E. J., & Sriraman, B. (Eds.) (2014). *Probabilistic thinking: Presenting plural perspectives* (Volume 7 of Advances in Mathematics Education Series). Berlin/Heidelberg: Springer Science.

Fischbein, E. (1975). *The intuitive sources of probabilistic of probabilistic thinking in children.* Dordrecht: Reidel.

Garfield, J. B., & Ahlgren, A. (1988). Difficulties in learning basic concepts in probability and statistics: Implication for research. *Journal for Research in Mathematics Education, 19*, 44–63.

Hawkins, A. S., & Kapadia, R. (1984). Children's conceptions of probability—A psychological and pedagogical review. *Educational Studies in Mathematics, 15*, 349–377.

Jones, G. A. (Ed.). (2005). *Exploring probability in school: Challenges for teaching and learning*. New York: Springer.

Jones, G. A., Langrall, C. W., & Mooney, E. S. (2007). Research in probability: Responding to classroom realties. In F. K. Lester (Ed.), *Second handbook of research on mathematics teaching and learning* (pp. 909–955). New York: Macmillan.

Jones, G. A., & Thornton, C. A. (2005). An overview of research into the learning and teaching of probability. In G. A. Jones (Ed.), *Exploring probability in school: Challenges for teaching and learning* (pp. 65–92). New York: Springer.

Kapadia, R., & Borovcnik, M. (Eds.). (1991). *Chance encounters: Probability in education.* Dordrecht: Kluwer.

Piaget, J., & Inhelder, B. (1975). *The origin of the idea of chance in students* (L. Leake, Jr., P. Burrell, & H. D. Fischbein, Trans.). New York: Norton. (Original work published 1951).

Shaughnessy, J. M. (1992). Research in probability and statistics. In D. A. Grouws (Ed.), *Handbook of research on mathematics teaching and learning* (pp. 465–494). New York: Macmillan.

Shaughnessy, J. M. (2003). Research on students' understanding of probability. In J. Kilpatrick, W. G. Martin, & D. Schifter (Eds.), *A research companion to principles and standards for school mathematics* (pp. 216–226). Reston: National Council of Teachers of Mathematics.

Shulte, A. P., & Smart, J. R. (Eds.). (1981). *Teaching statistics and probability: 1981 yearbook.* Reston: National Council of Teachers of Mathematics.

Index

C. Batanero and E. J. Chernoff (eds.), *Teaching and Learning Stochastics*, ICME-13 Monographs, https://doi.org/10.1007/978-3-319-72871-1

The manufacturer's authorised representative in the EU is Springer Nature Customer Service Centre GmbH, Europaplatz 3, 69115 Heidelberg, Germany. If you have any concerns regarding our products, please contact ProductSafety@springernature.com

Printed and bound by CPI Group (UK) Ltd, Croydon, CR0 4YY
30/09/2026
02225659-0012